PropTech Innovations in Real Estate Investment and Finance

The real estate industry is undergoing unprecedented transformations driven by emerging technologies and sustainability practices. This textbook explores the intersection of PropTech, blockchain, and sustainability, shaping the future landscape of real estate investment and finance. The book is a straightforward guide to understanding how new technologies are changing the real estate industry. It explains in clear, everyday language the key digital innovations like artificial intelligence, virtual reality, and blockchain, transforming how homes and buildings are designed, constructed, bought, sold, financed, and managed.

The book shows how these new PropTech tools make real estate transactions faster, easier, and more efficient for buyers, sellers, agents, investors, developers, and property managers through real-world examples and case studies. Readers will learn how smart sensors and automated systems make buildings and homes smarter, greener, and more sustainable. They will discover how innovations like digital listing platforms and virtual tours are improving marketing and connectivity between agents and clients.

The book covers property search, online transactions, crowdfunding, blockchain for fractional ownership, automated valuations, and sustainability in real estate. In a simple, modular format, readers can focus on the real estate tech topics most relevant to their needs and test their knowledge and understanding using the end-of-chapter engagement exercises. Whether you are a real estate professional, student, investor, or just a technology enthusiast, this book provides the essential knowledge you need to understand the digital transformation of the real estate sector in simple language anyone can understand.

The book equips readers with the necessary information to leverage these PropTech tools and innovations for smarter, more efficient, future-oriented real estate.

Yishuang Xu is currently a senior lecturer (associate professor) in real estate at the University of Manchester, UK. She completed her PhD in real estate finance at the University of Hong Kong. She worked as an assistant professor at the Capital University of Economics and Business, China, and as a researcher at the Chinese University of Hong Kong before joining the University of Manchester. Her recent research focuses on real estate finance with a core interest in digitalisation, innovation, and sustainability in real estate.

Ming Yan Lo is an estate surveyor and possesses over 20 years' experience in the real estate and capital market industry. By leveraging his digital assets and investment in an industry-leading portfolio of companies, Lo is developing the real estate business as a futurist. He is the business owner of a regional real estate services firm, CHFT, and a real estate investment house, Fonto Real Estate, plus the founder of Asia PropTech (https://www.asiaprop.tech/)

and his own real estate technology company, which is working on AI and blockchain applications in the real estate sector, with solutions in its automated valuation model and tech-enabled real estate data analysis. Lo is in high demand as a speaker for his thought leadership at international conferences and is regularly invited to share his thoughts on real estate, PropTech, and capital markets in several TV channels and business magazines, such as Cable TV, HKIBC, TVB, Now TV, and *Bloomberg Businessweek*. Lo teaches at the Chinese University of Hong Kong (Business School) as an adjunct assistant professor and HKU Institute for China Business as a guest lecturer.

King Man Yiu is currently the head of sustainable investing with Cheung On Securities, an asset management firm in Hong Kong. Prior to that, he was head of corporate finance with a multinational FinTech company, chief financial officer of an international real estate investment firm, and investment director of a Hong Kong–listed real estate investment company. Andy has over 15 years of professional experience in corporate finance with a global deal record of real estate investment of circa USD 1.4 billion. He is frequently invited to interviews and events as a lecturer, speaker, and moderator on topics related to tokenisation, ESG, and sustainability. Andy is a co-founding executive committee member of PropTech Institute, a non-profit association registered in Hong Kong promoting property technology for a more sustainable built world in Asia.

PropTech Innovations in Real Estate Investment and Finance

Sustainable Futures

Yishuang Xu, Ming Yan Lo, and
King Man Yiu

Routledge
Taylor & Francis Group
LONDON AND NEW YORK

Designed cover image: Getty Images

First published 2026
by Routledge
4 Park Square, Milton Park, Abingdon, Oxon, OX14 4RN

and by Routledge
605 Third Avenue, New York, NY 10158

Routledge is an imprint of the Taylor & Francis Group, an informa business

British Library Cataloguing-in-Publication Data
A catalogue record for this book is available from the British Library

ISBN: 978-1-032-82424-6 (hbk)
ISBN: 978-1-032-82423-9 (pbk)
ISBN: 978-1-003-50443-6 (ebk)

DOI: 10.1201/9781003504436

Typeset in Times New Roman
by SPi Technologies India Pvt Ltd (Straive)

Contents

Acknowledgement

We extend our sincere gratitude to Simon Yu, a leading real estate professional in Hong Kong, currently with CBRE, and a key contributor to this book. His invaluable insights, gleaned from 15 years of experience in transaction advisory services and his work interviewing PropTech founders across the APAC region, were crucial in shaping our understanding of the region's landscape. Simon's dedication extends beyond his professional career, as demonstrated by his roles as Co-Chair of the Property Committee at the Canadian Chamber of Commerce and President & Co-Founder of the PropTech Institute.

Simon's commitment to fostering innovation in the built environment is truly commendable. His work with the PropTech Institute perfectly embodies the need to bridge the gap between the traditional "bricks" and the evolving digital world. By championing the adoption of technology in real estate, Simon, along with others, is helping create a more sustainable, competitive, and globally appealing industry, ultimately attracting a new generation of talent crucial for future growth.

Chapter 1

Background and Introduction

1.1 Overview of the Real Estate Industry

The real estate industry is one of the most significant sectors in the global economy, its importance stemming from its substantial market size, its role in providing essential living and working spaces, and its interconnectedness with numerous other industries. Looking ahead, the industry is poised for substantial growth. Statista (2024) forecasts a projected compound annual growth rate (CAGR) of 2.77% from 2024 to 2029, and the total market value could reach USD 727.8 trillion by 2029 for global real estate.

The real estate industry has undergone significant transformations since its inception, continually adapting to societal changes. From the agrarian societies where land ownership denoted wealth and power, to the industrial revolution that spurred demand for urban housing and commercial spaces, the industry has evolved in tandem with human progress. Key milestones in this evolution include the rise of modern urban planning in the late 19th and early 20th centuries, the post–World War II housing boom and suburban expansion, the emergence of commercial real estate as a distinct asset class in the mid-20th century, and the globalisation of real estate markets in the late 20th and early 21st centuries.

In recent years, several trends have been reshaping the real estate landscape. The advent of PropTech is revolutionising how properties are invested, financed, managed, and developed, bringing unprecedented efficiency and transparency to the market. Meanwhile, growing awareness of climate change has put sustainability at the forefront of real estate development and investment decisions, driving innovations in green building technologies and sustainable urban planning. Demographic shifts, including rapid urbanisation and ageing populations in developed countries, are changing demand patterns and necessitating new approaches to property development and urban design.

The global COVID-19 pandemic has further accelerated certain trends, particularly the shift towards remote work, potentially reshaping demand for office and residential spaces in ways that may persist long after the immediate health crisis has passed. Moreover, new categories of real estate, such as data centres, logistics facilities, and co-living spaces, are gaining prominence, reflecting broader technological and social changes.

The real estate industry operates within a complex ecosystem of stakeholders, each playing a distinct role in its functioning. This ecosystem includes developers and construction companies who create new properties, property management firms that maintain and operate existing ones, real estate agents and brokers who facilitate transactions, investors (both individual and institutional) who provide capital, lenders and financial institutions that offer financing, and government bodies and regulators who shape the legal and policy

DOI: 10.1201/9781003504436-1

environment. Each player contributes to the industry's dynamism, influencing everything from property valuation to urban development policies.

1.2 Introduction to Innovation and Digital Transformation

1.2.1 Innovation and Digital Transformation

Innovation is a driving force in the evolution of industries, acting as a catalyst for growth, efficiency, and competitive advantage. At its core, innovation is creating value through the implementation of new ideas, methods, or technologies. It can range from incremental improvements to radical changes that disrupt entire markets.

Digital transformation, a significant subset of innovation in the modern era, refers to the integration of digital technology into all areas of business. This process fundamentally changes how operations are conducted and value is delivered to customers. It is not only about adopting new technologies but also about reimagining business processes, culture, and customer experiences in the digital age.

In traditional industries such as real estate, innovation and digital transformation are becoming increasingly crucial. They offer solutions to long-standing challenges, create new opportunities, and help businesses adapt to ever-changing market conditions and customer expectations.

1.2.2 Fundamental Innovation

Fundamental innovation goes beyond small improvements. It causes major shifts in how industries operate. These innovations are based on deep scientific and technological advances. They can disrupt existing ways of doing things and sometimes create entirely new industries. Fundamental innovations can change existing markets completely or create new ones. They are often risky and require big investments, but they can also bring huge rewards. These innovations do not happen in isolation. They rely on networks of researchers, innovators, investors, and partners working together. While fundamental innovations might start in one industry, they often end up affecting many others. This wide-ranging impact shows how powerful these innovations can be in reshaping how we do business and live our lives.

1.2.3 The Reasons for Innovation and Digital Transformation

Innovation and digital transformation are driven by a few factors across industries. Businesses are seeking to increase productivity and efficiency through digital tools and innovative practices that streamline processes, reduce waste, and speed up operations, resulting in significant time and cost savings. At the same time, they are working to improve customer experience, recognising that modern consumers expect seamless, personalised, digital-first interactions. Sustainability is another key driver, as growing concerns about climate change and resource scarcity push for innovations in sustainable practices and technologies. The need to manage complex data is also crucial, with digital transformation enabling more effective data collection, analysis, and utilisation for decision-making. These innovations help industries adapt to changing market dynamics, including shifts in demographics, global economic patterns, and consumer behaviours. Ultimately, in a rapidly changing business landscape, the ability to innovate and digitally transform has become a key competitive advantage, often separating market leaders from laggards.

1.2.4 *Business Model Innovation*

Innovation not only involves technology but also includes new business models. Business model innovation changes how companies create, deliver, and capture value, often allowing them to disrupt established markets or create new ones. This powerful form of innovation involves several key aspects. It redefines the value proposition, changing what is offered to customers and how it meets their needs. It also involves altering the revenue model and finding new ways to generate income from the value proposition. Companies may need to adjust their cost structure to deliver value more efficiently. Identifying and using new resources or processes to deliver value is another important aspect. While successful business model innovations often use new technologies, their real strength comes from rethinking how a business fundamentally operates and provides value to its customers. This comprehensive approach to innovation can lead to significant competitive advantages and industry transformations.

1.3 Theoretical Framework: Principles of Diffusion and Innovation and Information Technology Adoption Model (ITAM)

Two key theories build the theoretical frameworks to understand the dynamics of PropTech adoption in the real estate industry. They are the diffusion of innovations theory and the information technology adoption model (ITAM), which incorporates elements of the technology acceptance model (TAM).

1.3.1 *Diffusion of Innovations*

Everett Rogers's (1962) **diffusion of innovations theory**, was first published in 1962 and refined over subsequent decades. It provides a comprehensive framework for understanding how new ideas and technologies spread through social systems.

Rogers (1962) identified five attributes that influence the adoption rate of innovation, including relative advantage, compatibility, complexity, trialability and observability; and categorised adopters into five groups based on their innovativeness: innovator, early adopters, early majority, late majority, and laggards.

A five-stage process that individuals go through when deciding to adopt or reject an innovation (Rogers, 1962), as shown in Figure 1.1.

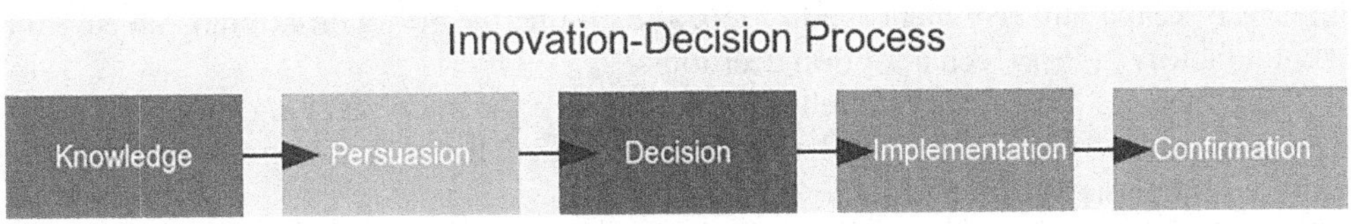

Figure 1.1 Five-Stage Innovation-Decision Process.

By author (2024) based on Rogers (1962).

Diffusion of innovations theory describes how innovations are communicated and adopted over time among members of a social system. In the context of real estate, this theory can explain (1) the characteristics of early adopters versus late adopters in the industry, (2) how PropTech innovations are communicated within real estate networks, and (3) the attributes of PropTech solutions that make them more likely to be adopted (like relative advantage, compatibility, and observability).

1.3.2 Information Technology Adoption Model (ITAM) and Technology Acceptance Model (TAM)

The **information technology adoption model (ITAM)** builds upon and extends several well-established theoretical frameworks, including Davis's (1989) **technology acceptance model (TAM)**. ITAM provides an approach to understanding the factors that influence an individual's decision to adopt and use new information technologies, making it particularly relevant to the study of PropTech adoption.

The ITAM comprises several key components. End-user sophistication, as described by Dixon (1999), refers to the user's depth and breadth of knowledge and skills in information technology (IT), including their ability to transfer this knowledge across contexts. Available Resources considers the technological, financial, and temporal resources at the user's disposal. IT Requirements include the specific demands of the technology in terms of hardware, software, skills, and knowledge. IT-User Fit measures the alignment between user capabilities and technology requirements. Perceived Usefulness, a concept from Davis's (1989) technology acceptance model (TAM), assesses how much a user believes a system will enhance their job performance, incorporating elements like relative advantage, subjective norms, compatibility, and feedback. Perceived Ease of Use, also from TAM, evaluates how effortless a user believes a system will be to use, including factors such as usability, perceived behavioural control, and available support. Intention to Use represents the user's stated plan to adopt and use the technology, while Actual Use refers to the observed adoption and continued usage of the technology. These components work together to provide a comprehensive framework for understanding technology adoption in various contexts.

ITAM provides a more comprehensive framework than TAM alone, incorporating elements that are particularly relevant to the complex ecosystem of PropTech. It acknowledges the importance of user characteristics, technological requirements, and the fit between the two, as well as the broader organisational and social context in which adoption occurs (Dixon, 1999). For PropTech in real estate, ITAM can help explain (1) how real estate professionals' perceptions of a PropTech solution's usefulness and ease of use affect their likelihood of adopting it, (2) the impact of an individual's technological sophistication on their propensity to adopt PropTech, (3) how the fit between a PropTech solution and a user's capabilities influences adoption, and (4) the role of organisational and environmental factors in PropTech adoption decisions.

These theories provide a comprehensive framework for understanding PropTech adoption at the industry and individual levels. This theoretical framework helps us to design the empirical analysis in Chapters 4–7.

1.4 Definition and Evolution of PropTech

PropTech, short for property technology, refers to the application of information technology and platform economics to real estate markets. This term refers to a range of digital innovations and technology applications designed to optimise the way people research, rent, buy, sell, and manage property.

The concept of PropTech emerged in the early 2000s, coinciding with the broader digital transformation across industries. Initially, PropTech mainly focused on online real estate listings and property management software. However, as technology advanced, the scope of PropTech expanded dramatically.

1.4.1 *Key Phases in PropTech Evolution*

As shown in Figure 1.2, PropTech has evolved over the past two decades. PropTech 1.0, from the early 2000s to the early 2010s, focused on digitising existing processes and information. This phase introduced online property listings and basic property management software, causing limited disruption to traditional real estate models. PropTech 2.0, from the mid-2010s to late 2010s, saw the integration of more advanced technologies. This period saw the emergence of virtual and augmented reality for property viewings, the adoption of big data analytics for market insights, and the rise of real estate crowdfunding platforms. The current phase, PropTech 3.0, which began in the late 2010s and continues to the present, is characterised by the adoption of cutting-edge technologies. This includes artificial intelligence and machine learning for predictive analytics, the Internet of Things (IoT) for smart building management, blockchain for transparent and secure transactions, and 3D printing and robotics in construction. Each phase in the diagram represents a significant leap in technological sophistication and its impact on the real estate industry (Andrew Baum et al., 2020; Shaw, 2020; Saiz, 2023). The COVID-19 pandemic (2020–2022) significantly accelerated digital adoption in the sector, particularly in remote property viewing technologies and contactless building management systems. More recently (2024–2025), the integration of generative AI has begun transforming property design, valuation models, and predictive maintenance.

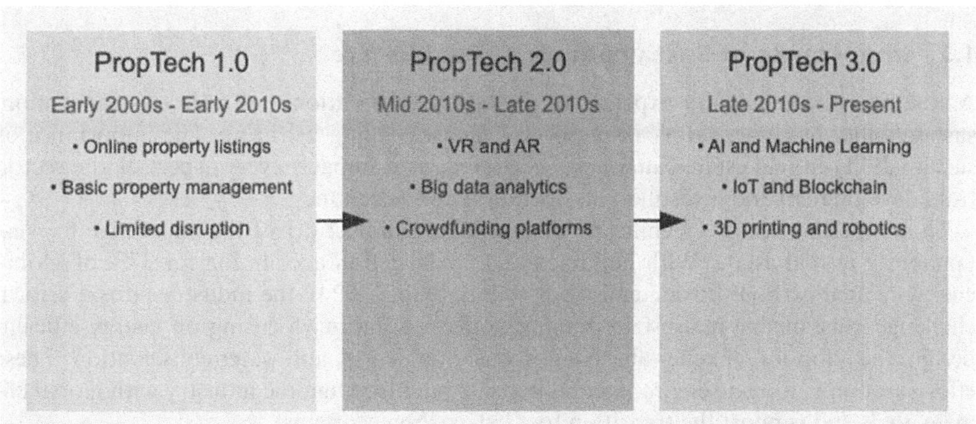

Figure 1.2 Key Phases in PropTech Evolution.
By author (2024).

1.4.2 *Current State and Future Trends*

PropTech has evolved from being just an enabler to a disruptor in the real estate industry. It is reshaping traditional business models and creating new opportunities.

The key trends shaping the current state and future of PropTech include:

1. *Space-as-a-Service:* The real estate industry is shifting from "space-as-a-product" to "space-as-a-service" or "real estate-as-a-service" (REaaS). This model prioritises flexibility and user experience over traditional ownership. This shift reflects a broader societal move towards a post-materialism era, where access is valued over ownership. Co-working spaces exemplify this trend, offering not just workplaces but also amenities and community-building opportunities. *While early pioneers like WeWork faced challenges, the model has proven resilient through PropTech integration.*
2. *Data-Driven Valuation:* Real estate value is increasingly determined by the data it generates. Properties that can extract and utilise more data per square meter are likely to be valued higher.
3. *Sustainability Focus:* PropTech plays a crucial role in making buildings more sustainable, offering solutions for energy efficiency, waste reduction, and smart resource management.
4. *Personalisation:* AI and big data enable highly personalised real estate experiences, from property searches to space utilisation and tenant management.
5. *Blockchain and Tokenisation:* These technologies are set to revolutionise property transactions, potentially increasing liquidity in real estate markets and supporting the democratisation of real estate investment.

As PropTech continues to evolve, it is blurring the lines between physical and digital real estate, creating new asset classes and investment opportunities. In future, it is likely to see deeper integration of AI, IoT, and blockchain, potentially leading to fully automated, self-managing properties and frictionless real estate transactions. Real estate may become analogous to a mobile device, with PropTech solutions acting as software applications. This shift suggests a need to redefine real estate from an innovative perspective, where value is increasingly determined by data generation capabilities.

1.5 Importance of Sustainability in Real Estate

As the real estate industry experiences digital transformation and embraces innovation, sustainability has emerged as an important factor shaping its future. The importance of sustainability in real estate cannot be overstated, as it impacts every aspect of the sector, from development and investment to operation and occupancy.

The real estate sector's significant environmental impact drives the imperative for sustainability in real estate. With buildings and construction accounting for 37% of global energy-related carbon dioxide emissions (GlobalABC, 2020), the industry faces a critical challenge. Sustainable real estate practices offer a solution, focusing on energy-efficient design, the adoption of renewable energy, waste reduction, and water conservation. These efforts not only mitigate environmental impacts but also align the industry with global climate goals and support the transition to a low-carbon economy.

Beyond environmental considerations, sustainability in real estate presents compelling economic benefits. Green buildings typically enjoy lower operating costs, higher property values, and increased tenant attraction and retention. Research indicates that sustainable buildings can command rental premiums of 6–11% and experience fewer void periods (JLL, 2021), demonstrating the tangible financial advantages of sustainability investments. This economic case is further strengthened by the enhanced long-term asset value and resilience against climate-related risks that sustainable properties offer.

Sustainable real estate also plays a critical role in promoting social responsibility and well-being. By creating healthier indoor environments, fostering community engagement, and improving urban liveability through green spaces and sustainable urban planning, the industry contributes to societal well-being. This focus aligns with growing consumer demand for responsible and community-oriented development, supporting the creation of more resilient and inclusive cities (WEF, 2021). As regulatory landscapes evolve, sustainable practices help real estate stakeholders navigate compliance challenges and mitigate risks. With governments worldwide implementing stricter environmental regulations, sustainability initiatives assist in complying with building codes and energy performance

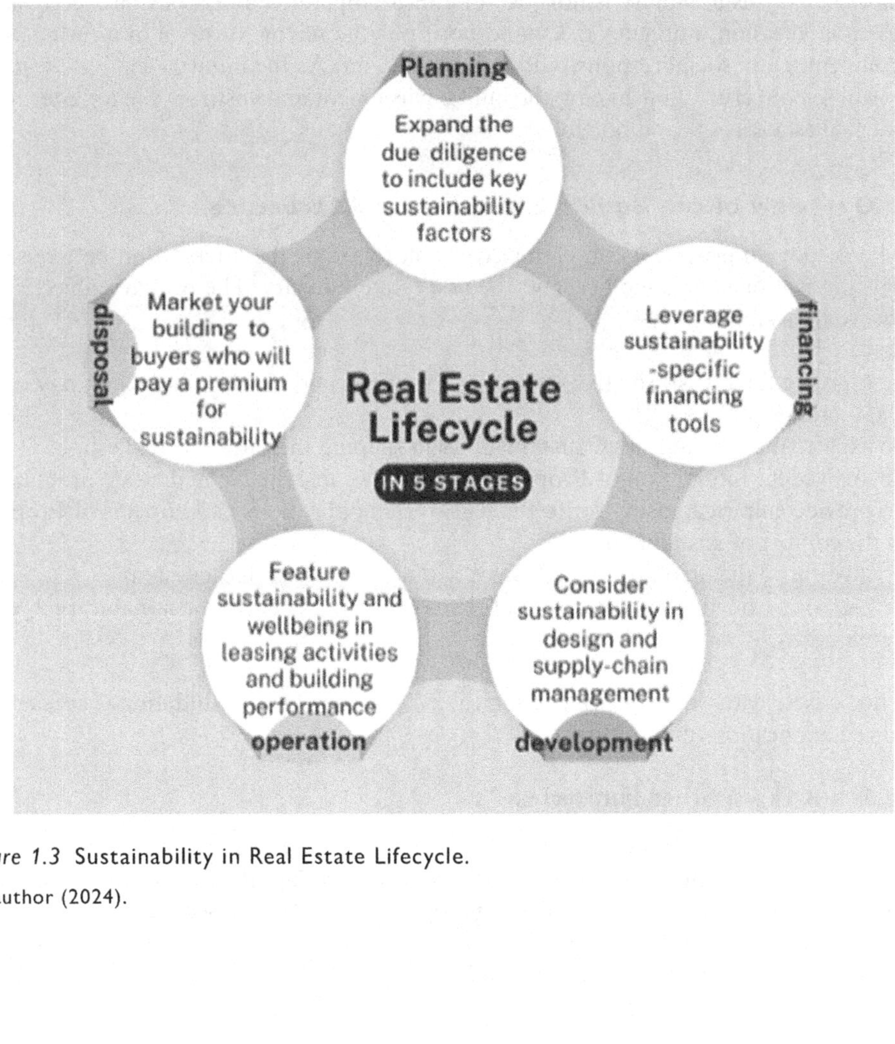

Figure 1.3 Sustainability in Real Estate Lifecycle.

By author (2024).

standards while also preparing for future regulatory changes (WEF, 2020). Investors and regulators' increasing focus on environmental, social, and governance (ESG) criteria further highlights the importance of sustainability in real estate.

As shown in Figure 1.3, sustainability can be integrated throughout the five stages of the modern real estate lifecycle. At the planning stage, due diligence can be expanded to include key sustainability factors, which involves a comprehensive approach considering environmental, social, and economic factors. This process begins with thoughtful site selection, prioritising locations that minimise ecological impact and leveraging existing infrastructure. Developers should focus on integrating renewable energy systems, implementing water conservation strategies, and designing for efficient waste management. Green spaces and biodiversity preservation should be prioritised, along with provisions for sustainable transportation options. Community engagement and mixed-use development concepts can enhance social sustainability.

As we progress through this book, we will explore how emerging technologies like PropTech and blockchain intersect with and enhance sustainability efforts in real estate. Sustainability is a foundational concept that underpins many of the technological advancements and innovations we will discuss in subsequent chapters. The real estate industry's commitment to sustainability is not just an ethical imperative but a business necessity. It drives value creation, mitigates risks, and positions the sector to thrive in a future where environmental and social responsibility are paramount. As the industry evolves, sustainability will remain critical in shaping the built environment and ensuring the long-term success of real estate assets and portfolios.

1.6 Overview of the Book's Objectives and Structure

This book aims to provide a comprehensive exploration of the intersection between technology, innovation, sustainability, and the real estate industry. The primary objectives of this book are as follows:

- To examine the transformative impact of PropTech and blockchain technology on the real estate sector
- To explore the critical role of sustainability in shaping the future of real estate
- To analyse the integration of PropTech strategies in sustainable real estate practices
- To provide empirical insights into the real-world applications and impacts of PropTech in the context of sustainability
- To investigate the drivers and challenges within the PropTech ecosystem
- To envision the future trajectory of PropTech in fostering sustainable real estate development

The book is structured to provide a logical progression from foundational concepts to advanced applications and outlook.

Chapter 1: Background and Introduction

This chapter provides a comprehensive overview of the real estate industry, introduces key concepts of innovation and digital transformation, and lays the theoretical groundwork for understanding the adoption of technology in the real estate sector.

Chapter 2: Transformations of Real Estate: PropTech, Blockchain, and Sustainability

Building upon the historical context, this chapter navigates through the transformative forces reshaping real estate, examining the symbiotic relationship between PropTech, blockchain, and sustainability.

Chapter 3: Sustainability in the Real Estate Industry: From Green Building to Sustainable Real Estate

This chapter explores the evolution of sustainable practices within real estate, from green building initiatives to the broader concept of sustainable real estate, including the emergence of ESG factors.

Chapter 4: PropTech in Real Estate Investment and Finance

Delving into the innovative realm of PropTech, this chapter examines its impact on real estate investment and finance, covering topics from digital platforms to blockchain integration and emerging technologies.

Chapter 5: PropTech Strategies in Sustainable Real Estate

This chapter explores strategies that leverage technology to drive sustainable real estate initiatives, examining data-driven decision-making, blockchain applications, and innovative financing models.

Chapter 6: PropTech and Sustainability

Through empirical analysis, this chapter examines the real-world impact of PropTech on sustainability within the real estate domain, presenting case studies, data-driven insights, and tangible examples.

Chapter 7: Drivers and Challenges for PropTech Ecosystem

This chapter provides an empirical examination of the factors driving the growth of the PropTech ecosystem and the challenges it faces, drawing on real-world scenarios and industry experiences.

Chapter 8: The Future of PropTech for Sustainable Real Estate

The final chapter envisions the future trajectory of PropTech in fostering sustainability, exploring emerging trends, potential disruptions, and the transformative possibilities that lie ahead.

Chapter 9: Conclusions

This chapter synthesises the insights gathered throughout the book, drawing comprehensive conclusions about the symbiotic relationship between PropTech, real estate investment, finance, and sustainability.

Throughout the book, we apply theoretical frameworks such as the diffusion of innovations theory and the information technology adoption model to understand the adoption and impact of these technologies in the real estate sector. Each chapter concludes with engaging activities designed to encourage readers to apply the knowledge gained.

By the end of this book, readers will have a comprehensive understanding of how PropTech. blockchain and AI are reshaping the real estate industry, the critical role of sustainability in this transformation, and the challenges and opportunities that lie ahead in this rapidly evolving landscape.

References

Baum, A., Saull, A., & Braesemann, F. 2020. *PropTech 2020: The future of real estate*. Oxford Future of Real Estate Initiative, University of Oxford.

Davis, F. D. 1989. Perceived usefulness, perceived ease of use, and user acceptance of information technology. *MIS Quarterly*, 13, 319–340.

Dixon, D. R. 1999. The behavioral side of information technology. *International Journal of Medical Informatics*, 56, 117–123.

GlobalABC. 2020. *Global status report for buildings and construction*. Global Alliance for Buildings and Construction.

JLL. 2021. *The impact of sustainability on value*. Jones Lang LaSalle IP.

Rogers, E. M. 1962. *Diffusion of innovations*. The Free Press of Glencoe.

Saiz, A. 2023. *The global housing affordability crisis: Policy options and strategies*. MIT Center for Real Estate Research Paper.

Shaw, J. 2020. Platform real estate: Theory and practice of new urban real estate markets. *Urban Geography*, 41, 1037–1064.

Statista. 2024. *Statista market insights*. Author.

WEF. 2020. *The real estate industry's transparency challenge—and how it can solve it*. World Economic Forum.

WEF. 2021. *A framework for the future of real estate*. World Economic Forum.

Chapter 2

Transformations of Real Estate

PropTech, Blockchain, and Sustainability

2.1 Evolving Real Estate Landscape: Digital Innovations, Blockchain, and Sustainability

2.1.1 The Evolution of the Real Estate Landscape

Real estate has evolved through history due to tech, economic, social, and regulatory changes (Figure 2.1). In pre-industrial times, land ownership meant agricultural productivity, wealth, and power. This land–agriculture connection persisted for centuries.

The industrial revolution triggered urban migration as people sought employment in cities. This population shift created housing demands that shaped construction methods and urban planning. Meanwhile, growing industries required factories, warehouses, and offices, establishing commercial real estate as a distinct category from residential properties and fostering more structured development approaches.

After World War II, housing and commercial space demands fueled a building boom. Innovations such as prefabrication and modern materials reduce construction time and costs, enabling large-scale housing and commercial developments that transform both landscapes and development practices.

The rise of PropTech in the past decade marked a new era in the real estate industry, introducing innovative solutions designed to enhance efficiency, improve customer experience, and streamline transactions. This technological revolution brought greater transparency, accessibility, and convenience to the real estate experience, reshaping the way properties are bought, sold, and managed.

The COVID-19 pandemic significantly accelerated the trends related to remote work and the wellness of living spaces. The widespread adoption of remote work practices led to a decrease in demand for traditional office spaces, while the need for comfortable and functional home offices increased. The pandemic accelerated the shift towards e-commerce, creating significant challenges for brick-and-mortar retail establishments. Many retailers faced closures due to reduced foot traffic and reliance on online shopping. Consequently, businesses are now focusing on omnichannel strategies, blending in-person and digital experiences to adapt to changing consumer behaviours. There has also been an increasing focus on creating multi-functional spaces that can easily adapt to different uses, such as co-working environments, co-living arrangements, and traditional residential setups. This flexibility allows properties to accommodate varying demands, making them more resilient to future market fluctuations.

DOI: 10.1201/9781003504436-2

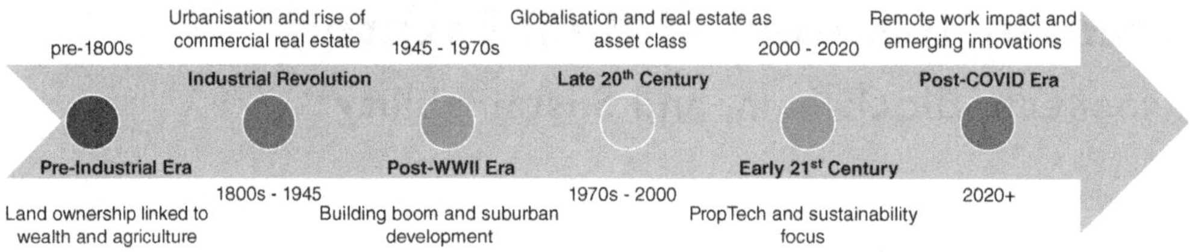

Figure 2.1 Evolution of the Real Estate Industry.

By author (2024).

The growing awareness of climate change and the urgency to address environmental concerns have placed a heightened emphasis on sustainability and green building practices in the real estate sector. Developers are increasingly incorporating sustainable materials, energy-efficient designs, and renewable energy sources into their projects.

2.1.2 *The Impact of Digital Innovations*

Digital innovations are reshaping the real estate landscape by creating a more efficient, transparent, and consumer-centric real estate industry. The industry has become increasingly data-driven, with technology playing a critical role in shaping the future of real estate.

In property search and marketing, virtual and augmented reality tours have revolutionised how potential buyers view properties, allowing for immersive remote experiences. AI-powered recommendation engines have enhanced property matching, making the search process more efficient and personalised. For instance, Zillow[1] reported a 600% increase in 3D home tours during the pandemic, highlighting the growing importance of these technologies.

The realm of transactions and financing has also seen significant changes. Digital platforms now streamline purchase processes and documentation, reducing paperwork and speeding up closings. Blockchain technology has emerged as a game-changer, enabling smart contracts and fractional ownership. In 2017, Propy[2] made history by completing the first blockchain-based real estate transaction, paving the way for more secure and transparent property transfers.

Property management has been revolutionised by the Internet of Things (IoT) and smart home technologies. These innovations optimise building operations, from energy usage to security systems. AI-driven predictive maintenance is reducing costs and improving efficiency by anticipating and addressing issues before they become significant problems. Honeywell's Forge platform[3] exemplifies this trend, using machine learning to reduce energy consumption by up to 25% in commercial buildings.

Data analytics and valuation have become more sophisticated with the advent of big data and machine learning. These technologies provide deeper insights into market trends and property values, enhancing decision-making for investors and developers. Automated valuation models (AVMs) have become increasingly accurate (Xiong et al., 2021), with companies like HouseCanary[4] claiming a median error rate of just 2.5% for their AI-powered valuation model.

In construction and development, building information modeling (BIM) has improved design and construction processes, allowing for better coordination and reduced errors. 3D printing and modular construction techniques are pushing the boundaries of what's possible, reducing construction time and costs. WinSun[5] demonstrated the potential of 3D printing in construction by printing 10 houses in just 24 hours using recycled materials (Mohr & Khan, 2015).

Energy management and sustainability have become key focus areas, with smart meters and AI-driven energy management systems optimising resource consumption in real time. Products like Nest's learning thermostat, which can reduce heating and cooling bills by 10–12%, showcase how technology can contribute to both cost savings and environmental sustainability (Ayan & Turkay, 2018).

The tenant experience has been enhanced through mobile apps that facilitate communication between tenants and landlords, streamlining service requests and providing real-time updates. AI chatbots now offer 24/7 customer service for property-related queries, improving responsiveness and tenant satisfaction. Platforms like Livly[6] have reported significant increases in resident engagement through their tenant experience apps.

Space utilization, particularly in commercial real estate, has been optimized through sensors and analytics tools. These technologies help property managers make data-driven decisions about space allocation and usage. The flexible workspaces also leverage technology to enable on-demand booking and efficient space management (Richardson, 2021).

Real estate investment has been democratized through online crowdfunding platforms, allowing smaller investors to participate in previously inaccessible opportunities. AI-driven analytics help investors make informed decisions based on comprehensive market data. Fundrise,[7] with over 150,000 active investors and USD 5 billion in real estate assets, demonstrates the growing popularity of these platforms.

Property insurance and risk assessment have been transformed by technologies like drones and satellite imagery, which improve property inspection processes. AI algorithms analyze historical data to predict and price insurance risks more accurately. Insurtech companies like Lemonade[8] are leveraging AI to process claims in record time.

Finally, compliance and regulatory tech are streamlining adherence to zoning laws and building codes. Automated systems and blockchain-based solutions are enhancing transparency in regulatory reporting. Companies like Buildrz[9] are using AI to automate zoning analysis, reducing processes that once took weeks to mere minutes.

These digital innovations enhance efficiency, reduce costs, and create new business models and revenue streams within the real estate industry. As technology continues to evolve, we can expect further disruption and transformation across all aspects of the real estate sector, from how properties are built and managed to how they are bought, sold, and experienced by occupants.

2.1.3 Blockchain's Potential in Real Estate

Blockchain has great potential to be applied in real estate (as shown in Figure 2.2).

Blockchain technology, also known as distributed ledger technology (DLT), has emerged as a potentially transformative force in the real estate industry. At its core, blockchain offers a secure, transparent method of recording and transferring data across a distributed network of computers.

The key features that make blockchain particularly relevant to real estate include its distributed structure, which eliminates single points of failure; enhanced transparency through immutable record-keeping; cryptographic security that protects against fraud; and the capability for smart contracts that can automate various processes.

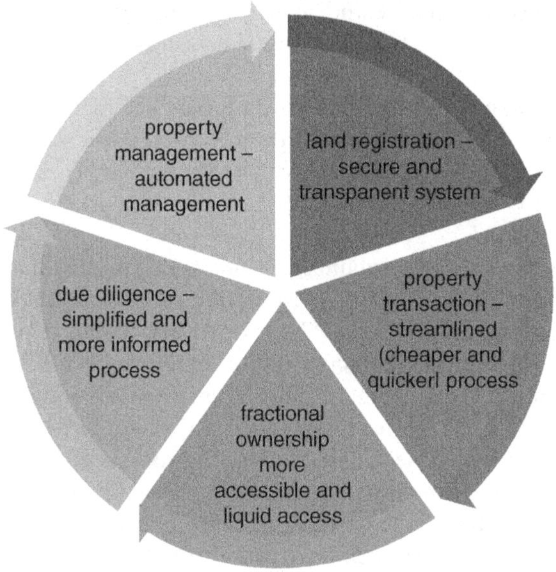

Figure 2.2 Potential Application of Blockchain in Real Estate.

By author (2024).

In real estate, blockchain has several promising applications. It could revolutionise land registration by creating more secure and accessible property records. The technology could streamline property transactions through automated processes and reduce paperwork. Tokenisation—the digital representation of real estate assets—could enable fractional ownership and potentially increase market liquidity. Additionally, blockchain could simplify due diligence processes and enhance property management through automated leases and maintenance records.

While these applications show significant promise, blockchain implementation in real estate faces challenges including regulatory uncertainty, integration with existing systems, and industry adoption hurdles. These applications and challenges will be explored in greater detail in Section 2.3, which provides a comprehensive examination of blockchain's role in transforming various aspects of the real estate industry.

2.1.4 The Growing Importance of Sustainability in Real Estate

Sustainability has become increasingly important in shaping the future of the real estate industry. As awareness of environmental challenges grows and regulatory pressures increase, the real estate sector is increasingly focusing on sustainable practices in development, management, and investment strategies. As shown in Figure 2.3, several key areas have been frequently discussed in relation to the sustainability of real estate.

1 *Green Building Practices:* Green building practices have become a cornerstone of sustainable real estate, emphasising energy efficiency, water conservation, and the use of sustainable materials in construction and renovation. New developments increasingly

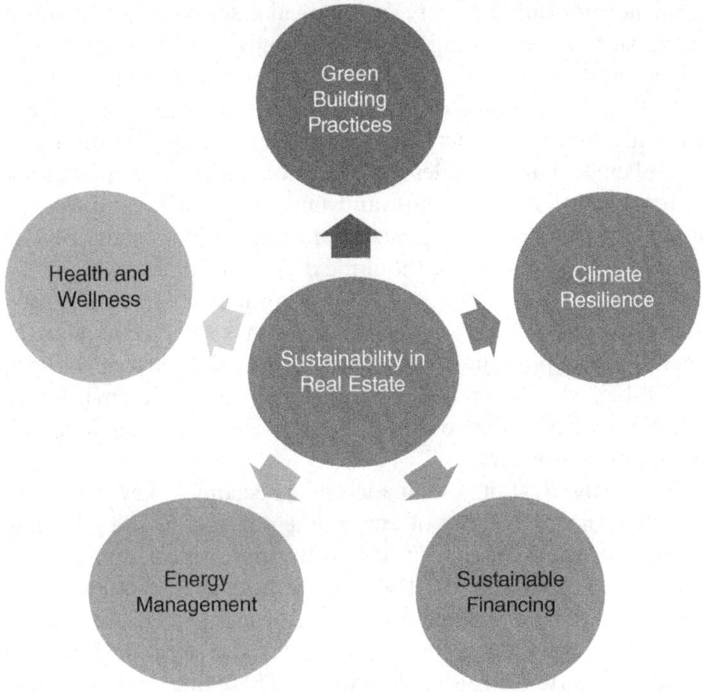

Figure 2.3 Key Areas of Focus in Sustainability of Real Estate.
By author (2024).

incorporate energy-efficient designs and technologies, such as improved insulation, high-performance windows, and advanced heating, ventilation, and air conditioning (HVAC) systems, which not only reduce energy consumption but also lower operational costs. The use of sustainable construction materials, including recycled, reclaimed, and eco-friendly options, is gaining popularity as builders opt for materials with lower environmental impact that contribute to healthier indoor environments. Governments are supporting this trend by implementing stricter building codes that mandate sustainable practices, including energy efficiency and reductions in emissions. Additionally, certification programs like LEED, BREEAM, and Green Star help property owners demonstrate their commitment to sustainability, often enhancing property values and marketability.

2 *Climate Resilience:* Climate resilience has emerged as a critical focus in sustainable real estate, addressing the need to design and retrofit buildings to withstand extreme weather events and the long-term impacts of climate change. This approach involves incorporating features that enhance a building's ability to adapt to and recover from climate-related stresses. Strategies may include elevated construction in flood-prone areas, reinforced structures to withstand high winds, and the use of fire-resistant materials in wildfire-prone regions. Climate-resilient designs also consider future climate scenarios, incorporating flexible systems that can adapt to changing conditions over time. This

focus on resilience not only protects the physical asset but also ensures business continuity and occupant safety, making it an increasingly important consideration for investors, developers, and property owners in the face of growing climate uncertainty.

3 *Health and Wellness*: The health and wellness aspect of sustainable real estate has gained significant attention, with an increasing focus on creating environments that promote occupant well-being. This includes emphasis on indoor air quality using non-toxic materials, proper ventilation systems, and maximization of natural lighting. Designs that encourage physical activity, such as prominent staircases and on-site fitness facilities, are becoming more common. Biophilic design principles, which incorporate natural elements into built environments, are also being adopted to reduce stress and improve cognitive function. The WELL Building Standard has emerged as a leading certification program in this area, providing guidelines for creating spaces that enhance human health and well-being. As awareness of the link between built environments and health grows, properties that prioritize wellness features are seeing increased value and attractiveness to tenants and buyers.
4 *Sustainable Financing*: Sustainable financing has become a key driver in the growth of green real estate projects. The rise of environmental, social, and governance (ESG) criteria in investment decision-making has led to increased demand for properties that meet high sustainability standards. These properties are often viewed as offering better returns and reduced risks, further promoting investment in green real estate. Green bonds and sustainability-linked loans have emerged as popular financing options for developers looking to fund sustainable projects. These financial instruments often offer favourable terms for projects that meet specific environmental criteria, encouraging investment in energy-efficient and environmentally friendly constructions. The growth of sustainable finance is not only providing the necessary capital for green development but also creating a positive feedback loop, where increased funding availability is driving more developers to incorporate sustainability features into their projects.
5 *Energy Management*: Energy management has become a crucial aspect of sustainable real estate, focusing on the implementation of smart technologies and renewable energy solutions to optimize resource consumption. Real estate developers are increasingly incorporating on-site renewable energy sources, such as solar panels and wind turbines, into their projects. This approach not only reduces reliance on fossil fuels but can also lead to substantial long-term savings on energy costs. Smart building technologies, including IoT devices and AI-driven systems, are being deployed to monitor and optimize energy usage in real time. These systems can adjust lighting, heating, and cooling based on occupancy and external conditions, significantly reducing energy waste. Furthermore, properties designed with energy resilience in mind are better equipped to withstand energy crises, natural disasters, and other disruptions. This resilience is becoming an important consideration for investors and developers, recognizing the long-term value and operational stability it provides to real estate assets.

These sustainability initiatives are not only addressing environmental concerns but are also creating new opportunities for innovation, cost savings, and value creation in the real estate sector. A more comprehensive discussion of sustainable development principles and their application in real estate will be presented in Section 2.4.

2.2 PropTech Development Phases

2.2.1 *The Evolution of PropTech*

The term "PropTech" initially emerged to describe the technological innovations aimed at improving the real estate industry in the early 2010s. While its exact origins are difficult to pinpoint, the term gained widespread usage around 2013–2014, marking the beginning of a new era in real estate.

PropTech's roots, however, stretch back much further. In the late 20th and early 21st centuries, several technological solutions laid the groundwork for what would later be recognized as PropTech. Online real estate listings, for instance, revolutionized property searches. Zillow,[10] launched in 2006, and Realtor.com,[11] dating back to 1996, were pioneers in this space, offering consumers unprecedented access to property information and market trends.

Property management also saw early technological adoption. Yardi,[12] founded in 1984, and Buildium,[13] established in 2004, introduced software solutions that streamlined operations for landlords and property managers. These platforms digitized tasks such as tenant tracking, maintenance management, and accounting, setting the stage for more advanced PropTech solutions.

The real estate analytics sector similarly experienced early tech integration. CoStar Group,[14] founded in 1987, and Real Capital Analytics (RCA),[15] established in 2000 and acquired by MSCI later, began providing crucial data and market insights for commercial real estate professionals. These platforms enabled more informed decision-making, a cornerstone of modern PropTech.

Other notable precursors to PropTech include virtual tour technologies, online transaction platforms like DocuSign,[16] and customer relationship management tools adapted for real estate use. Even crowdfunding platforms like Fundrise[17] emerged as the concept of PropTech was taking shape, illustrating the industry's readiness for technological disruption.

One particularly significant development was building information modeling (BIM). The concept of BIM, which allows for the creation of comprehensive digital representations of physical and functional characteristics of buildings, began to take form in the 1970s and 1980s. By the 1990s, software solutions like ArchiCAD[18] and Revit[19] were incorporating BIM principles, revolutionizing architectural design and construction processes.

2.2.2 *The PropTech Wave: From 1.0 to 3.0 and Beyond*

The evolution of PropTech can be divided into phases (shown in Figure 2.4). Andrew Baum et al. (2020)'s influential work provides a framework to understand this evolution.

2.2.2.1 PropTech 1.0

This phase marked the initial digitisation of traditional processes in the real estate industry. PropTech 1.0 primarily focused on automating existing workflows rather than fundamentally changing business models. It laid the groundwork for efficiency gains but was limited by the technological capabilities of the time.

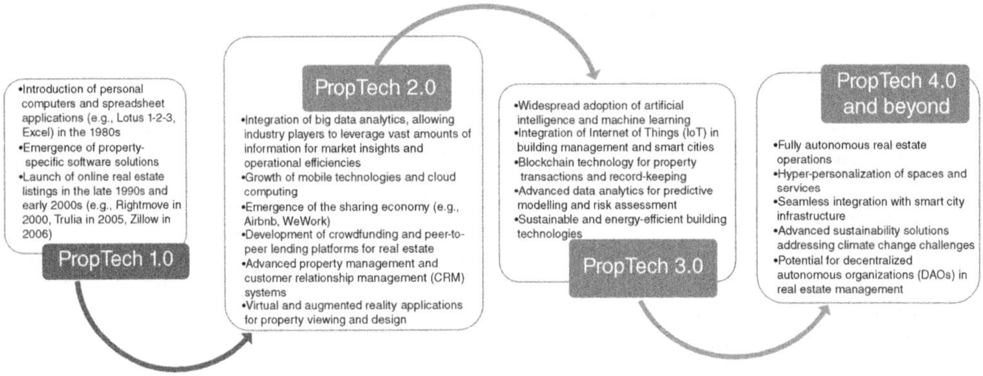

Figure 2.4 Evolution of PropTech Wave.

By author (2024).

2.2.2.2 PropTech 2.0

PropTech 2.0 saw a leap in innovation and the introduction of more advanced tools. Big data analytics became central, allowing industry players to leverage vast amounts of information for market insights and operational efficiencies. Customer relationship management systems and sophisticated property management software reshaped how real estate professionals interacted with clients and managed properties. This phase is marked by innovation and the introduction of more sophisticated tools. PropTech 2.0 is transforming the way real estate professionals interact with clients, manage properties, and make informed investment decisions. It is characterised by a more disruptive approach to traditional real estate practices.

2.2.2.3 PropTech 3.0

Andrew Baum et al. (2020) imagines PropTech 3.0 as a deeper transformation characterised by the convergence of various advanced technologies. This phase is expected to fundamentally change investment strategies, property management, and tenant experiences. Smart buildings are emerging as a significant trend, integrating digital technology to optimize energy usage, enhance security, and improve occupant comfort.

2.2.2.4 PropTech 4.0 and Beyond

While not explicitly defined in Andrew Baum et al. (2020)'s work, the concept of PropTech 4.0 and beyond suggests a future of fully autonomous real estate operations, hyper-personalisation, and seamless integration with smart city infrastructure. This vision points to a future where technology not only drives efficiency but also creates sustainable and adaptive urban environments.

The future of PropTech suggests a pathway where technology drives efficiency and fosters sustainable, adaptable urban environments.

2.2.3 *The Post-COVID Landscape: Consolidation and Co-Creation*

The COVID-19 pandemic has accelerated certain PropTech trends and reshaped the industry's trajectory. In the post-pandemic era, the PropTech landscape is entering a consolidation phase, with a shift from rapid expansion to a focus on quality and effectiveness.

Companies are now prioritising the refinement of their core solutions, investing in reliability and features that truly address client needs. This trend is accompanied by increased consolidation within the market, as larger firms acquire smaller startups to integrate innovative technologies and broaden their service offerings. This consolidation is driven by the desire for increased market reach and the ability to offer more comprehensive, integrated solutions.

An interesting trend in this new phase is the emphasis on developing integrated systems that operate seamlessly across various platforms and technologies. These integrated solutions aim to provide holistic approaches to property management, tenant experiences, and building operations, enhancing user experience and usability.

Simultaneously, there is a growing focus on co-creation in PropTech development. This approach involves collaboration among PropTech companies, end-users (such as tenants and property managers), and real estate corporations in the development of the solution. By directly incorporating insights from those who will use these technologies, companies can better address real-world challenges and user preferences.

This co-creation model encourages the sharing of data and insights among stakeholders, leading to more informed decision-making and effective solutions. It also prioritises real-world testing and experimentation, allowing for continuous refinement based on user feedback.

The shift towards consolidation and co-creation in the PropTech sector represents a maturation of the industry. By leveraging diverse expertise and fostering cooperative relationships, stakeholders are driving innovation, enhancing service delivery, and contributing to a more sustainable and user-centric future in real estate.

2.3 Blockchain's Application in Real Estate

2.3.1 *Blockchain (Distributed Ledger Technology)*

Blockchain technology, which is also known as distributed ledger technology (DLT), has emerged as a potentially transformative force in various industries, including real estate. DLT refers to a system of recording information across multiple locations, ensuring accuracy, security, and transparency (Crosby et al., 2016). Unlike traditional centralised systems, DLT allows multiple parties to access and update a shared database without relying on a single authority (Farahani et al., 2021). Five core characteristics of DLT are relevant to real estate.

- *Decentralisation:* The ledger is distributed across a network of computers (nodes), eliminating single points of control or failure. This structure enhances security and resilience, as there is no central point of vulnerability (Nakamoto, 2008).
- *Transparency:* All transactions or data entries are visible to authorised participants in the network, fostering an environment of trust and accountability (Tapscott & Tapscott, 2016).

- *Immutability:* Once recorded, it's extremely difficult to alter or delete information without detection. This feature creates a reliable, tamper-proof audit trail (Crosby et al., 2016).
- *Consensus Mechanisms:* The network uses various methods to reach agreement on the state of the ledger among participants, maintaining data integrity across the network (Viriyasitavat & Hoonsopon, 2019).
- *Cryptographic Security:* DLT uses advanced cryptographic techniques to secure data and transactions, enhancing protection against fraud and unauthorised access (Crosby et al., 2016).

By utilising these characteristics, DLT has a few potential impacts on the real estate industry. First, the enhanced security and reduced fraud risk could significantly benefit the real estate sector, which has historically been vulnerable to various forms of fraud (Deloitte, 2017). The immutable and cryptographically secure nature of DLT could provide a more reliable system for property records and transaction histories. Second, DLT could increase efficiency in real estate transactions. By streamlining processes and potentially reducing the need for intermediaries, DLT could speed up transactions and reduce associated costs (Veuger, 2018). Smart contracts, self-executing contracts with terms directly written into code, could automate various real estate processes such as property transfers, rental agreements, and escrow services (Dutta, 2020). Third, the transparent nature of DLT could increase trust and accountability in real estate dealings. All authorised participants would have access to the same information, potentially reducing disputes and increasing market confidence (Tapscott & Tapscott, 2016). Finally, blockchain technology enables the tokenisation of real estate assets, potentially allowing for fractional ownership and increased market liquidity (Veuger, 2018). This could democratise real estate investment, making it accessible to a broader range of investors.

The potential applications of blockchain technology are briefly outlined in Section 2.1.3. In Sections 2.3.2–2.3.6, the detailed scenarios of applying blockchain technology in real estate will be explained and demonstrated.

2.3.2 Blockchain for Real Estate Titles

Blockchain technology offers a revolutionary approach to managing real estate titles and land registries. Several jurisdictions have applied blockchain in real estate title management or land registry. These initiatives aim to enhance transparency, security, and efficiency in land and property registration processes. Before showing specific examples, we want to demonstrate the mechanism and process of applying blockchain in real estate titles.

As illustrated in Figure 2.5, the mechanism of using blockchain technology for land registry started from the technology foundation, the blockchain, which serves as a decentralised ledger. Each property is represented by a unique digital identifier linked to all relevant information. Smart contracts, depicted as a key component, automate processes like property transfers and lease agreements. The system ensures immutable record-keeping, with each transaction or status change recorded as a new block, creating a tamper-proof history. Decentralised verification, represented in the diagram, replaces central authorities with collective validation by the network, enhancing security. Public key cryptography provides secure access and transaction authorisation for property owners. The integration layer at the bottom shows how blockchain systems connect with existing land registry databases, ensuring a smooth transition and maintaining compatibility with current

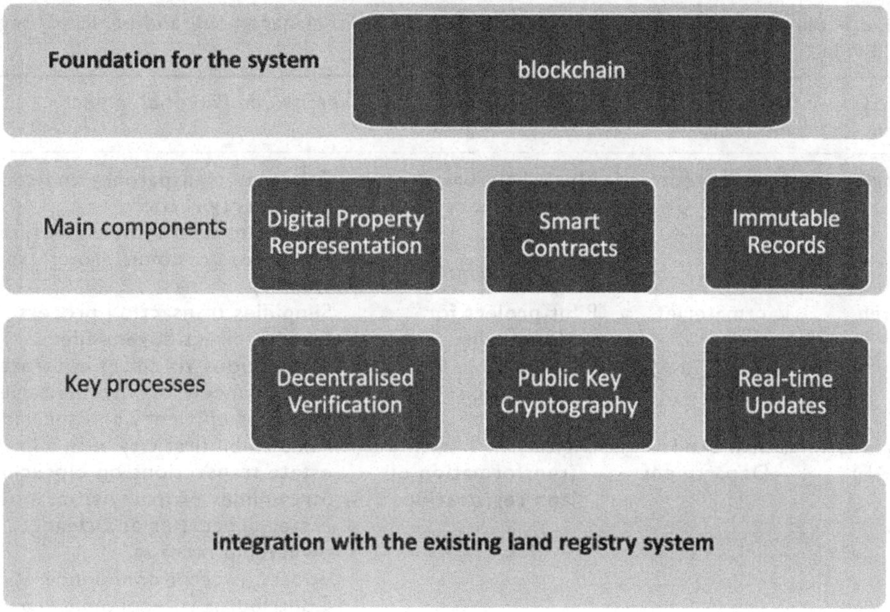

Figure 2.5 Blockchain Mechanism for Land Registry.

By author (2024).

frameworks. While not explicitly shown, the system incorporates permission access to balance transparency with privacy. Finally, the real-time updates component illustrates how changes in property status are immediately reflected across the entire network, eliminating delays and reducing conflicts. This comprehensive approach transforms traditional land registry systems, offering enhanced security, efficiency, and transparency in property management.

The implementation of a blockchain-based land registry system follows a structured process. It begins with the digitisation of existing records, converting paper-based documents into a digital format and ensuring data accuracy through verification and cleaning. Next, an appropriate blockchain platform is selected based on specific needs such as scalability, security, and functionality. Smart contracts, which govern property transactions and record-keeping, are then developed and rigorously tested. The blockchain system is integrated with existing land registry databases and relevant government systems to ensure seamless operation. User-friendly interfaces are created to allow property owners, government officials, and other stakeholders to interact effectively with the system. Before full deployment, thorough testing and pilot implementations are conducted to identify and address any issues. Finally, relevant personnel are trained, and the system is gradually rolled out, often starting with a specific region or property type. This systematic approach ensures a smooth transition from traditional to blockchain-based land registry systems, maximising the benefits of the new technology while minimising disruption to existing processes.

The benefits of using blockchain in land registries are clear: It helps to reduce the risk of fraud and unauthorised alterations. All stakeholders can access and verify property

Table 2.1 Blockchain's Application in Real Estate Title Management and/or Land Registry Worldwide

Country/ Region	*Implementing Authority*	*Initiative*	*Benefits and Potential Impact*
Georgia	Government and Bitfury Group	Blockchain-based land registry system	- Enhanced transparency and security of property records - Addresses land disputes and fraud - Promotes economic growth and stability
Sweden	Lantmäteriet	Pilot project for blockchain property transactions	- Simplifies transaction process - Enables direct buyer-seller transactions via smart contracts - Reduces need for intermediaries - Improves efficiency and security
Dubai (UAE)	Dubai Land Department	Digital transformation of land registration	- Aims to be first city with all real estate transactions on blockchain - Streamlines property transactions - Faster processing and clear ownership records - Boosts investor confidence - Contributes to economic growth
British Columbia (Canada)	Land Title and Survey Authority	Pilot projects for blockchain in land title management	- Enhances efficiency and security of property transfers - Improves title verification processes - Benefits both buyers and sellers
India	Various state governments	Pilot projects (including Andhra Pradesh)	- Addresses land title disputes and fraud - Creates transparent and secure registry system - Streamlines transaction processes - Potential to transform real estate sector

records. It offers faster processing of property transactions and record updates, which reduces administrative costs. The clear, verifiable ownership records minimise property disputes. Ultimately, the enhanced trust in the property market can attract more investment.

Despite the challenges in the regulatory framework, digital literacy, data privacy, and scalability issues, there are a handful of jurisdictions piloting blockchain's application in real estate title management or land registry (Table 2.1).

2.3.3 Smart Contracts for Real Estate Transactions

Smart contracts are self-executing contracts with the terms of the agreement directly written into code and stored on a blockchain. They automatically enforce and execute actions when predefined conditions are met. In the real estate sector, smart contracts provide innovative solutions that can streamline processes, enhance transparency, and lower transaction costs in various types of real estate transactions (as shown in Figure 2.6).

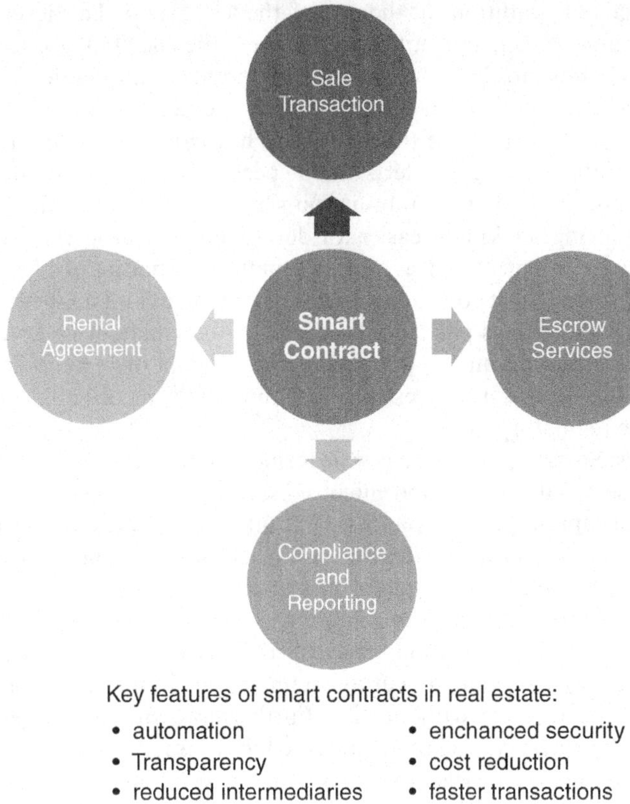

Figure 2.6 Smart Contract Application in Real Estate Transactions.

By author (2024).

1 *Property Sale Transactions:* Smart contracts have revolutionised property sale transactions by streamlining the process and significantly reducing costs for both buyers and sellers. By automating many aspects of the transaction, smart contracts eliminate the need for costly intermediaries, thereby reducing expenses associated with traditional real estate deals. This automation also minimises the potential for errors or fraud that can occur due to human intervention, making the entire process more reliable and efficient. Furthermore, smart contracts leverage cryptographic security to ensure that transactions are tamper-proof. This enhanced security builds trust and confidence in the process, as all parties can be assured that the terms of the agreement will be executed exactly as specified, without the risk of manipulation or unauthorised changes.

2 Rental Agreements: Rental agreements, as another form of transaction in real estate, can also utilise smart contracts to automate processes such as rent collection and security deposit handling. This automation simplifies the complex interactions between tenants and landlords, reducing the likelihood of misunderstandings or disputes. By clearly defining and automatically executing the terms of the lease, smart contracts promote transparency in the rental relationship. For instance, rent payments can be automatically deducted on specified dates, and security deposits can be automatically returned

based on predefined conditions at the end of the lease term. This level of automation and clarity in rental agreements not only improves efficiency but also fosters better tenant–landlord relationships by reducing common points of contention.

3 *Compliance and Reporting:* Smart contracts play a crucial role in simplifying compliance and reporting in real estate transactions. They can be programmed to automatically generate reports or trigger alerts when specific conditions are met or regulatory thresholds are approached. This automation significantly reduces the risk of errors in compliance reporting, making it easier for developers and property managers to meet their regulatory requirements. Moreover, the immutable nature of blockchain technology, which underpins smart contracts, creates a permanent and unalterable record of all transactions. This feature is particularly valuable for regulatory compliance, as it provides a transparent and tamper-proof audit trail. Regulators can easily verify compliance and conduct audits by accessing this immutable record, increasing trust and efficiency in the regulatory process.
4 Escrow Services: Smart contracts have transformed escrow services in real estate transactions by introducing automation and enhanced security. Traditional escrow services often involve time-consuming processes and significant costs. However, smart contracts can automate these services, ensuring that funds are held securely until all predefined conditions of the sale are met. For example, a smart contract can be programmed to automatically release funds to the seller once conditions such as successful property inspections, title clearance, and fund verification are satisfied. This automation builds trust between buyers and sellers by providing a neutral, technology-driven intermediary that executes the terms of the agreement without bias. Furthermore, automating escrow processes through smart contracts can significantly expedite transactions. By eliminating manual processing and reducing the need for human intervention, smart contracts can complete escrow processes much faster than traditional methods. This increased speed and reduced human involvement also lead to lower costs than traditional escrow services, making real estate transactions more efficient and economical for all parties involved.

Being in the early stage, the applications of blockchain in real estate transactions are not widely used yet; however, there are some cases that show the potential. In this section, we will use Propy [20] as an example.

Case Study—Propy: Betting on Blockchain to Revolutionise Global Real Estate Transactions

Natalia Karayaneva, CEO of Propy, saw an opportunity to transform the industry using blockchain technology. In 2016, she founded Propy to make real estate transactions faster, easier, and more secure, not just domestically but on a global scale.

PROPY AND ITS BUSINESS MODEL

Propy was founded as a real estate transaction platform that leverages blockchain technology to streamline property purchases. The company's core offerings include:

1 *Offer Management:* A system for agents to field and manage offers efficiently.
2 *Transaction Platform:* A streamlined process for handling real estate transactions from offer to closing.

3 *Title and Escrow:* Services to facilitate fast and secure closings.
4 *Blockchain Integration*: Recording transactions on blockchain for enhanced security and transparency.
5 *Cryptocurrency and NFT (Non-Fungible Token) Capabilities:* Enabling property purchases using cryptocurrency and selling properties as NFTs.

Propy's business model is based on subscription fees for real estate professionals, with different tiers offering various features and benefits. The company also generates revenue through its involvement in NFT property sales.

Propy began rolling out its platform to real estate professionals in 2018. The company's innovative approach attracted a diverse range of early adopters, from individual agents to large brokerages, each finding unique value in the platform.

Propy's success can be attributed to its innovative approach in addressing real market needs within the real estate industry. By leveraging blockchain technology, Propy created a secure, transparent, and efficient platform for real estate transactions. This innovation directly addressed key issues, including privacy concerns, fraud prevention, and streamlined documentation processes, which were particularly significant for international and high-net-worth clients.

The adaptability of Propy's solution to changing market conditions further contributed to its success. The platform proved especially valuable during the COVID-19 pandemic when in-person transactions became challenging. Propy's system allowed for secure, remote closings of deals, maintaining business continuity in a disrupted market. This adaptability, coupled with the platform's enhanced security and privacy features, made it particularly attractive to high-net-worth international clients who required a secure environment for exchanging sensitive documents and conducting transactions.

By effectively addressing these key areas and demonstrating scalability in handling both residential and commercial transactions of varying scales, Propy has positioned itself as a pioneer in applying blockchain technology to real estate transactions. Its pioneering success stems not just from having innovative technology but from effectively applying this technology to solve real-world problems in the real estate industry, particularly for international and high-value transactions.

2.3.4 Real Estate Tokenisation

Real estate tokenisation refers to the process of converting rights to a real estate asset into digital tokens on a blockchain. This innovative approach is reshaping how real estate assets are owned, traded, and managed, offering potential solutions to long-standing issues in the market, such as illiquidity, high transaction costs, and limited accessibility (Baum, 2021; Chow & Tan, 2022).

2.3.4.1 Concept, Mechanism, and Proces

Tokenisation involves creating digital representations of real estate assets, typically through security tokens or asset-backed tokens. These tokens can represent full ownership of a property, fractional ownership, or even specific cash flows generated by the property (Mottaghi et al., 2024). The underlying blockchain technology ensures transparency, immutability of records, and potential for programmable features through smart contracts. The process of real estate tokenisation typically involves several key steps (shown in

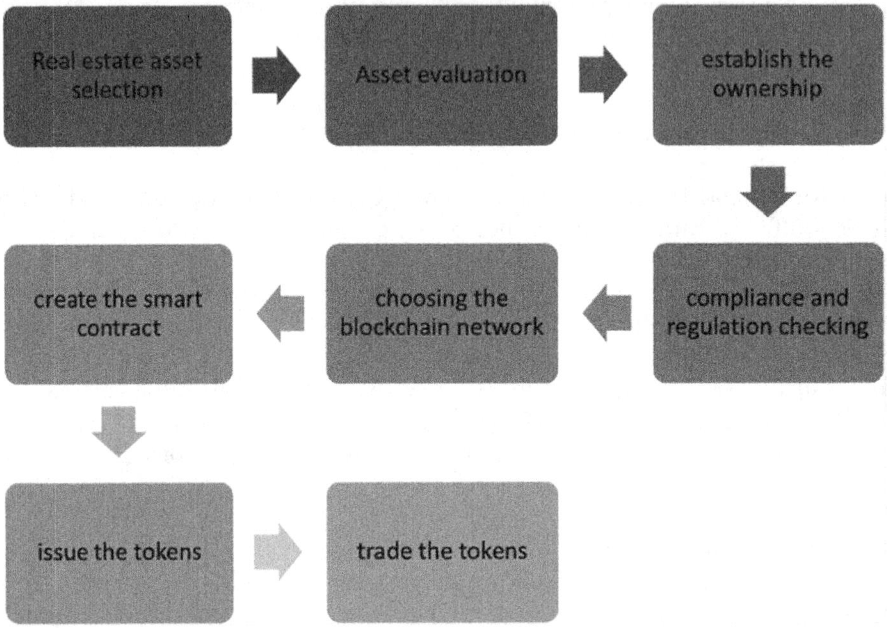

Figure 2.7 Process of Real Estate Tokenisation.

By author (2024).

Figure 2.7), throughout which various technological and legal considerations come into play, including ensuring the security of the blockchain platform, maintaining regulatory compliance, and establishing clear mechanisms for managing the underlying real estate asset.

2.3.4.2 Benefits

Real estate tokenisation offers a multitude of benefits that are reshaping the landscape of property investment. One of the most significant advantages is increased accessibility, as the lowering of minimum investment amounts allows a broader range of investors to participate in real estate markets that were previously out of reach (Chow & Tan, 2022). This democratisation of real estate investment is closely tied to enhanced liquidity, as tokenised real estate assets are potentially easier to buy and sell compared to traditional methods, reducing the illiquidity premium typically associated with property investments (Baum, 2021; Chow & Tan, 2022). The use of blockchain technology in tokenisation also enables faster transactions by streamlining processes and reducing transaction times, a significant improvement over the often-lengthy procedures involved in traditional real estate deals (Chow & Tan, 2022). Cost efficiency is another key benefit, as automation and disintermediation can potentially lower transaction costs, making real estate investment more economical for all parties involved (Chow & Tan, 2022; Mottaghi et al., 2024). Transparency is greatly enhanced through blockchain technology, which provides a clear, immutable record of ownership and transactions, increasing trust and reducing the potential for fraud in real estate dealings (Chow & Tan, 2022). Finally, tokenisation facilitates global market access, allowing real estate tokens to be traded internationally and opening up a world of investment opportunities that were previously limited by geographical constraints (Chow & Tan, 2022; Mottaghi et al., 2024). Collectively, these benefits have the potential to transform real estate into a more accessible, efficient, and globally connected asset class.

The real estate tokens generally fall into three categories. Equity tokens represent ownership shares in the property, similar to traditional equity investments. Debt tokens, on the other hand, represent a debt instrument secured by the property, offering a different risk–return profile for investors. Lastly, revenue tokens entitle holders to a share of the property's income, providing a way for investors to benefit from the cash flows generated by the asset without necessarily owning a share of the property itself (Baum, 2021; Chow & Tan, 2022). These varied token types allow for a range of investment strategies and risk profiles, catering to diverse investor needs within the real estate market.

The real estate tokenisation market has seen several successful experimental implementations across various regions, demonstrating the growing adoption and potential of this innovative approach. One of the earliest and most high-profile examples is the Aspen Coin, which represents fractional ownership in the St. Regis Aspen Resort, setting a precedent for luxury property tokenisation (Baum, 2021). In Asia, the SiriHub Token, launched in Thailand in September 2021, marked a significant milestone as the first regulated real estate–backed public initial coin offering (ICO) on the Tezos blockchain network (Chow & Tan, 2022). South Korea has also embraced this technology, with the Kasa platform successfully issuing digital asset-backed securities (DABS) for commercial real estate in Seoul, showcasing the potential for tokenisation in the commercial property sector (Chow & Tan, 2022). The trend is not limited to these countries; platforms like BrickX in Australia or ADDX and Minterest in Singapore, have been attracting strong interest from both institutional and retail investors, indicating a growing appetite for tokenised real estate investments across the Asia-Pacific (APAC) region (Chow & Tan, 2022). These diverse examples illustrate the global reach and versatility of real estate tokenisation, spanning different property types, regulatory environments, and investor profiles. A more detailed case study will be provided in Chapter 4.

2.3.4.3 *Regulatory Landscape*

The regulatory environment for real estate tokenisation varies by jurisdiction. In the United States, tokens are generally treated as securities, regulated by the Securities and Exchange Commission (SEC) under existing securities laws (Mottaghi et al., 2024). Europe is more advanced in this area. Table 2.2 summarises the regulatory frameworks for digital assets in Europe. There are four major legislation bodies across Europe that are exploring the regulations for digital assets, namely eWPG from Germany, MiCA for the whole EU, TVTG from Liechtenstein, and FCA from the UK. The Markets in Crypto-Assets (MiCA) regulation, which started to be enforced in December 2024, will provide a comprehensive framework for crypto-assets, including tokenised real estate.

In emerging markets, various economies have set up regulatory frameworks for tokenised real-world assets (RWA), including real estate. For example, in Thailand, the Securities and Exchange Commission (SEC) has approved real estate-backed ICOs (Chow & Tan, 2022). In Singapore and Hong Kong, a few sandbox programs for digital token offerings have been launched, and regulatory stances have been clarified (Chow & Tan, 2022). In Japan, the Financial Services Agency (FSA) regulates tokenised real estate under the Financial Instruments and Exchange Act (FIEA) (Mottaghi et al., 2024).

Table 2.2 Regulatory Frameworks for Digital Assets in Europe

	eWPG	*MiCA*	*TVTG*	*FCA*
Passed into law on	Jun 2021	Around 2022–2024	Jan 2020	Jan 2020
Financial products/ tokens	Bonds, mortgage bonds, certain funds, crypto asset	Securities, currencies, utilities, asset-backed, crypto	Securities, currencies, utilities, asset-backed, crypto	All tokens/ products
Not covered	Shares, equity	NFTs	NFTs	NFTs
Licensing requirements	Issuance, custody, exchange	Issuance, custody, exchange	Exchange, issuer, generator, validator etc.	Exchange, custody (AML only)
Passporting	Germany only	EEA	EEA	UK only
Challenges	Only a few selected instruments currently are covered	Draft status – requires adoption by all members	Local entity requirements and no automatic passporting	Unregulated, but stricter stance expected by FCA

By author (2024)

2.3.4.4 Challenges

Real estate tokenisation faces significant challenges, including regulatory uncertainty across jurisdictions, technology risks like smart contract vulnerabilities, and limited market adoption (Chow & Tan, 2022; Mottaghi et al., 2024). Valuation complexities and legal integration issues present additional hurdles (Baum, 2021). The lack of a centralised marketplace for real estate tokens can lead to liquidity problems and higher search costs for investors (Chow & Tan, 2022). The future of real estate tokenisation appears promising, with particularly strong potential in the emerging markets. As the technology matures, we can expect to see integration with other cutting-edge technologies such as artificial intelligence, the Internet of Things, and big data, which will enhance property management and investment analysis capabilities (Mottaghi et al., 2024). Standardization is likely to play a crucial role in the industry's development, with efforts focused on creating industry-wide standards for tokenisation processes and improving interoperability between platforms.

The emerging markets, in particular, present a fertile ground for the growth of real estate tokenisation. The relatively undercapitalised financial systems in many emerging markets create demand for alternative capital-raising platforms among developers and asset owners (Chow & Tan, 2022). Furthermore, the ongoing development of central bank digital currencies (CBDCs) in the region could contribute to lowering administration and transaction costs in post-tokenisation processes (Chow & Tan, 2022).

Real estate tokenisation represents a significant shift in how property can be owned and traded. While it offers numerous benefits, including increased accessibility, liquidity, and efficiency, its widespread adoption depends on overcoming regulatory, technological, and market challenges. As the technology matures and regulatory frameworks evolve, tokenisation has the potential to revolutionise the real estate industry. This transformation is particularly vibrant in the APAC region, where digital adoption rates are high and regulatory clarity is improving.

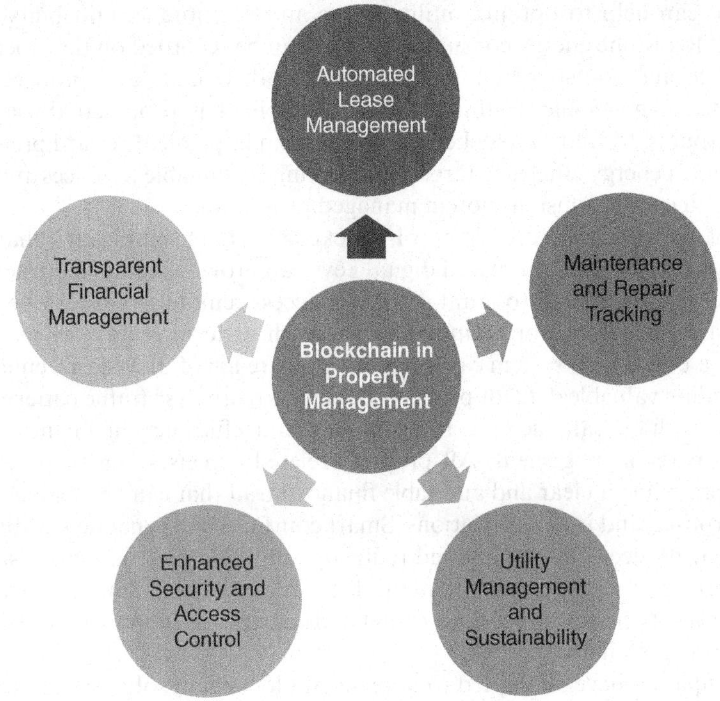

Figure 2.8 Applying Blockchain in Property Management.

By author (2024).

2.3.5 *Blockchain in Property Management*

By leveraging the transparency, security, and efficiency of blockchain, property managers can streamline operations, enhance tenant experiences, and improve overall asset performance. Blockchain technology offers several innovative solutions to many challenges faced in property management (shown in Figure 2.8).

Smart contracts on blockchain platforms can automate many aspects of lease management. These digital agreements can automatically collect rent payments on predetermined dates, reducing late payments and administrative overhead. Additionally, the system can initiate lease renewal processes automatically, sending timely notifications to tenants and property managers when lease terms are nearing expiration. Blockchain can also securely hold and manage security deposits, automatically releasing funds based on predefined conditions at the end of a lease term, thus simplifying this often contentious aspect of property management.

Blockchain can create an immutable record of all the maintenance and repair activities, enhancing transparency and accountability. All maintenance requests, approvals, and completions can be recorded on the blockchain, creating a transparent and auditable trail of property upkeep. Smart contracts can automate payments to service providers upon completion of work, enhancing efficiency and reducing disputes. Furthermore, by integrating with IoT devices, blockchain systems can track equipment performance and trigger maintenance requests automatically when needed, enabling a predictive maintenance approach that can significantly reduce downtime and repair costs.

Blockchain can help to optimise utility usage and promote sustainability in property management. Real-time energy consumption data can be recorded on the blockchain, providing accurate and transparent information for both tenants and property managers. Smart contracts can calculate and distribute utility bills based on actual usage, reducing errors and disputes. Moreover, blockchain systems can implement reward programmes for tenants who meet energy efficiency targets, promoting sustainable practices and potentially reducing overall energy consumption in managed properties.

In terms of security and access control, blockchain offers significant enhancements to traditional systems. Blockchain-based digital keys can provide secure and trackable access to properties, with the ability to grant or revoke access remotely. This can be particularly useful for managing access for maintenance staff, short-term rentals, or in emergencies. Additionally, a blockchain system can maintain a secure log of all visitors, enhancing security and providing valuable data for property managers to analyse traffic patterns and usage.

Finally, blockchain can increase transparency and efficiency in financial operations related to property management. All property-related expenses can be recorded on the blockchain, providing a clear and auditable financial trail that can be invaluable for budgeting, tax purposes, and investor relations. Smart contracts can generate real-time financial reports, improving decision-making and reducing administrative burdens associated with periodic reporting. For properties with multiple investors, blockchain can provide real-time updates on property performance and financial distributions, enhancing transparency and trust in investment scenarios.

Several companies have attempted to implement blockchain solutions in property management over the past few years; however, the situation remains challenging. A primary concern is the integration of blockchain solutions with existing software and processes, which may require substantial time and resources to ensure seamless operation. Data privacy presents another crucial challenge; while blockchain enhances transparency, it is vital to strike a balance with tenants' privacy rights, ensuring that sensitive information is protected appropriately. Finally, the adoption of this new technology requires careful consideration of the human element. Both property management staff and tenants will need comprehensive education and ongoing support to effectively use and benefit from blockchain-based systems.

2.3.6 *Blockchain for Real Estate Financing*

Blockchain technology also has the potential to revolutionise real estate finance. As shown in Figure 2.9, it can potentially streamline the mortgage process, create opportunities for mortgage securitisation and trading, and provide alternative financing models for real estate investment. By utilising blockchain technology, the real estate financing sector can benefit from enhanced risk assessment and management as well as improved mortgage servicing and administration.

Blockchain technology is set to significantly reduce inefficiencies in the mortgage application and approval process. Through the implementation of smart contracts, much of the underwriting process can be automated, including credit checks, income verification, and property appraisals. This automation has the potential to dramatically reduce both the time and cost associated with mortgage origination (Wouda & Opdenakker, 2019). Some innovative platforms have the capability to reduce mortgage closing times from weeks to mere days, thereby improving efficiency for both lenders and borrowers. This streamlining

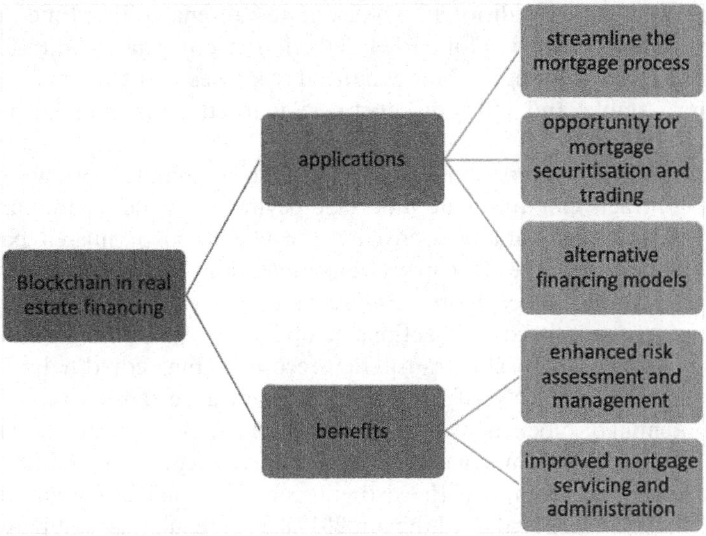

Figure 2.9 Blockchain for Real Estate Financing: Potential Application and Benefits.

By author (2024).

of the mortgage process represents a significant step forward in making real estate financing more accessible and user-friendly.

Blockchain offers exciting new possibilities for mortgage securitization and secondary market trading. By representing mortgage-backed securities as tokens on a blockchain, the process of securitization can become more transparent and efficient. This tokenisation could potentially increase liquidity in the secondary mortgage market, making it easier for investors to buy and sell these securities (Baum, 2021). Furthermore, blockchain-based platforms could enable real-time trading of mortgage-backed securities, significantly reducing settlement times and increasing overall market efficiency (Kejriwal & Mahajan, 2017). This enhanced liquidity and efficiency could lead to better pricing for mortgage-backed securities and potentially lower costs for borrowers.

The advent of blockchain technology is enabling new models of real estate financing that were previously impractical or impossible. Blockchain platforms can facilitate direct peer-to-peer real estate lending, potentially offering better rates for borrowers and new investment opportunities for lenders. Moreover, blockchain could enable the creation of fractional mortgages, where multiple lenders contribute to a single mortgage. This innovative approach could potentially make homeownership more accessible by allowing borrowers to piece together funding from various sources. These alternative financing models have the potential to democratize real estate investment and financing, opening new opportunities for both small investors and borrowers who might struggle in the traditional mortgage market.

Blockchain technology has the potential to significantly improve risk assessment in mortgage lending. By leveraging blockchain's ability to securely aggregate data from multiple sources, lenders can potentially create more comprehensive and accurate credit scoring models. This could lead to fairer lending practices and better risk management, potentially opening mortgage opportunities to borrowers who traditional credit scoring methods might overlook. Additionally, blockchain can enable real-time monitoring of mortgage performance, allowing for earlier detection of potential defaults and more proactive risk management strategies. This enhanced risk assessment and management could lead to a more stable and efficient mortgage market, benefiting both lenders and borrowers.

Blockchain technology offers significant potential to streamline mortgage servicing processes. Smart contracts can automate mortgage payments, reducing administrative overhead for servicers and potentially decreasing the likelihood of missed payments. This automation could lead to more efficient servicing operations and potentially lower costs for borrowers. Furthermore, blockchain's immutable ledger can provide a clear, auditable record of all mortgage-related transactions, simplifying reporting and compliance processes (Mulligan et al., 2018). This transparent record-keeping could reduce disputes and make it easier for all parties to verify the status of a mortgage at any given time.

While the potential of blockchain in real estate financing is significant, several challenges remain to be addressed. The mortgage industry is heavily regulated, and blockchain solutions must be designed to comply with existing regulations while also adapting to evolving regulatory landscapes (Kejriwal & Mahajan, 2017). This regulatory complexity presents a significant hurdle for widespread adoption. Additionally, many financial institutions and mortgage companies rely on legacy systems, and integrating blockchain technology with these existing systems presents a significant technical challenge (Ferreira Santana et al., 2021). Moreover, while blockchain offers enhanced security in many ways, ensuring the privacy of sensitive financial data on a distributed ledger remains a challenge that must be addressed. Overcoming these challenges will require concerted effort from technologists, regulators, and industry stakeholders.

2.4 Sustainable Development Principles

Sustainable development is defined as "the use of environment and resources to meet the needs of the present without compromising the ability of future generations to meet their own needs" by the World Commission on Environment and Development (WCED, 1987). It is important to translate this generic ideal into practice. The question of how to translate this generic ideal into practice has been addressed in various ways over the years. The 2002 World Summit on Sustainable Development (WSDS, 2002) introduced the notion of the three pillars of sustainable development: the environmental, social, and economic, which are considered "interdependent and mutually reinforcing."

As described in the reports by the World Green Building Council (WGBC, 2024) and Urban Land Institute (ULI, 2024), sustainable development in real estate refers to the practice of creating and managing built environments that meet the needs of the present without compromising the ability of future generations to meet their own needs. In the context of real estate, this involves designing, constructing, operating, and renovating buildings and infrastructure in ways that:

- Minimise negative environmental impacts,
- Maximise resource efficiency
- Enhance human health and well-being
- Support social equity and community development
- Ensure long-term economic viability

This includes the entire lifecycle of real estate assets, from site selection and design through construction, operation, renovation, and eventual decommissioning or repurposing.

2.4.1 Practising Sustainable Development in Real Estate

1 *Define and Align Sustainability Strategy:* A critical first step in sustainable real estate management is to define and align the sustainability strategy. As Rogmans and Ghunaim (2016) pointed out, it is important to establish a clear, unambiguous definition of sustainability for the real estate portfolio. This definition should not exist in isolation but must be responsive to local conditions and contexts, recognising that sustainability challenges and opportunities can vary significantly across different locations and property types. Warren-Myers (2012) further stresses the importance of aligning real estate sustainability targets with the overall corporate strategy, ensuring that sustainability efforts in real estate contribute meaningfully to broader organisational goals. In this alignment process, Warren-Myers (2012) advises balancing exchange value strategies, which focus on cost reduction and financial performance, with value strategies that prioritise user benefits and experience. This balanced approach ensures that sustainability initiatives deliver economic benefits and enhanced value for building occupants and the wider community, creating a more holistic and effective sustainability strategy for the real estate portfolio.

2 *Focus on Comprehensive Sustainability Performance:* A comprehensive approach to sustainability performance in real estate requires attention to multiple dimensions. ULI (2024) highlights the importance of addressing operational and embodied carbon in buildings, recognising that a building's environmental impact extends beyond its day-to-day operations to include the carbon footprint of its materials and construction. The implementation of energy efficiency measures and renewable energy solutions is one of the key strategies for reducing operational carbon emissions (Deloitte, 2024e). However, Warren-Myers (2012) argues that sustainability performance is influenced by physical and non-physical factors, suggesting that managers must look beyond the building's physical characteristics to consider aspects such as occupant behaviour and management practices. Meanwhile, both academics and practitioners acknowledge that it is critical to measure actual performance rather than relying solely on design intentions or certifications (Warren-Myers, 2012; ULI, 2024). This focus on real-world outcomes ensures that sustainability initiatives deliver tangible benefits and allows for continuous improvement based on empirical data rather than theoretical projections.

3 Implement Robust Performance Measurement and Reporting: Implementing robust performance measurement and reporting systems is critical for sustainable real estate management. Collecting and monitoring sustainability data across the entire real estate portfolio is the first step (ULI, 2024). This data-driven approach allows for informed decision-making and continuous improvement. Adopting established frameworks and standards for sustainability reporting offers multiple benefits (Deloitte, 2024f). This approach facilitates comparison across organisations, boosts stakeholder confidence, highlights data transparency and verifiability, encourages trust, and helps meet growing regulatory demands. Rogmans and Ghunaim (2016) add a critical dimension to it by stressing the importance of systematically evaluating sustainability rating methodologies. This evaluation process helps organisations select the most appropriate metrics

and methodologies for their specific context, ensuring their sustainability performance measurement captures the most relevant and impactful aspects of their real estate operations. By combining these elements, organisations can create a robust framework for measuring, reporting, and continuously improving their real estate sustainability performance.

4 *Enhance Climate Resilience and Adaptation:* Climate resilience is showing its importance and has become an emerging focus in sustainable real estate management. ULI (2024) recommends thoroughly assessing physical climate risks across the entire real estate portfolio, providing a foundation for informed decision-making. Building on this assessment, ULI (2024) further advises developing resilience strategies at asset and portfolio levels, ensuring a multi-tiered approach to climate adaptation. Climate change has emerged as a critical factor for long-term investors when making location decisions for new acquisitions. Both acute events, like flooding and hurricanes, and chronic risks, such as sea level rise and temperature increases, are increasingly impacting economically important cities and regions. Recent research by Clayton et al. (2021) suggests that areas previously less exposed to extreme weather may now face long-lasting declines in property values or liquidity due to climate events. This has led to growing awareness and concern among global real estate investors, with many now incorporating climate risk assessments into their investment strategies. Commercial property investors increasingly apply higher risk premiums to entire metro areas affected by climate events, regardless of whether individual properties were directly impacted. Some are even excluding high-risk locations from investment consideration altogether. As a result, investors are reassessing risks, allocations, and potential mitigation actions when making location decisions for new acquisitions. This shift is particularly notable in the commercial sector, where investor-owners tend to factor in long-term climate impacts more heavily than residential owner-occupiers. More recently, some high-profile campaigns, such as 100RC (2024), started calling for integrating resilience planning with social equity considerations, arguing that climate impacts often disproportionately affect vulnerable communities. Mohabat Doost et al. (2023) examined how European cities address equity concerns in their resilience strategies, focusing on three dimensions of social equity—distributional, procedural, and recognitional—and found that while some resilience plans effectively incorporate equity considerations, others fall short. They recommend several strategies to enhance equity in future resilience planning, including promoting structural transformations, considering political processes, adopting participatory approaches, fostering trust between citizens and governing bodies, and prioritising upstream inequality factors.

5 *Adopt Various Approaches to Sustainability:* The approach to sustainability in real estate extends beyond environmental considerations to encompass a broader range of factors. This multi-faceted perspective recognises the interconnectedness of sustainability issues and their impact on various stakeholders. It is important to consider health, well-being, and social equity impacts of real estate decisions, acknowledging the significant role buildings play in shaping occupant experiences and community dynamics (ULI, 2024). ULI (2024) further recommends incorporating nature-based solutions and biodiversity enhancements into real estate projects, recognising the potential for buildings to contribute to local ecosystems positively. Similarly, Deloitte (2024d) suggests adopting a circular economy approach to real estate, which involves designing out waste and pollution,

keeping materials in use, and regenerating natural systems. This wider range of approaches to sustainability can minimise negative impacts and create positive value across environmental, social, and economic dimensions.

6 *Engage Stakeholders and Drive Collaboration:* Engaging stakeholders and driving collaboration are essential to a successful, sustainable real estate strategy. For example, partnering with tenants on sustainability initiatives and data sharing shows its importance, as occupier behaviour significantly impacts a building's environmental performance. This collaborative approach extends to the supply chain, and real estate companies work closely with suppliers to reduce emissions throughout the value chain (ULI, 2024). The industry-wide collaboration to develop standards, share best practices, and create a collective approach to sustainability challenges is helpful in the sustainable development of real estate. New York City's Green Codes Task Force showcased the effectiveness of involving industry stakeholders in the city government's effort to promote green building in New York City (Scheib et al., 2014). Deloitte (2024b) pointed out that the engagement with local communities and governments ensures that real estate sustainability initiatives align with broader community goals and regulatory requirements.

7 *Navigate the Evolving Regulatory Landscape:* The legal environment and regulatory landscape are becoming increasingly complex. Real estate companies need to proactively monitor emerging sustainability regulations and disclosure requirements to stay ahead of compliance challenges. Meeting minimum requirements is not sufficient; to maintain a competitive advantage, companies should strive to exceed regulatory standards (ULI, 2024). Through engagement with policymakers and active participation in future regulations and standards, real estate companies can contribute their expertise to developing practical, effective sustainability regulations (Deloitte, 2024c). This helps create more informed policies and allows companies to anticipate and prepare for future regulatory trends, turning potential challenges into strategic opportunities.

8 *Integrate Sustainability into All Real Estate Decisions:* To create a sustainable real estate portfolio, sustainability should be considered at every stage of the real estate lifecycle, from acquisitions and developments to refurbishments and disposals. Sustainability criteria should be incorporated into leasing and property management practices, and sustainable principles should be embedded into day-to-day operations. This approach ensures that sustainability is not an afterthought but a core consideration in all property-related decisions (Sinke, 2015; ULI, 2024). Taking green leases as an example, Deloitte (2024d) finds it practical and effective to align landlord and tenant sustainability goals, developing collaboration and shared responsibility for sustainable building performance. By integrating sustainability considerations into every aspect of real estate decision-making and management, companies can create a coherent, effective approach to sustainability that permeates their entire real estate strategy and operations.

9 *Drive Continuous Improvement and Innovation:* Innovation and sustainability are closely intertwined in the context of urban real estate and property management. While distinct concepts, they share several common features and mutually reinforce each other. Innovation, particularly in technology and processes, often drives sustainability improvements in buildings and urban development, such as through green building technologies or smart systems that optimise resource use (Bradley et al., 2013; Kauko, 2018).

Conversely, the sustainability agenda and environmental regulations push the industry to innovate, developing new products and services to meet evolving goals and requirements (Hellström, 2007). Both innovation and sustainability focus on effective resource management, require a long-term perspective, and demand supportive organisational cultures (Mieg, 2012; Thompson, 2015). They can drive economic benefits through cost savings and new market opportunities (Cainelli et al., 2004), while also having the potential to disrupt existing industry practices. Both face challenges in measurement and implementation (Hildén & Rosenström, 2008; Mooya, 2011), often requiring multi-stakeholder approaches and coordination across various levels of governance (Gualini, 2006). As cities and companies strive to create more efficient, resilient, and valuable urban environments, the synergy between innovation and sustainability becomes increasingly critical, with each concept feeding into and strengthening the other in a virtuous cycle of urban development and management (Kauko, 2019; Thompson, 2015). Moreover, Deloitte (2024a) highlights the importance of continuously educating teams on sustainability best practices. This ongoing learning ensures that all team members have the latest knowledge and skills to implement and improve sustainability initiatives effectively.

10 *Create and Measure Long-Term Value*: Creating and measuring long-term value is essential for sustainable real estate management that delivers lasting benefits. ULI (2024) highlights balancing short-term costs with long-term value creation and risk mitigation, recognising that sustainability investments often yield returns over extended periods.

While taking a long-term perspective can reconcile financial and sustainability objectives (RICS, 2015; Sullivan, 2015; Haran et al., 2019), the lack of convincing market data and valuation methodologies makes it difficult to quantify long-term value creation (Jackson & Orr, 2021). The heterogeneous nature of real estate further complicates isolating sustainability's value contribution. However, improved data standards and analysis gradually enhance sustainability performance measurement (Jackson & Orr, 2021). Sustainability strategies can create long-term value by future-proofing assets, reducing obsolescence risks, and attracting desirable tenants (Ellison et al., 2007; Lorenz & Lützkendorf, 2008). Collaborative approaches between landlords and tenants can also enhance relationships and potentially influence values over time (Jackson & Orr, 2021). Nevertheless, short-term financial pressures, misalignment between payback periods and lease lengths, and slow market reactions in some sectors pose barriers to long-term value creation through sustainability (Jackson & Orr, 2021). Overcoming these challenges requires improved data, valuation methodologies, and potential policy interventions to fully capitalise on the long-term value potential of sustainable management practices in real estate.

2.4.2 *Conceptual Framework: PropTech Dynamics in Real Estate Sustainability*

We propose a conceptual framework (shown in Figure 2.10) that integrates several key theories and models to understand the complex interplay between PropTech and sustainability in the real estate sector. This framework provides a theoretical mapping for PropTech dynamics in real estate sustainability. It integrates key theories from technological innovation, sustainability science, and real estate economics to offer a comprehensive view of how PropTech influences sustainability outcomes in the built environment.

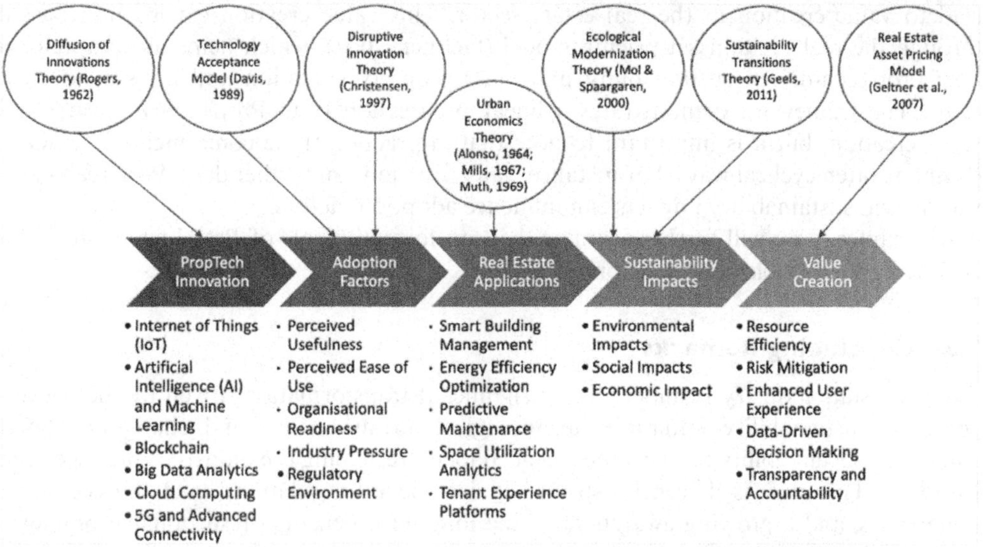

Figure 2.10 Conceptual Framework: PropTech Dynamics in Real Estate Sustainability. By author (2024).

At the core of our framework lies PropTech innovation, encompassing technologies like the IoT, artificial intelligence (AI), blockchain, big data analytics, cloud computing, and advanced connectivity solutions like 5G. These innovations serve as the driving force for change in the real estate industry. The diffusion of innovation theory (Rogers, 1962), 1962) helps us understand how these PropTech solutions spread through the real estate sector. Simultaneously, the disruptive innovation theory (Christensen, 1997)frames how PropTech can fundamentally alter traditional real estate practices.

The adoption of PropTech solutions is influenced by several key factors: perceived usefulness, perceived ease of use, organisational readiness, industry pressure, and regulatory environment. These factors are grounded in the technology acceptance model (Davis, 1989), which helps explain the conditions under which new technologies are embraced by real estate stakeholders.

When adopted, PropTech innovations manifest in various real estate applications, such as smart building management, energy efficiency optimisation, predictive maintenance, space utilisation analytics, and tenant experience platforms, among others. These applications can be understood through the lens of urban economics theory (Alonso, 1964, Mills, 1967, Muth, 1969), which provides insights into how PropTech affects spatial distribution and land use in urban areas.

Implementing PropTech in real estate leads to sustainability impacts across three key dimensions: environmental, social, and economic impacts. This view of sustainability aligns with the triple-bottom-line concept (Elkington, 1994). The ecological modernisation theory (Mol and Spaargaren, 2000) further explains how PropTech can drive environmental improvements in the built environment. The sustainability transitions theory (Geels, 2011) offers a multi-level perspective on how these PropTech applications can facilitate systemic changes towards sustainability in real estate. Sustainability impacts ultimately lead to value creation in the real estate sector. This value creation can be understood through the real estate asset pricing model (Geltner, 2007), which helps us comprehend how PropTech-driven sustainability influences property valuation and investment decisions. The framework demonstrates a linear progression from PropTech innovation to value creation, but it is important to note that in practice, these components interact in complex, often cyclical ways. For instance, value creation can further drive PropTech innovation, and sustainability impacts can influence adoption factors.

In Chapter 5, we will further examine the transformative role of PropTech strategies in advancing sustainable real estate practices.

2.5 Concluding Remarks

The real estate industry is undergoing a significant transformation driven by the convergence of sustainability initiatives, technological advancements, and changing market dynamics. Sustainability has become a central focus, reshaping the industry's practices and priorities. This shift is driven by stricter environmental regulations, evolving consumer preferences, and a growing awareness of the long-term benefits of sustainable practices. Governments worldwide are implementing more stringent environmental standards and offering incentives for green development, while investors and occupiers increasingly demand properties with strong sustainability credentials. This trend is driving the integration of smart technologies to enhance building performance and energy efficiency, as well as a heightened focus on climate resilience in design and construction.

The rise of ESG factors in investment decision-making is further accelerating this transformation. Real estate companies are under increasing pressure to demonstrate their commitment to sustainability, not just in their developments but across their entire operations. This shift is also driving innovation in financing models, with green bonds and impact investing gaining traction in the real estate sector. As a result, the industry is seeing the emergence of new value-creation opportunities, with sustainable properties often commanding premium values and attracting a wider pool of investors and tenants.

Parallel to the sustainability revolution, distributed ledger technology (DLT), including blockchain, is poised to fundamentally change the real estate landscape. DLT offers the potential to lower barriers to entry through fractional ownership models, enhancing market accessibility for a broader range of investors. It also promises to increase liquidity in the traditionally illiquid real estate market through tokenisation and near-instantaneous transaction capabilities. Furthermore, DLT could facilitate the globalisation of real estate markets by enabling standardised, cross-border transactions and creating unified protocols for property deals across various jurisdictions.

The advent of DLT in real estate is likely to drive significant regulatory evolution as governments and industry bodies grapple with the implications of these new technologies and transaction methods. It is also reshaping traditional roles within the industry, potentially disintermediating some functions while creating new specialised positions in areas such as blockchain consultancy and digital asset management. However, the widespread adoption of DLT in real estate faces several challenges, including regulatory uncertainty, the need for integration with existing systems, and the requirement for broad industry acceptance.

The PropTech sector, at the intersection of real estate and technology, is moving towards consolidation and the creation of holistic ecosystems. Integrated platforms are emerging that combine property search, transaction management, and financing options, streamlining the entire real estate journey for users. This consolidation is reducing friction in navigating multiple systems and services, leading to increased user engagement and retention. Moreover, these comprehensive platforms are driving innovation by combining various technologies and data sources, offering unprecedented insights and efficiencies.

As the PropTech ecosystem matures, it has the potential to fundamentally disrupt traditional real estate practices, offering more efficient, transparent, and user-centric solutions. The convergence of sustainability initiatives, DLT applications, and PropTech innovations is set to reshape the real estate industry significantly. While challenges remain in terms of technology adoption, regulatory alignment, and market acceptance, these developments offer unprecedented opportunities for increased efficiency, accessibility, and value creation in the sector. The future of real estate lies in embracing these changes, adapting to new technologies, and prioritising sustainable practices to create resilient, efficient, and valuable assets for the long term.

2.6 Post-Chapter Engagement

To help reinforce your understanding and apply the concepts covered in this chapter, complete the following activities:

1 *Reflection Questions*

- Identify three key ways in which digital innovations are reshaping the real estate landscape. Discuss how these changes might impact traditional real estate practices in the next decade.
- Compare and contrast the potential benefits and challenges of implementing blockchain technology in real estate transactions. Consider aspects such as security, efficiency, and regulatory compliance.

2 *Design Challenge*

- Imagine you are a real estate developer planning a new mixed-use project in your city. Create a basic outline (one to two pages) for a sustainable development plan that incorporates at least five key sustainability principles discussed in this chapter. Explain how each principle would be implemented and the potential benefits for stakeholders.

3 *Regulatory Framework Assessment*

- Select a country or region, and research its current regulatory framework for real estate tokenization or blockchain-based property transactions. Identify potential gaps or areas for improvement in the regulations. Suggest how the framework could be enhanced to better support innovation while protecting stakeholders.

Notes

1 https://www.zillow.com/.
2 https://propy.com/home/.
3 https://www.honeywellforge.ai/.
4 https://www.housecanary.com/.

5 https://www.winsun3d.com/En/News/news_inner/id/465.
6 https://www.livly.io/.
7 https://fundrise.com/.
8 https://www.lemonade.com/uk?f=1.
9 https://www.buildrz.io/?lang=en.
10 www.zillow.com.
11 https://www.realtor.com/.
12 https://www.yardi.co.uk/.
13 https://www.buildium.com/.
14 https://www.costar.com/.
15 https://www.msci.com/our-solutions/real-assets/real-capital-analytics.
16 https://www.docusign.com/en-gb.
17 https://fundrise.com/.
18 https://myarchicad.graphisoft.com/.
19 https://www.autodesk.com/uk/products/revit/overview?term=1-YEAR&tab=subscription.
20 https://propy.com/home/.

References

100RC. 2024. 100 resilient cities—Pioneered by The Rockefeller Foundation. https://www.rockefellerfoundation.org/wp-content/uploads/d631b84b1efe41685a_3gm6b18ip1.pdf

Alonso, W. 1964. *Location and land use: Toward a general theory of land rent*, Harvard University Press.

Baum, A., Saull, A., & Braesemann, F. 2020. *PropTech 2020: The future of real estate*. Oxford Future of Real Estate Initiative, University of Oxford.

Ayan, O., & Turkay, B. Smart thermostats for home automation systems and energy savings from smart thermostats. *2018 6th International Conference on Control Engineering & Information Technology (CEIT)*, 2018. IEEE, 1–6.

Baum, A. 2021. Tokenization—The future of real estate investment. *The Journal of Portfolio Management*, 47, 41–61.

Bradley, J., Loucks, J., Noronha, A., Macaulay, J., & Buckalew, L. 2013. Internet of everything (IoE). *Survey Report*. https://www.cisco.com/c/dam/en_us/about/business-insights/docs/ioe-value-index-whitepaper.pdf

Cainelli, G., Evangelista, R., & Savona, M. 2004. The impact of innovation on economic performance in services. *The Service Industries Journal*, 24, 116–130.

Chow, Y. L., & Tan, K. K. 2022. Real estate insights: Is tokenization of real estate ready for lift off in APAC? *Journal of Property Investment & Finance*, 40, 284–290.

Christensen, C. M. 1997. *The innovator's dilemma: When new technologies cause great firms to fail*. Harvard Business Review Press.

Clayton, J., Devaney, S., Sayce, S., & Van De Wetering, J. 2021. Climate risk and real estate prices: What do we know? *Journal of Portfolio Management*, 47. https://doi.org/10.3905/jpm.2021.1.278

Crosby, M., Pattanayak, P., Verma, S., & Kalyanaraman, V. 2016. Blockchain technology: Beyond bitcoin. *Applied Innovation*, 2, 71.

Davis, F. D. 1989. Perceived usefulness, perceived ease of use, and user acceptance of information technology. *MIS Quarterly*, 319–340.

Deloitte. 2017. *Blockchain in commercial real estate: The future is here*.

Deloitte. 2024a. *Business solutions spread from real estate ESG certifications*.

Deloitte. 2024b. *ESG in giga-cities: Navigating trends, challenges, and opportunities*.

Deloitte. 2024c. *ESG—A driving force behind investor intentions*.

Deloitte. 2024d. *Green leases are coming of age*.

Deloitte. 2024e. *Sustainable portfolio transformation*.

Deloitte. 2024f. *Unravelling the CSRD: Navigating sustainability challenges and opportunities in the real estate sector*.

Dutta, S. K. 2020. Smart contracts. In *The definitive guide to blockchain for accounting and business: Understanding the revolutionary technology*. Emerald Publishing Limited.

Ellison, L., Sayce, S., & Smith, J. 2007. Socially responsible property investment: Quantifying the relationship between sustainability and investment property worth. *Journal of Property Research*, 24, 191–219.

Elkington, J. 1994. Towards the sustainable corporation: Win-win-win business strategies for sustainable development. *California Management Review*, 36, 90–100.

Farahani, B., Firouzi, F., & Luecking, M. 2021. The convergence of IoT and distributed ledger technologies (DLT): Opportunities, challenges, and solutions. *Journal of Network and Computer Applications*, 177, 102936.

Ferreira Santana, F., Mira da Silva, M., & Galvão da Cunha, F. 2021. Blockchain for real estate: A systematic literature review. In E. Insfran, F. González, S. Abrahão, M. Fernández, C. Barry, H. Linger, M. Lang, & C. Schneider (Eds.), *Information systems development: Crossing boundaries between development and operations (DevOps) in information systems (ISD2021 Proceedings)*. Universitat Politècnica de València.

Geels, F. W. 2011. The multi-level perspective on sustainability transitions: Responses to seven criticisms. *Environmental Innovation and Societal Transitions*, 1, 24–40.

Geltner, D. 2007. *Commercial real estate analysis and investments*. Thompson Higher Education Mason.

Gualini, E. 2006. The rescaling of governance in Europe: New spatial and institutional rationales. *European Planning Studies*, 14, 881–904.

Haran, M., McGreal, W. S., Davis, P., Berry, J., McCord, M., Lo, D., Griffiths, P., Hirsch, J., Spanner, M., & Lafuente, J. J. 2019. Carbon risk integration in corporate strategies within the real estate sector. *CRREM Report No. 2*.

Hellström, T. 2007. Dimensions of environmentally sustainable innovation: The structure of eco-innovation concepts. *Sustainable Development*, 15, 148–159.

Hildén, M., & Rosenström, U. 2008. *The use of indicators for sustainable development*. Wiley Online Library.

Jackson, C., & Orr, A. 2021. The embeddedness of sustainability in real estate investment decision-making. *Journal of European Real Estate Research*, 14, 362–380.

Kauko, T. 2018. *Pricing and sustainability of urban real estate*. Routledge.

Kauko, T. 2019. Innovation in urban real estate: The role of sustainability. *Property Management*, 37, 197–214.

Kejriwal, S., & Mahajan, S. 2017. *Blockchain in commercial real estate: The future is here*. Deloitte Center for Financial Services.

Lorenz, D., & Lützkendorf, T. 2008. Sustainability in property valuation: Theory and practice. *Journal of Property Investment & Finance*, 26, 482–521.

Mieg, H. A. 2012. Sustainability and innovation in urban development: Concept and case. *Sustainable Development*, 20, 251–263.

Mills, E. S. 1967. An aggregative model of resource allocation in a metropolitan area. *The American Economic Review*, 57, 197–210.

Mohabat Doost, D., Brunetta, G., & Caldarice, O. 2023. In search of equitable resilience: Unravelling the links between urban resilience planning and social equity. *Sustainability*, 15, 13818.

Mohr, S., & Khan, O. 2015. 3D printing and its disruptive impacts on supply chains of the future. *Technology Innovation Management Review*, 5, 20.

Mol, A. P., & Spaargaren, G. 2000. Ecological modernisation theory in debate: A review. *Environmental Politics*, 9, 17–49.

Mooya, M. 2011. Of mice and men: Automated valuation models and the valuation profession. *Urban Studies*, 48, 2265–2281.

Mottaghi, S. F., Steininger, B. I., & Yanagawa, N. 2024. Real estate insights: The current state and the new future of tokenization in real estate. *Journal of Property Investment & Finance*, 42(6), 614–620.

Mulligan, C., Scott, J. Z., Warren, S., & Rangaswami, J. 2018. *Blockchain beyond the hype: A practical framework for business leaders*. White paper of the World Economic Forum.

Muth, R. F. 1969. *Cities and housing; The spatial pattern of urban residential land use*. University of Chicago Press.

Nakamoto, S. 2008. *Bitcoin: A peer-to-peer electronic cash system*. Satoshi Nakamoto.

Richardson, L. 2021. Coordinating office space: Digital technologies and the platformization of work. *Environment and Planning D: Society and Space*, 39, 347–365.

RICS. 2015. *Advancing responsible business practices in land, construction and real estate use and investment*. Royal Institute of Chartered Surveyors/UN Global Compact.

Rogers, E. M. 1962. *Diffusion of innovations*. The Free Press of Glencoe.

Rogmans, T., & Ghunaim, M. 2016. A framework for evaluating sustainability indicators in the real estate industry. *Ecological Indicators*, 66, 603–611.

Scheib, C., Unger, R., Leigh, R., & Council, U. G. 2014. Greening a city's building codes: The NYC Green Codes Task Force. *ACEEE Summer Study on Energy Efficiency in Buildings*, 6–261.

Sinke, G. 2015. *Corporate real estate sustainability management: A strategic management framework for implementing a sustainable corporate real estate strategy*. Eindhoven University of Technology.

Sullivan, R. 2015. *Fiduciary duty in the 21st century*. SSRN 2724866.

Tapscott, D., & Tapscott, A. 2016. *Blockchain revolution: How the technology behind bitcoin is changing money, business, and the world*. Penguin.

Thompson, B. 2015. Innovation in property management. *Journal of Property Investment & Finance*, 33, 436–445.

ULI. 2024. *Global sustainability outlook 2024*.

Veuger, J. 2018. Trust in a viable real estate economy with disruption and blockchain. *Facilities*, 36, 103–120.

Viriyasitavat, W., & Hoonsopon, D. 2019. Blockchain characteristics and consensus in modern business processes. *Journal of Industrial Information Integration*, 13, 32–39.

Warren-Myers, G. 2012. Sustainable management of real estate: Is it really sustainability? *Journal of Sustainable Real Estate*, 4, 177–197.

WCED. 1987. *Our common future*. World Commission on Environment and Development.

WGBC. 2024. *What is a sustainable built environment?* World Green Building Council. https://worldgbc.org/what-is-a-sustainable-built-environment/

Wouda, H. P., & Opdenakker, R. 2019. Blockchain technology in commercial real estate transactions. *Journal of Property Investment & Finance*, 37, 570–579.

WSDS. 2002. *Ensuring sustainable livelihoods: Challenges for government, corporates, and the civil society*. https://wsds.teriin.org/2002/

Xiong, Q., Graham, L., & Baum, A. 2021. *The future of automated real estate valuations*. SSRN 4927480.

Chapter 3

Sustainability in the Real Estate Industry

From Green Building to Sustainable Real Estate

3.1 Introduction of Green Building

The concept of green buildings (GB) has emerged as a critical response to growing global challenges, including climate change, energy shortages, increasing pollution, population growth, and rapid urbanisation. This chapter explores the historical development, economic implications, and evolving landscape of green buildings and sustainable real estate.

3.1.1 *Origins of Green Building Practices*

Recent data indicates the urgency of addressing these challenges. According to NASA (2023), Earth's average surface temperature in 2023 was approximately 1.36°C warmer than the late 19th-century pre-industrial average, marking the warmest year on record since 1880. Concurrently, the International Energy Agency (IEA, 2023) reported that global energy-related CO_2 emissions grew by 1.1% in 2023, reaching a new record high of 37.4 billion tonnes.

The built environment plays a significant role in these environmental concerns. Buildings are responsible for approximately 42% of annual global CO_2 emissions (Architecture2030, n.d.). Of this, about 27% comes from building operations, while the embodied carbon of key construction materials accounts for an additional 15%. Moreover, the building sector generates 40–50% of global water pollution and contributes 40% of total solid waste in developed countries (Ade & Rehm, 2020).

The concept of green buildings emerged in the 1960s and 1970s, spurred by the energy crisis and growing environmental awareness. This period saw increased research into improving energy efficiency and reducing environmental pollution in construction (Mao et al., 2009). Early pioneers focused on passive solar design, natural ventilation, and the use of local, sustainable materials. The 1980s and 1990s saw the formalisation of green building practices, culminating in the development of the first green building rating tool, the Building Research Establishment Environmental Assessment Method (BREEAM), in the UK in 1990. This marked the beginning of a global movement towards standardising and promoting sustainable building practices. The 1992 Earth Summit (United Nations Conference on Environment and Development) further stimulated efforts to build environmental protection.

Green buildings are structures designed, constructed, and operated to minimise negative environmental impacts while maximising resource efficiency and occupant well-being. This concept emerged in response to growing global challenges, including climate change, energy

DOI: 10.1201/9781003504436-3

shortages, increasing pollution, population growth, and rapid urbanisation. The World Green Building Council (WGBC) acknowledges that countries and regions have unique characteristics such as history, culture, traditions, climatic conditions, building types, and environmental, economic, and social priorities that shape their approaches to green building. As a result, green building definitions and practices are not uniform across the globe. WGBC defines green building as aiming to reduce or eliminate negative impacts on the environment during the whole building lifecycle, creating positive impacts on the climate and environment (Council, 2013). The U.S. Environmental Protection Agency (EPA) describes green building as "the practice of creating structures and using processes that are environmentally responsible and resource-efficient throughout a building's life cycle, from siting to design, construction, operation, maintenance, renovation, and deconstruction." The U.S. Green Building Council (USGBC) focuses on energy use, water use, indoor environmental quality, material use, and the building's effects on its site as central considerations in green building. In the UK and European Union, green building is often associated with preserving the environment while also considering occupant well-being in terms of space usage and air quality. The Japanese Architectural Institute defines sustainable (green) buildings as those designed to save energy and resources, recycle materials, minimise toxic emissions, harmonise with local climate and surroundings, and sustain and improve human life quality while maintaining ecosystem capacity. China's Assessment Standard of Green Buildings describes green buildings as those that save resources (including energy, land, water, and materials) throughout their lifecycle, protect the environment, reduce pollution, and provide healthy, comfortable, and efficient use of spaces that work in harmony with nature. Table 3.1 exhibits the definitions for green buildings from the key organisations around the world.

There are many factors influencing the green building development. For example, government policies and economic incentives are key drivers. Certification systems help set professional standards to benchmark and follow. In addition, the development and adoption of green building technology increases operational efficiency. Sufficient training and education for the stakeholders would enhance their usage of green buildings. Green buildings aim to mitigate these negative impacts throughout a building's lifecycle, from design and construction to operation and demolition. While green building definitions vary by country, they generally focus on energy efficiency, resource conservation, and occupant well-being. The concept of green buildings continues to evolve alongside the development of the construction industry.

3.1.2 Key Milestones of Green Building Development

The evolution of green building initiatives over the past three decades reflects a growing global commitment to sustainable development in the real estate sector. Since the launch of BREEAM in 1990, the industry has witnessed a proliferation of certification systems, standards, and initiatives aimed at improving the environmental performance of buildings. These milestones not only mark the introduction of new tools and frameworks but also highlight shifting priorities within the field. From its initial focus on energy efficiency and resource conservation, the industry has expanded its scope to encompass broader sustainability concerns, including occupant health, climate resilience, and achieving net-zero carbon emissions. The following timeline (Figure 3.1) illustrates key developments from 1990

Table 3.1 Definitions for Green Buildings Around the World

Country	*Organization*	*Key Aspects of Green Building Definition*
USA	World Green Building Council	• Reduces or eliminates negative impacts • Creates positive impacts on climate and environment • Considers design, construction, and operation
USA	U.S. Environmental Protection Agency (EPA)	• Environmentally responsible and resource-efficient • Covers entire building lifecycle • Includes siting, design, construction, operation, maintenance, renovation, and deconstruction
USA	U.S. Green Building Council (USGBC)	• Focuses on energy use, water use, indoor environmental quality • Considers material use and building's effects on its site
UK	Building Research Establishment	• Enhances well-being of occupants • Helps protect natural resources • Makes more attractive property investments
Europe	European Commission Delegation	• Preserves the environment • Considers occupant well-being • Focuses on space usage and air quality
Germany	German Sustainable Building Council (DGNB)	• Consciously uses available resources • Minimizes energy consumption • Preserves the environment
France	Haute Qualite Environment (HQE)	• Considers overall building performance • Focuses on energy, environment, health, and comfort
Australia	Green Building Council Australia	• Incorporates sustainable development principles • Meets present needs without compromising the future
Japan	Architectural Institute of Japan (AIJ)	• Saves energy and resources • Recycles materials and minimizes toxic emissions • Harmonizes with local climate, traditions, and culture • Sustains and improves human life quality • Maintains ecosystem capacity at local and global levels
China	Assessment Standard of GBs	• Saves resources throughout building lifecycle • Protects environment and reduces pollution • Provides healthy, comfortable, efficient use space • Works in harmony with nature
Singapore	Inter-Ministerial Committee on Sustainable Development (IMCSD)	• Energy and water efficient • High quality and healthy indoor environment • Integrated with green spaces • Constructed from eco-friendly materials

Documented by author (2024)

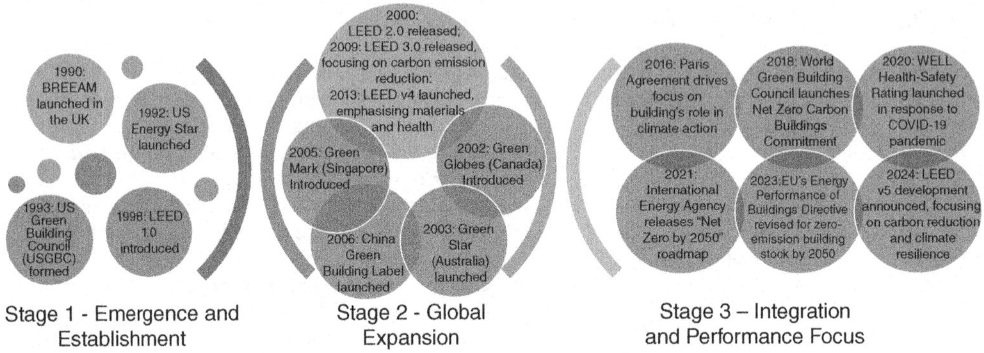

Figure 3.1 Key Milestones of Green Building Development (1990–2024).
By author (2024).

to 2024, showcasing the dynamic nature of green building practices and their increasing integration with global sustainability goals.

The evolution of green building initiatives can be divided into three key stages, each representing a shift in focus and approach (Table 3.2):

- *Stage 1: Emergence and Establishment (1990–2000)*: This stage marks the birth of formal green building certification systems and the initial efforts to standardise sustainable building practices.
- *Stage 2: Global Expansion and Diversification (2001–2015)*: During this period, green building concepts spread globally, with many countries developing their own certification systems. The focus expanded beyond energy efficiency to include broader environmental and health considerations.
- *Stage 3: Integration and Performance Focus (2016–Present)*: This stage is characterised by a shift towards performance-based standards, integration with broader sustainability goals, and an increased focus on carbon neutrality and occupant health.

This three-stage framework demonstrates the evolution from initial standardisation efforts to global adoption and diversification, and finally to a more integrated, performance-driven approach aligned with broader sustainability and health goals.

3.2 Green Building Certification Systems: Catalysts for Sustainable Real Estate

3.2.1 *Major Green Building Certification Programs*

It has been generally accepted that the world's first green building rating tool was BREEAM. This tool was created by the BRE (Building Research Establishment) Group in 1990. The project to develop this rating tool was contracted to BRE by Ron German, a senior project manager from Stanhope Developers in London. The key motivation for this project was Stanhope's belief that their buildings offered superior environmental quality compared to others in the booming market of the 1990s. However, they could not prove this to investors or tenants, and they thus desired to create a rating system that would allow them to demonstrate this conclusively (Zhang et al., 2019).

Table 3.2 Timeline of Green Building Development Events (1990–2024)

Year	*Event*
1990	BREEAM launched
1992	U.S. ENERGY STAR launched
1993	U.S. Green Building Council (GBC) formed
1995	U.S. ENERGY STAR Homes launched
1996	BEAM PLUS (formerly HK-BEAM) launched
1998	LEED 1.0; Spain GBC formed; BREEAM 98; NABERS launched
2000	LEED 2.0; Green Globes (Canada) debut
2001	Japan CASBEE created; Indian GBC formed
2002	Canada GBC and World GBC formed
2003	Green Star, GBC Australia formed
2005	Green Mark (Singapore) launched
2006	U.S. federal GSA adopts LEED; CGBL launched
2007	German Sustainable Building Council (DGNB) formed
2008	BREEAM registrations top 1 million buildings and homes
2009	LEED 2009; BREEAM International launched
2010	Green Globes ANSI Standard approved
2012	National Green Building Standard ICC-700 updated
2013	U.S. federal GSA accepts both LEED and Green Globes; LEED v4 launched
2014	LEED used in more than 140 countries
2015	EDGE launched; World GBCs in 100 countries
2016	Green Globes ANSI Standard updated
2017	LEED v4.1 introduced, focusing on building performance and user experience
2018	World Green Building Council launches Net Zero Carbon Buildings Commitment
2019	Global Alliance for Buildings and Construction reports buildings and construction responsible for 39% of global energy-related carbon emissions
2020	WELL Health-Safety Rating launched in response to COVID-19 pandemic
2021	International Energy Agency releases "Net Zero by 2050" roadmap, emphasizing the role of buildings in achieving climate goals
2022	GRESB assessment covers USD 6.9 trillion of real estate and infrastructure value
2023	EU's Energy Performance of Buildings Directive revised, aiming for a zero-emission building stock by 2050
2024	LEED v5 development announced, focusing on carbon reduction and climate resilience

Having examined why and how BREEAM originated, it is now time to explore how it has influenced the construction of more environmentally friendly buildings worldwide. We will compare it with other important green building programs like LEED. Since the first GB assessment BREEAM was issued in 1990, the development of GB aligns with the development of the green building certification programs or green building rating system (GBRS). Over 40 GBRSs have been developed by governments or third parties to promote sustainable buildings. These programs have come up in different places, each with its own rules and goals to help the environment, economy, and society. By looking at these different programs, we can learn more about the different ways building practices are made more sustainable and improved.

3.2.1.1 BREEAM

BREEAM, launched in 1990 by BRE Group, provides social and economic benefits while mitigating the built environment's environmental impacts. It is now a voluntary international standard operated and applied through a network of international operators, assessors, and industry professionals.

BREEAM promotes energy-efficient building design and systems, as well as equipment that supports the sustainable use of energy in buildings. It also emphasises sustainable management of the building's operation. Credits in this aspect assess measures to improve the building's inherent energy efficiency, encourage carbon emission reduction, and support efficient management throughout the building's operational phase.

3.2.1.2 LEED

LEED was developed by the U.S. Green Building Council (USGBC), an international and widely used GBRS. It is a voluntary, market-driven, consensus-based tool that serves as a guideline and assessment mechanism for designing, constructing, and operating high-performance green buildings and neighbourhoods. LEED is for all building types and phases, including new construction, interior fit-outs, operations and maintenance, and core and shell.

LEED aims to reduce the environmental and economic harm of excessive energy use. The requirements for building energy performance are stated in the prerequisite (minimum energy performance) and the credit (optimise energy performance) for both new and existing buildings. The former requires a minimum level of energy efficiency for a building and its system, while the latter intends to achieve increasing levels of energy performance beyond the prerequisite standard.

3.2.1.3 CGBL

The CGBL, namely, the China Green Building Label, was launched by the Ministry of Housing and Urban-Rural Development of the People's Republic of China in 2006 as a voluntary local standard. It has been widely accepted and applied throughout the whole country. Some local governments make it compulsory for new buildings to obtain a one-star level of certification as a minimum requirement. The latest version of this local standard has been officially implemented since August 2019. This local standard addresses building performance in terms of safety and durability, health and comfort, occupant convenience, resource saving, and environmental liveability. The CGBL aims to improve buildings' energy efficiency and curb consumption growth by setting savings and efficient use requirements. It focuses on the performance of each building service system by setting minimum prescriptive requirements and standard credits. Relatively, it allocates lower scores to the overall energy saving of the whole building compared to other rating systems.

3.2.1.4 BEAM Plus

BEAM Plus, launched in 1996 and formerly HK-BEAM, is Hong Kong's local standard for offering independent assessments of building sustainability performance. It is a voluntary scheme that is tailor-made for the high-rise, high-density built environment and the subtropical climate of Hong Kong. Recognised and certified by the Hong Kong Green Building Council (HKGBC), BEAM Plus offers a comprehensive set of performance criteria for a wide range of sustainability issues relating to building planning, design, construction, commissioning, management, operation, and maintenance. BEAM Plus encourages building owners to achieve better energy performance beyond the minimum requirements of local building energy codes. BEAM Plus focuses on reducing energy consumption in building operations for new buildings. For existing buildings, BEAM Plus stimulates a thorough evaluation of building and service system design performance and provides more significant investments for improving energy performance. This is a way to reduce energy consumption, environmental impacts, and peak electricity demand.

3.2.1.5 Green Mark

The Green Mark scheme was launched by BCA in 2005 as an initiative to drive Singapore's construction industry towards more environmentally friendly buildings. It is intended to promote sustainability in the built environment and raise environmental awareness among developers, designers, and builders when they start project conceptualisation and designs, as well as during construction. Green Mark comprises several rating tools that collectively assess the built environment's environmental performance. Green Mark aims to achieve high energy efficiency in buildings and sets requirements to guide the design, construction, and operation of buildings energy-efficiently. Apart from a prerequisite for total building energy reduction set for GoldPlus and Platinum ratings, several credits also address the energy efficiency of different building service systems based on prescriptive criteria.

3.2.1.6 Green Star

Green Star, launched by the Green Building Council of Australia in 2003, is a voluntary local sustainability rating system for buildings and communities. It assesses the sustainability of projects at all stages of the built environment lifecycle. It improves environmental efficiencies in buildings while also considering users' health and productivity and operational cost savings. Green Star aims to reward projects designed and constructed to reduce overall greenhouse gas (GHG) emissions by energy demand reduction, use efficiency, and power generation from alternative sources. Two integrated energy-related credits are set for new and existing buildings: the Greenhouse Gas Emissions credit and the Peak Electricity Demand Reduction credit. Under these two credits, several sub-items are set for different aspects of building service systems.

3.2.1.7 EDGE

EDGE, an innovation of the International Finance Corporation (IFC), comprises free web-based software, a universal standard, and a certification system for more than 160 countries. It focuses on making buildings more resource-efficient for energy, water, and materials. Rather than suggesting a perfect or prescribed scenario, EDGE provides users with best-practice options to explore and identify an optimum design solution. EDGE provides energy-efficiency measures for users to select and apply to their new and existing buildings. It is a performance-based system that requires a minimum 20% reduction in energy consumption compared to local practices. With EDGE online software, the overall energy-simulated performance of each project building will be assessed based on actual parameters against 45 energy-efficiency measures adopted in the building.

3.2.1.8 NABERS

NABERS is a performance-based rating system for buildings initiated by the Australian government. NABERS provides separate standalone rating schemes to measure the energy efficiency, water usage, waste management, indoor environmental quality of a building or tenancy, and environmental impact. (These are NABERS Energy, NABERS Water, NABERS Waste, and NABERS Indoor Environment.) It assesses the actual environmental impact of a building by using third-party verified data, such as utility bills, and compares its performance to similar buildings in the market. The NABERS Energy rating is calculated by benchmarking the actual energy consumption data and associated GHG emissions within 12 months for the rated premises and comparing them against buildings of the same category. The consumption figures are adjusted based on factors such as building area, hours of use, climate, equipment density (tenancy and overall building ratings), and the greenhouse intensity of the energy source to ensure fair comparison. The NABERS Energy rating is mandatory for office buildings with areas exceeding 1,000 m^2.

Table 3.3 provides a concise overview of the major green building certification systems, highlighting their key strengths, weaknesses, and how they've been adapted to different regional contexts.

3.2.1.9 Challenges and Criticisms of Certification Systems

The expense of certification can be prohibitive, especially for smaller projects or in developing regions. Many systems are criticised for being overly complex, requiring specialised knowledge and potentially discouraging wider adoption. Some critics argue that the point-based systems can lead to a focus on achieving certification rather than truly sustainable outcomes. There is often a discrepancy between the predicted performance at the design stage and the actual operational performance. Many systems focus primarily on environmental aspects, potentially overlooking social and economic sustainability. The proliferation of different systems globally can lead to confusion and difficulty in comparing buildings across regions. While some systems have regional variations, critics argue that they still may not adequately address local environmental, social, and economic priorities. Many systems are critiqued for focusing more on new buildings rather than the existing building stock, which represents a larger portion of the built environment.

Despite these challenges, green building certification systems continue to evolve, addressing criticisms and adapting to new priorities such as carbon neutrality, resilience to climate change, and occupant health and well-being. Their role in driving market transformation towards more sustainable building practices remains significant, even as the industry grapples with how to best measure and incentivise truly sustainable outcomes.

3.2.2 Green Building Certifications' Impact on Real Estate

Green building certifications and associated technologies have had a profound impact on the real estate industry, influencing economic decisions and market trends and addressing key market failures. The following points highlight the major impacts and developments.

Table 3.3 Overview of the Major Green Building Certification Systems

Certification System	*Strengths*	*Weaknesses*	*Regional Adaptation*
BREEAM (UK)	- Comprehensive - Adaptable to different building types - Strong focus on sustainable procurement	- Complex assessment process - Potentially high certification costs	BREEAM International adapts to local conditions while maintaining global comparability
LEED (U.S.)	- Widely recognized globally - Covers various building types and phases	- Criticized for point-chasing behaviour - May not always result in better building performance	Country-specific versions (e.g., LEED India, LEED Italia) to address local priorities
CGBL (China)	- Tailored to Chinese construction practices and regulations - Government-backed	- Less recognized internationally - Focuses more on design than actual performance	Specifically designed for China's rapid urbanization and unique environmental challenges
BEAM Plus (Hong Kong)	- Addresses Hong Kong's high-density urban environment - Includes site aspects often overlooked in other systems	- Limited international recognition - Complex assessment process	Tailored for Hong Kong's subtropical climate and compact city layout
Green Mark (Singapore)	- Strong focus on tropical climate solutions - Emphasis on energy efficiency	- Limited international application - May overlook some broader sustainability issues	Designed to address Singapore's specific urban and climatic conditions
EDGE (International)	- Simple, fast, and affordable - Suitable for developing markets	- Less comprehensive than some other systems - Focuses primarily on resource efficiency	Universally applicable but with country-specific baselines
NABERS (Australia)	- Performance-based - Focuses on actual rather than predicted performance	- Limited to specific building types - Primarily used in Australia	Tailored to Australian climate zones and building practices

By author (2024)

3.2.2.1 *Economic Impacts*

As shown in Figure 3.2, green building certification can bring various economic benefits to the property. Meta-analysis by Dalton and Fuerst (2018) shows an overall significant rent premium of 6% for green buildings; commercial properties see a 5.4% premium, while residential properties enjoy an 8.2% premium; occupancy rates for commercial green real estate are 5–9% higher. They also found an overall price premium of 7.6% for green buildings. Commercial properties command an 11.5% premium, while residential properties see a 5.5% premium.

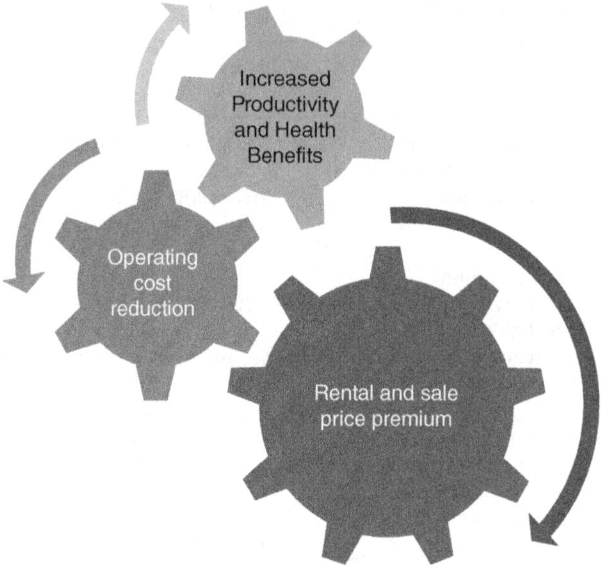

Figure 3.2 Economic Impact of Green Building Certifications.
By author (2024).

Green buildings generally have lower operating costs due to energy and resource savings, as well as reduced maintenance costs. The World Green Building Council (2013)'s analysis of 33 LEED-certified buildings showed significant net present value benefits from operational cost savings.

While harder to quantify, green buildings are associated with improved productivity, health, and well-being of occupants. These benefits can translate into significant value for commercial tenants, potentially outweighing increased rental costs.

3.2.2.2 *Addressing Market Failures*

Green building certifications also help to address a few market failures in the real estate sector. Green building certifications help address the information asymmetry problem by providing credible third-party verification of a building's green features. They help differentiate truly green buildings from those making unsubstantiated claims (greenwashing). The issue of split incentives between landlords and tenants is being addressed through innovative lease structures. Green leases, which include clauses for cost pass-through and shared benefits, are becoming more common. As of 2021, 34% of commercial leases already have some green lease conditions, with another 40% planning to sign by 2025 (Torres et al., 2023). Certifications and associated technologies help internalise positive externalities associated with green buildings. They provide a framework for valuing benefits that might otherwise be overlooked, such as reduced carbon emissions and improved public health.

3.2.2.3 *Market Trends and Technological Advancements*

- *Growing Adoption:* Dodge Construction Network (2018) indicates that close to 50% of participants expect over 60% of their building projects to be certified under a recognised green building system by 2021.
- *Technological Integration:* The introduction of platforms like Arc by Green Business Certification Inc. (GBCI) in 2016 has facilitated performance tracking and certification. As of 2019, Arc was certified for 1.5 billion square meters across 80 countries (Zhang et al., 2019).
- *Evolution of Standards:* LEED v4.1, introduced in 2019, offers a new approach to enhancing green building performance. Over 94,000 LEED-certified commercial projects globally, with an average of 2.2 million square meters being LEED-certified daily.
- *Passive House Technology:* Passive house technology offers significant energy savings, with up to 90% reduction in heating energy, and provides better resilience to power outages during climate disasters. It improves occupant health through controlled ventilation and filtered fresh air. The initial cost premium of 3–10% decreases over time due to learning effects and market adaptation (Council, 2013).
- *Building Electrification:* Features such as heat pump technology are gaining traction as a pathway to net-zero emissions. Building electrification offers benefits such as highly efficient heating and cooling, elimination of separate systems, and improved occupant comfort. The adoption varies by climate and local conditions, with both favourable (e.g., technological improvements, subsidies) and unfavourable (e.g., high upfront costs, electricity prices) factors influencing uptake.

3.2.2.4 *Policy Pressures and Decarbonization Strategies*

The global real estate sector is facing increasing pressure to decarbonise, with policies being implemented across various regions to encourage or mandate sustainable building practices. In response, the industry has developed several key decarbonisation strategies. These include improving energy efficiency through better insulation and more efficient operations, switching to renewable energy sources both onsite (e.g., solar panels) and offsite, and electrifying heating systems by transitioning from gas-based to electricity-based solutions. Some stakeholders are also exploring the purchase of carbon offsets, although this strategy faces challenges related to additionality and effectiveness. Despite these efforts, significant performance gaps persist in the green building sector. Actual energy consumption often exceeds projected consumption, highlighting the need for better modelling and implementation practices. These discrepancies can be attributed to various factors, including behavioural aspects, inconsistencies between design intentions and operational realities, and the persistent issue of split incentives between landlords and tenants. Addressing these challenges will be crucial for the industry to meet its decarbonisation goals and fully realise the potential of green building technologies and practices.

The green building industry faces several significant challenges as it continues to evolve. The high cost of obtaining certifications and implementing advanced technologies can be a deterrent, particularly for smaller projects. Additionally, the complexity of some systems makes them difficult to understand and implement, especially for those with limited resources or expertise. There is also an increasing focus on actual performance verification rather than just design elements, driving the development of more sophisticated monitoring and reporting tools. This shift reflects a growing recognition that a building's real-world performance is often more important than its theoretical design. Another emerging challenge is the expanding scope of emissions reporting, with growing pressure on both landlords and tenants to report on their Scope 1, 2, and 3 emissions. This trend is fostering more collaborative approaches to building management and energy use. Furthermore, as climate change impacts become more severe, green building strategies are increasingly incorporating resilience measures. These include features such as improved insulation for extreme temperatures and enhanced water management systems. Despite these challenges, green building certifications and associated technologies continue to drive the adoption of sustainable building practices.

3.3 Shifted Focus from Green Building to Sustainability

3.3.1 *Rise of Sustainability*

The concept of sustainability has evolved to encompass a broader perspective than just green building practices. According to the United Nations, sustainability is defined as "meeting the needs of the present without compromising the ability of future generations to meet their own needs" (UN, 2015). This definition emphasises the need for an integrated approach that considers both environmental concerns and economic development. As discussed in Chapter 2, the United Nations' 17 Sustainable Development Goals (SDGs) are shaping this broader perspective on sustainability. The World Green Building Council has prioritised 11 of these SDGs, recognising the building and construction sector's potential to contribute significantly to sustainable development (WGBC, 2024).

RICS (2023) suggests that sustainability encompasses many environmental, economic, and social aspects, often referred to as the "three dimensions" in the UN's SDGs. In corporate and investment circles, these factors are frequently considered under the umbrella of environmental, social, and governance (ESG) considerations.

3.3.2 *Financial Incentives for Sustainable Real Estate*

One of the most important factors influencing stakeholders' green building implementation decisions is the economic benefits. Ofek et al. (2018) found that potential energy, maintenance savings, and increases in real estate values are the main forces driving consumers' decisions. Green building premium size is positively correlated to the level of maintenance savings. Meanwhile, developers are more concerned about the energy price and the pursuit of innovation.

Despite the generally extra initial construction costs of 1–10% compared to conventional buildings (Zhang et al., 2019), from a maintenance perspective, some researchers suggest that green buildings perform better than conventional buildings in terms of energy and water efficiency, improving cost efficiency. (Vyas and Jha, 2017) suggests that savings from green buildings can cover the incremental cost of green buildings. Zhao et al. (2018) studied the impact of the green building policy over time on energy efficiency in low-income housing. As a result of decreased energy consumption in green buildings, there was an annual financial saving of USD 648.

Ormond (2021)'s research shows that green building ratings positively impact building values in both hemispheres. An 8–18% sales price premium was found for green-rated buildings compared to equivalent buildings without a BREEAM or NABERS rating across London, Sydney, and Melbourne markets, depending on the level of green rating. The research illustrates that the properties in London with a BREEAM Excellent certification command a 10.5% higher sales price than comparable uncertified buildings. Similarly, properties with a NABERS rating of 5 or above in Sydney and Melbourne enjoy a 17.9% premium in sales price. The study also indicates that rental income and yield influence these green rating-related price premiums. On the rental side, the most ESG-driven tenants are willing to pay a premium for buildings that align with their values.

In the United States, office spaces with LEED certification in general enjoy rents 5.6% higher than their non-certified counterparts. This premium is especially noticeable for buildings with LEED Platinum certification, which are expected to experience annual rental growth rates 0.51% higher than those of non-certified properties (CBRE, 2022).

Deepki's survey of 250 European pension fund managers reveals that 79% anticipate commercial properties with strong ESG profiles yielding superior returns or 'green value' in the coming five years (Jones, 2022). Conversely, properties lacking ESG credentials are expected to face substantial value decline through "brown discounting." 67% of the interviewees expect a significant increase in the impact of brown discounting over the next three years. The survey reports that 40% of pension funds in the past year reported a 21–30% depreciation due to brown discounting. Moreover, buildings with poor environmental performance often face limited returns due to capital expenditure (CAPEX) requirements. Investors increasingly incorporate these factors into their cash flow projections, anticipating legislative changes during their expected holding periods.

Green-rated buildings may reduce risk premiums due to improved liquidity and lower legislation and obsolescence risks, supporting downward pressure on yields. Demand and liquidity for green-rated buildings are likely to increase as more investors utilise ESG criteria as a screening tool. Green-rated buildings enjoy greater futureproofing against potential legislative changes, reducing the risk of assets becoming "stranded" due to obsolescence. Lenders are increasingly incentivised to weigh lending towards green assets, which can lower the risk premium and reduce the cap rate.

The research adds evidence that green building ratings positively correlate with building prices, resulting in higher financial returns. This pattern was observed in different global markets, suggesting a strong relationship between sustainability and value.

3.3.3 *Socially Responsible Real Estate Development*

Real estate stands out among asset classes because it is deeply woven into the fabric of neighbourhoods and has a huge social impact on the communities. It is beyond buildings. Rather, it is about shaping the places where people live, work, and interact. The shift to sustainability has also emphasised the social aspects of real estate development.

The housing affordability issue is pressing worldwide. However, it could create opportunities for impact investing, which is about generating positive social and environmental outcomes without sacrificing financial returns.

Some investors find they can achieve attractive returns from their impact portfolios, challenging the common belief that there's always a trade-off between positive social impact and financial performance. Take PGIM Real Estate Impact's Value Partners Fund VI, for example. It's raised USD 363 million and aims for 12–14% gross internal rates of return by focusing on sustainable real estate development and affordable housing in growing U.S. markets. Then there's Turner Impact Capital's TMIF II, which has raised USD 357 million and targets a 10% net return. It's concentrating on community infrastructure, healthcare facilities, and affordable rentals. With USD 175 million raised and a 15% annual return goal, J.P. Morgan's Urban Renaissance Property Fund is all about urban renewal projects, including affordable housing (CBRE, 2022).

Another angle from which to look at the social perspective of properties is the well-being of occupants, particularly in the workplace, where people spend significant hours a day. The COVID-19 pandemic has intensified the general public's attention to it, which is already a priority for office investors and landlords. There are several key areas of focus. For example, people install ventilation and filtration systems for better air quality. More frequent cleaning is demanded for better hygiene, and more creative space design for more natural light and enough personal space.

According to the International Well Building Institute (2024), companies that invest in health strategies can benefit from improved performance and better financial outcomes. For example, certified healthy buildings can command rents that are 7.7% higher per square foot. High-performing, health-oriented buildings can generate USD 115 per square foot in 10-year net present value. Moreover, enhancing air quality alone can boost employee productivity by 8%. As a result, landlords are increasingly adopting the WELL Building Standard.

This shift from a narrow focus on green building to a broader sustainability perspective reflects the real estate industry's growing recognition of its role in addressing global environmental, social, and economic challenges. It also highlights the potential for sustainable practices to create value for investors, occupants, and communities alike.

3.4 The Emergence of ESG Factors in Real Estate

3.4.1 *ESG Introduction and Integration*

2020 marked a turning point for ESG investment in real estate. Disruptions related to the pandemic, climate change, and increased awareness of social inequalities led investors to adopt more comprehensive approaches to sustainability-related risks.

According to GRESB's definitions of ESG data in the real estate context, ESG stands for:

- *E (Environmental):* Environmental indicators relate to actions and efficiency measures undertaken in order to monitor and decrease the environmental footprint of a portfolio.
- *S (Social):* Indicators relate to an entity's relationship with and impact on its stakeholders and the direct social impact of its activities.
- *G (Governance):* Indicators relate to the governance of ESG, policies, and procedures, as well as the approach to ESG at the entity level.

CBRE (2022) revealed that 60% of participants had already integrated ESG criteria into their investment strategies. This change was observed across the world, with each region demonstrating increased focus on ESG issues relative to previous years.

As ESG principles become increasingly central to corporate operations, investors are now incorporating ESG considerations throughout the entire property lifecycle, from initial due diligence and acquisitions to leasing and ongoing asset management.

3.4.2 ESG Frameworks and Standards

Since sustainability performance reporting is voluntary, real estate companies often need more standardised approaches concerning the content of sustainability reports. While real estate companies are increasingly interested in reporting non-financial information, it is noted that the real estate sector lags behind other sectors in sustainable reporting. Furthermore, there is no uniform approach to the standardised publication of non-financial information, including details on auditing, stakeholder participation, and environmental performance metrics, which European real estate companies most frequently report.

To address this issue in real estate reporting, organisations such as the Carbon Disclosure Project (CDP), Global Reporting Initiative (GRI), and Global Real Estate Sustainability Benchmark (GRESB) offer standardised yet voluntary frameworks for presenting sustainability information (Ionaşcu et al., 2020).

In Europe, with the enactment of Directive 2014/95/EU, the EU member states have mandated non-financial reporting for all public interest entities with over 500 employees, including publicly traded real estate companies. The European Commission recommends the reporting framework of the UN Global Compact initiative, urging companies to align their strategies with universal principles of human rights, labour, the environment, and anti-corruption and to devise creative strategies to achieve the SDGs. Consequently, the 17 SDGs provide a novel framework for businesses to convert global needs into business solutions and offer new management tools for addressing environmental, social, and economic challenges. Table 3.4 illustrates the key sustainability reporting frameworks globally that apply to the real estate sector.

3.4.3 Transparency in ESG Reporting

For more than two decades, green building has provided a framework for market transformation, creating a sizable industry. The result is many certified green building projects that became more efficient, less polluting, and healthier for their occupants. For example, in Australia, the NABERS (2024) rating system is mandatory for office buildings over 1,000 square m. On average, participating offices have reduced their energy use by 35% and halve their carbon dioxide emissions. Some buildings have even slashed their energy consumption by a whopping 80%.

Today, the demand for "green" is rising among institutional investors, property companies and investment funds. This standard includes new targets for greenhouse gas emissions, new requirements for supply chain management, and an increasing focus on social factors such as justice and equity. In addition, increasingly companies, investment funds, and individual assets are required to show their aspirations and plans for confirmed, measured performance. As a result, considering regulations on ESG and sustainability issues are expected to continue to increase, investors and lenders are motivated to seek tools to guarantee "ESG-proof" businesses.

As the real estate industry increasingly focuses on ESG performance, tools like the Global Real Estate Sustainability Benchmark (GRESB) are becoming more critical. GRESB evaluates and compares ESG performance across real estate and infrastructure assets. Participants receive a GRESB score, showing their absolute ESG performance, and a GRESB rating that ranks them against peers. They also get a detailed analysis highlighting strengths and weaknesses in leadership, policies, risk management, health and safety, emissions, certifications, and stakeholder engagement.

Table 3.4 Key Sustainability Reporting Frameworks for Real Estate

Framework	*Who Reports*	*Key Metrics Reported*	*Reporting Period*	*Region*	*Objective*
Global Real Estate Sustainability Benchmark (GRESB)	Real estate companies (asset managers, investors, developers)	Energy, GHG emissions, water, waste, social indicators	Annual (April 1–July 1)	Global	Assess sustainability performance and benchmark against peers
EU Sustainable Finance Disclosure Regulation (SFDR)	Financial market participants, financial advisers	Principal adverse impacts (PAIs) on sustainability factors, sustainability risks, promotion of environmental or social characteristics, and sustainable investment objectives	Continuous (regular updates required)	EU and global funds marketing to EU	Enhance transparency in sustainable investment products and protect against greenwashing
EU Corporate Sustainability Reporting Directive (CSRD)	EU companies, large companies that are public interest entities (PIEs)	ESG factors, climate change impact, board diversity, supply chain due diligence, others	Annual	European Union	Ensure consistency and comparability of sustainability information provided by companies
International Financial Reporting Standards (IFRS) S1 and S2	Any company of any size, public or private	S1: General sustainability-related disclosures; S2: Climate-related disclosures aligned with TCFD	Annual	Global	Set a global baseline for sustainability-related disclosures for capital markets
U.S. SEC Climate Disclosure Rule[a]	SEC-registered public companies	Climate-related risks, greenhouse gas emissions (Scope 1 and 2), climate-related financial statement metrics, progress towards climate-related goals	Annual	United States	Enhance and standardize public companies' disclosures regarding climate-related risks and impacts
UK Sustainability Disclosure Requirements (UK SDR)	UK companies, banks, insurers, and asset managers	ESG impacts, including climate change, waste and resource use, social factors, and governance structures	Annual	United Kingdom	Improve transparency around companies' sustainability practices and impacts

Global Reporting Initiative (GRI)	All organizations	Comprehensive ESG metrics across economic, environmental, and social domains	Annual	Global	Promote transparency and accountability in sustainability reporting
Carbon Disclosure Project (CDP)	Companies, cities, states, and regions	Climate change data, GHG emissions, water security, forest risks	Annual	Global	Encourage environmental transparency and action towards a sustainable economy
Dow Jones Sustainability Indexes (DJSI) and S&P Global Corporate Sustainability Assessment (CSA)	Publicly listed companies	ESG performance in terms of economic, environmental, and social criteria	Annual for CSA, varies for DJSI inclusion	Global	Assess and rank companies based on their sustainability performance

[a] As of early October 2025, the SEC's climate disclosure rule remains stayed and is not in effect, and the consolidated litigation is being held in abeyance by the Eighth Circuit pending further action by the SEC to either recommit to defending the rule or to reconsider it through notice-and-comment rulemaking. In March 2025, the SEC voted to end its defense of the 2024 climate disclosure rule, leading the court to pause proceedings; multiple legal and practitioner updates throughout late September and October 2025 confirm the rule is still paused with no compliance obligations in force. There is no federal SEC climate disclosure compliance requirement in effect at this time (Oct-2025); any future obligation depends on whether the SEC reopens rulemaking or resumes defense and how the Eighth Circuit ultimately rules thereafter. https://www.mintz.com/insights-center/viewpoints/2151/2025-09-30-sustainable-energy-infrastructure-litigation-updates; https://www.velaw.com/insights/federal-climate-change-disclosures-no-more

Documented by author; source: Measurabl (2024)

Another key initiative is the Principles for Responsible Investment (PRI). Signatories must integrate ESG criteria into at least 50% of their assets under management (AUM). They're also required to develop plans for incorporating ESG factors across all portfolios and creatively aggregate fund-level data to the company level.

While certifications like LEED, BREEAM, and NABERS will remain crucial, stakeholders will increasingly expect granular ESG data and information on building performance.

ESG ratings or certifications are set to become essential criteria for investment fund managers. As a result, property managers and investors will need to enhance their data-gathering, monitoring, and analysis capabilities across their portfolios.

3.4.4 ESG in Real Estate Investment

2020 marked a turning point for ESG investment in real estate. Disruptions related to the pandemic, climate change, and increased awareness of social inequalities led investors to adopt more comprehensive approaches to sustainability-related risks.

CBRE (2022) revealed that 60% of participants had already integrated ESG criteria into their investment strategies. This change was observed across the world, with each region demonstrating increased focus on ESG issues relative to previous years. As ESG principles become increasingly central to corporate operations, investors are now incorporating ESG considerations throughout the entire property lifecycle, from initial due diligence and acquisitions to leasing and ongoing asset management.

While decarbonisation might not directly increase investment returns, it will be important in maintaining asset value as tenants increasingly avoid buildings with poor environmental performance. Properties with better sustainability credentials will likely become more valuable as they attract more occupiers and investors. Devine and Kok (2015) found that GRESB participation and performance are both significant predictors of cross-sectional fund returns. They are also associated with the price appreciation component of fund total returns.

Meanwhile, building certifications like LEED, BREAM, and NABERS will stay relevant for measuring environmental performance. The focus is shifting towards initiatives like the World Green Building Council's Net Zero Carbon Buildings Commitment, aiming for all buildings to have zero net carbon emissions by 2050.

CBRE (2022) uses City Square Mall in Singapore as an example of a building where the tenants agree to adopt eco-friendly practices in the mall. Using motion-sensor lighting on escalators and parking areas saves over 50,000 kilowatt hours (kWh) of electricity each year. High-efficiency air conditioning and variable air volume (VAV) systems are installed to control the mall's temperature. As a result, energy is saved, and the mall's annual carbon dioxide emissions are under 4,000 tons.

Property owners and occupants are increasingly turning to "green leases" to monitor and improve the environmental performance of the buildings. The common measurements include power usage per worker, water use by square footage, and the percentage of waste in landfills.

While office tenants may find reducing energy use and targeting net-zero emissions easier, investors need help with retail tenants, who often need energy-intensive lighting for displays. Nevertheless, property owners and managers can still drive significant improvements through green leases and the adoption of smart technology.

3.5 Concluding Remarks

The evolution of green building initiatives has significantly impacted the real estate industry over the past few decades. Since the introduction of BREEAM in 1990, green building certification systems have proliferated globally, driving sustainable construction and property management practices. These systems, including LEED, CGBL, BEAM Plus, Green Mark, Green Star, EDGE, and NABERS, have played a crucial role in setting standards for energy efficiency, resource conservation, and occupant well-being.

The focus has gradually shifted from green building to broader sustainability concerns, encompassing environmental, social, and governance (ESG) factors. This shift has been driven by increasing awareness of climate change, social inequalities, and the need for responsible corporate governance. Given its significant environmental impact and social influence, the real estate sector has become a key player in addressing these sustainability challenges.

Financial incentives have emerged as a powerful driver for sustainable real estate development. Green-certified buildings often command premium rents and sales prices while benefiting from lower operating costs and increased tenant satisfaction. Moreover, socially responsible real estate development has gained traction, with impact investing in affordable housing and community infrastructure offering attractive returns alongside positive social outcomes.

Integrating ESG principles into real estate investment strategies has become increasingly prevalent, with investors considering these factors throughout the property lifecycle. Various ESG frameworks and reporting standards have been developed to improve industry transparency and comparability. Tools like GRESB and initiatives like the PRI have become essential for evaluating and benchmarking ESG performance in real estate portfolios.

As the industry moves forward, the focus will likely shift towards more granular ESG data and performance metrics, emphasising achieving net-zero carbon emissions. Adopting green leases and intelligent technologies will be crucial in driving further improvements in building performance and sustainability.

In conclusion, the real estate industry has made significant strides in embracing sustainability and ESG principles, driven by regulatory pressures, market demands, and the recognition of long-term value creation. Sustainability is set to remain a core consideration in real estate development, investment, and management for the foreseeable future.

3.6 Post-Chapter Engagement

To help reinforce your understanding and apply the concepts covered in this chapter, complete the following activities:

- How has the concept of green buildings evolved into broader sustainability and ESG considerations in real estate?
- Compare and contrast two major green building certification systems. What are their strengths and weaknesses?
- Think about the potential conflicts between environmental sustainability and affordability in real estate development. How might these be reconciled?

References

Ade, R., & Rehm, M. 2020. The unwritten history of green building rating tools: A personal view from some of the "founding fathers". *Building Research & Information*, 48, 1–17.

Architecture2030. n.d. *Why the built environment?* https://www.architecture2030.org/

CBRE. 2022. *ESG and real estate: The top 10 things investors need to know*.

Council, W. G. B. 2013. *The business case for green building: A review of the costs and benefits for developers, investors and occupants*. WGBC.

Dalton, B., & Fuerst, F. 2018. The "green value" proposition in real estate: A meta-analysis. In S. Wilkinson, T. Dixon, N. Miller, & S. Sayce (eds.), *Routledge handbook of sustainable real estate*. Routledge.

Devine, A., & Kok, N. 2015. Green certification and building performance: Implications for tangibles and intangibles. *Journal of Portfolio Management*, 41, 151–163.

Dodge Construction Network. 2018. *World green building and construction trends 2018*. SmartMarket report. Dodge Construction Network.

IEA. 2023. *CO_2 emissions in 2023*.

International Well Building Institute. 2024. *WELL v2*. International WELL Building Institute. https://www.wellcertified.com/certification/v2/

Ionașcu, E., Mironiuc, M., Anghel, I. & Huian, M. C. 2020. The involvement of real estate companies in sustainable development—An analysis from the SDGs reporting perspective. *Sustainability*, 12, 798.

Jones, M. G. 2022. *How are ESG matters affecting the investment market and how does this tie in with the repurposing agenda? RE: Imagining Retail*. Savills.

Mao, X., Lu, H., & Li, Q. A comparison study of mainstream sustainable/green building rating tools in the world. In *2009 International Conference on Management and Service Science*, 2009. IEEE, 1–5.

Measurabl. 2024. *The top 10 sustainability reporting frameworks you should know*. https://www.measurabl.com/sustainability-reporting-frameworks-you-should-know/

NABERS. 2024. *Ratings*. National Australian Built Environment Rating System. https://www.nabers.gov.au/ratings

NASA. 2023. *Global temperature*. https://climate.nasa.gov/vital-signs/global-temperature/?intent=121

Ofek, S., Akron, S., & Portnov, B. A. 2018. Stimulating green construction by influencing the decision-making of main players. *Sustainable Cities and Society*, 40, 165–173.

Ormond, V. 2021. *Green building value: Do green-rated buildings add a premium to sales price?* https://www.knightfrank.com/research/article/2021-09-29-green-building-value-do-greenrated-buildings-add-a-premium-to-sales-price

RICS. 2023. *Sustainability and ESG in commercial property valuation and strategic advice*. Royal Institution of Chartered Surveyors.

Torres, P., Miglani, K., & Lewis, S. 2023. *Green leasing 2.0: Bridging the owner-occupier divide to deliver shared ESG value*. JLL.

UN. 2015. *Sustainable development goals*. United Nations Department of Global Communications. https://sdgs.un.org/goals

Vyas, G., & Jha, K. 2017. Benchmarking green building attributes to achieve cost effectiveness using a data envelopment analysis. *Sustainable Cities and Society*, 28, 127–134.

WGBC. 2024. *What is a sustainable built environment?*. World Green Building Council. https://worldgbc.org/what-is-a-sustainable-built-environment/

Zhang, Y., Wang, H., Gao, W., Wang, F., Zhou, N., Kammen, D. M., & Ying, X. 2019. A survey of the status and challenges of green building development in various countries. *Sustainability*, 11, 5385.

Zhao, D., McCoy, A. P., Agee, P., Mo, Y., Reichard, G., & Paige, F. 2018. Time effects of green buildings on energy use for low-income households: A longitudinal study in the United States. *Sustainable Cities and Society*, 40, 559–568.

Chapter 4

PropTech in Real Estate Investment and Finance

4.1 Introduction to PropTech in Real Estate Investment and Finance

4.1.1 *An Overview of PropTech in Real Estate Investment and Finance*

PropTech, short for property technology, has emerged as a transformative force in the real estate industry, particularly in investment and finance. It includes various technological innovations designed to optimise and streamline real estate value chain processes. PropTech refers to applying cutting-edge technologies, such as artificial intelligence (AI), blockchain, big data analytics, and virtual reality, to traditional real estate practices in real estate investment and finance. Figure 4.1 shows the key areas where PropTech is making significant impacts in real estate investment and finance.

PropTech is revolutionising real estate investment and finance through several key innovations. Digital investment platforms, including online marketplaces and crowdfunding platforms, are reshaping how real estate investments are bought, sold, and funded, democratising access to investment opportunities. Advanced data analytics and predictive modelling tools transform market analysis, risk assessment, and investment decision-making processes. In contrast, automated financial analysis streamlines the modelling and evaluation of real estate projects, enhancing both efficiency and accuracy. Implementing smart contracts and blockchain technology brings unprecedented security, transparency, and efficiency to property transactions and financing.

Furthermore, real estate tokenisation enables fractional ownership and increased liquidity by converting real estate assets into digital tokens, thus reshaping traditional investment structures. This chapter will discuss these key areas in detail.

4.1.2 *Development of PropTech in Real Estate Investment and Finance*

The development of PropTech in real estate investment and finance has unfolded over several phases since the early 2000s. The journey began with the emergence of online property listing platforms, marking the first significant step towards digitisation in the industry. From 2008 to 2015, the sector witnessed the rise of real estate crowdfunding platforms, which democratised access to real estate investments by allowing smaller investors to participate in larger projects. The period from 2015 to 2020 saw a pivotal shift in integrating big data and AI in real estate analytics, enabling more sophisticated market analysis and

DOI: 10.1201/9781003504436-4

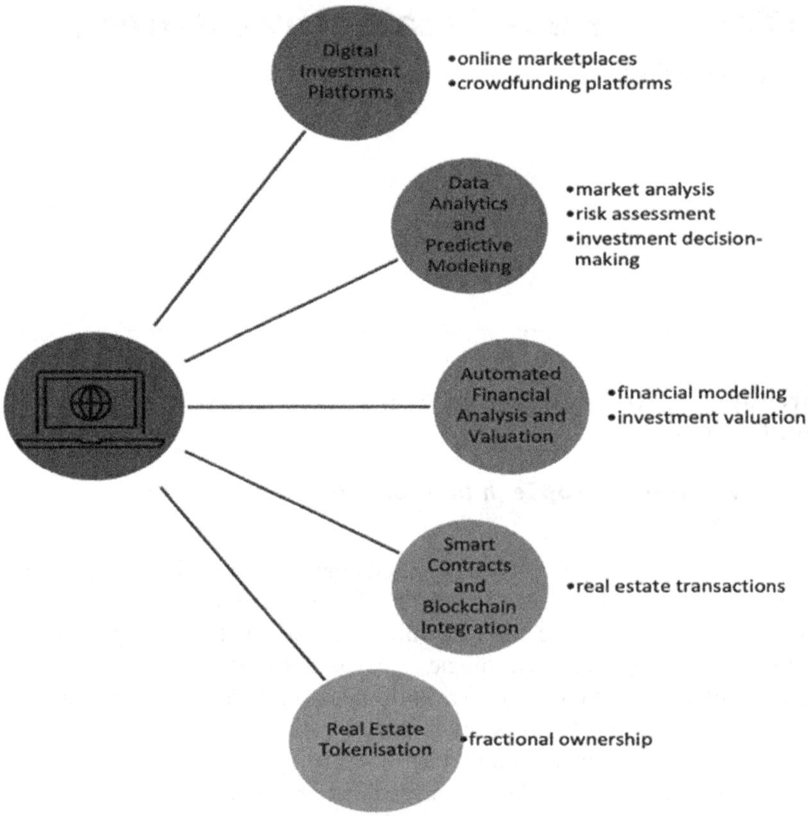

Figure 4.1 PropTech in Real Estate Investment and Finance.
By author (2024).

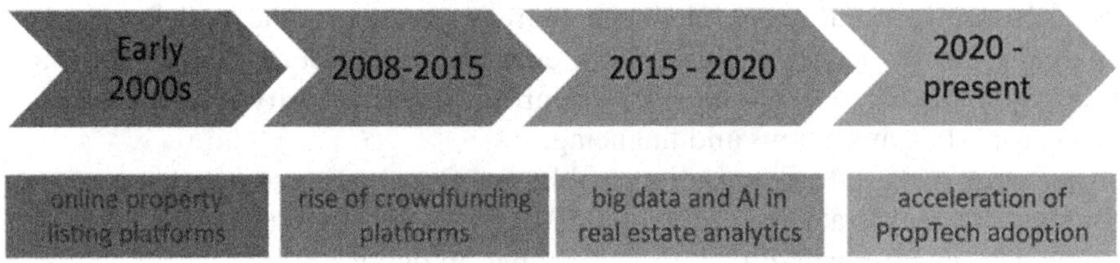

Figure 4.2 The Evolution of PropTech in Real Estate Investment and Finance.
By author (2024).

predictive modelling. This technological leap significantly enhanced decision-making processes for investors and developers alike. From 2020 to the present, the most recent phase has been characterised by an acceleration of PropTech adoption, driven mainly by the global pandemic. This period has seen an increased focus on virtual property tours, blockchain-based transactions, and AI-driven investment strategies, fundamentally transforming how real estate investment and finance operate in the digital age. Figure 4.2 illustrates the evolution of PropTech in real estate investment and finance from the 2000s to the present.

The evolution of PropTech has been punctuated by several significant milestones that have reshaped the real estate investment and finance landscape. In the mid-2000s, the launch of online marketplaces such as Redfin[1] in 2004 and Zillow[2] in 2006 revolutionised property search and valuation, making real estate information more accessible to the general public. This was followed by the introduction of real estate crowdfunding platforms, exemplified by Fundrise[3] in 2010, which opened up new avenues for smaller investors to participate in large-scale real estate projects. As technology advanced, sophisticated AI-powered analytics tools emerged, with companies like HouseCanary[4] leading the way in enhancing property valuation and market forecasting capabilities. More recently, blockchain technology has given rise to platforms like Propy,[5] facilitating secure and transparent real estate transactions. These milestones collectively illustrate the rapid pace of innovation in PropTech and its transformative impact on real estate investment and finance practices.

4.1.3 Regulatory Landscape of PropTech in Real Estate Investment and Finance

The regulatory landscape for PropTech in real estate investment and finance is complex and evolving as regulators work to keep pace with technological advancements (Aghion et al., 2023). Regulations can significantly impact innovation and growth in various sectors, including the real estate industry. Figure 4.3 summarises the key areas where PropTech in real estate investment and finance can be or have been regulated.

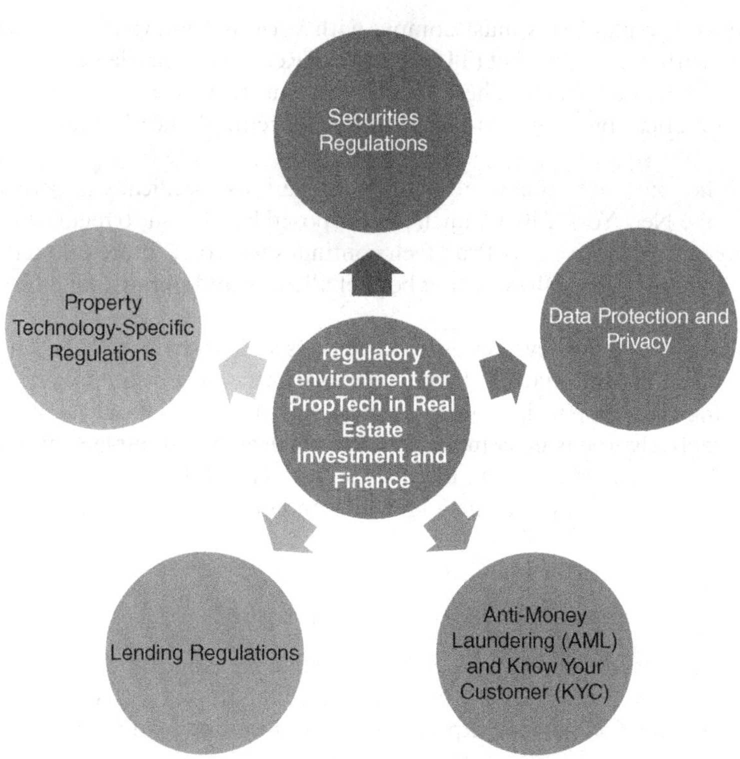

Figure 4.3 Regulatory Environment for PropTech in Real Estate Investment and Finance. By author (2024).

Securities regulations are crucial in shaping the PropTech landscape, particularly for platforms that offer tokenised assets or crowdfunding opportunities. These companies often must comply with securities laws, which in the United States typically involve registration with the Securities and Exchange Commission (SEC) or qualifying for specific exemptions. The regulatory requirements can vary significantly across jurisdictions, adding complexity for companies operating in multiple markets.

Data protection and privacy regulations have become increasingly important as PropTech companies collect and analyse vast amounts of data. Compliance with regulations such as the General Data Protection Regulation (GDPR) in Europe or the California Consumer Privacy Act (CCPA) in the United States is crucial. These regulations impose strict requirements on how companies handle personal data, affecting everything from data collection practices to user consent mechanisms.

Anti-Money Laundering (AML) and Know Your Customer (KYC) regulations are particularly relevant for PropTech platforms involved in financial transactions. These regulations require companies to implement robust procedures to prevent fraud and ensure compliance with financial regulations. The complexity of these requirements can be substantial, particularly for platforms that facilitate cross-border transactions.

As blockchain technology and cryptocurrencies become more prevalent in real estate transactions, companies must navigate evolving regulations surrounding digital assets and blockchain-based transactions. The regulatory approach to these technologies varies widely between jurisdictions, creating a complex landscape for PropTech companies operating in this space.

Digital mortgage platforms must comply with various lending regulations, including laws like the Truth in Lending Act (TILA) and the Real Estate Settlement Procedures Act (RESPA) in the United States. These regulations aim to protect consumers and ensure fair lending practices but can also create significant compliance burdens for innovative platforms.

Some jurisdictions are beginning to develop regulations specifically targeting PropTech. For example, the New York City Council has proposed legislation to regulate the use of AI in tenant screening processes. As PropTech continues to evolve, more tailored regulations are likely to emerge, potentially creating both challenges and opportunities for companies in this space.

The regulatory landscape evolves as policymakers strike a balance between innovation, consumer protection, and market stability. Regulations can substantially impact innovation, suggesting that PropTech companies must stay informed about regulatory changes and work proactively to ensure compliance as they develop and implement new technologies in the real estate investment and finance sector. The ability to navigate this complex regulatory environment may become a key differentiator for successful PropTech companies in the future.

4.2 Digital Platforms Transforming Investment

Digital platforms are revolutionising how real estate investments are bought, sold, and managed. This section will explore two key digital platforms reshaping the real estate investment landscape: online marketplaces and crowdfunding platforms.

4.2.1 *Online Real Estate Marketplaces*

4.2.1.1 *The Rise of Online Shopping and Marketplaces*

The concept of online marketplaces has revolutionised commerce across various sectors since the late 1990s, with the launch of platforms like Amazon and eBay leading to exponential growth in online shopping and fundamentally changing consumer behaviour and business models. These marketplaces are characterised by several key features: 24/7 accessibility, allowing consumers to shop anytime from anywhere with internet access; a vast selection of products often surpassing that of physical stores; price transparency enabling easy comparison across multiple sellers; user reviews providing feedback from previous buyers to inform purchase decisions; personalisation through AI-driven recommendations based on browsing and purchase history; and efficient logistics with integrated shipping and delivery services. The success of this model in retail has inspired similar approaches in other industries, including real estate.

4.2.1.2 *Online Real Estate Marketplaces*

Online real estate marketplaces have emerged as powerful tools for connecting buyers, sellers, and investors in the real estate market. These platforms leverage technology to streamline the property discovery, evaluation, and transaction process.

Online real estate marketplaces have extended the features of online shopping marketplaces to suit the asset. At the core of these platforms are comprehensive property listings, offering extensive databases of properties for sale or investment. These listings often include detailed information, high-quality photos, and virtual tours. Unlike traditional real estate listings, online platforms can update this information in real time, ensuring that users always access the most up-to-date property data.

Platforms provide advanced search and filter tools to help users navigate these vast databases. These allow users to narrow down properties based on specific criteria such as location, price, property type, and investment potential. Many platforms have taken this a step further by incorporating AI to provide personalised property recommendations based on user preferences and behaviour, enhancing the user experience and potentially uncovering opportunities that might be overlooked.

Market analytics and insights have become a crucial component of these platforms. Integrated tools provide market data, trends, and predictive analytics to help investors make informed decisions. This often includes historical price data, neighbourhood statistics, and future growth projections. For investors, these platforms frequently offer tools to calculate potential returns, estimate renovation costs, and analyse rental market conditions, providing a comprehensive view of a property's investment potential.

Virtual and augmented reality features have dramatically enhanced the visual aspect of property hunting. Many platforms now offer virtual property tours and augmented reality tools that allow potential buyers to visualise properties and even potential renovations without physical visits. This technology has proven particularly valuable when in-person viewings are challenging or impossible.

Communication and transaction management have also been streamlined. Platforms often facilitate direct communication channels between buyers, sellers, and real estate professionals, simplifying the negotiation process. Some platforms have taken it even further, offering comprehensive transaction management tools that cover the entire process, from offer submission to closing, including document signing and escrow services. These features make the property transaction process smoother and more efficient for all parties involved.

4.2.1.3 Global Online Real Estate Marketplaces

Real estate online marketplaces have become increasingly important information sources for house purchases (NAR, 2024). In 2023, they were the second most popular channel for purchasing a house after real estate agents. Especially for young people, over 80% of home buyers under 42 use a mobile or tablet search device (NAR, 2024).

Figure 4.4 showcases the representative online real estate marketplaces globally. Each platform has evolved to meet the specific needs of its local market, while often sharing common features such as property listings, search tools, and market analytics. The dominance of national platforms in many markets, as opposed to global players, highlights the localised nature of real estate markets and the importance of understanding local regulations, customs, and market dynamics.

Rightmove is the largest property portal in the United Kingdom, offering comprehensive listings and market analytics. It's complemented by Zoopla, another major player known for its property listings and automated valuation model (AVM). These platforms have become integral to the UK property search process.

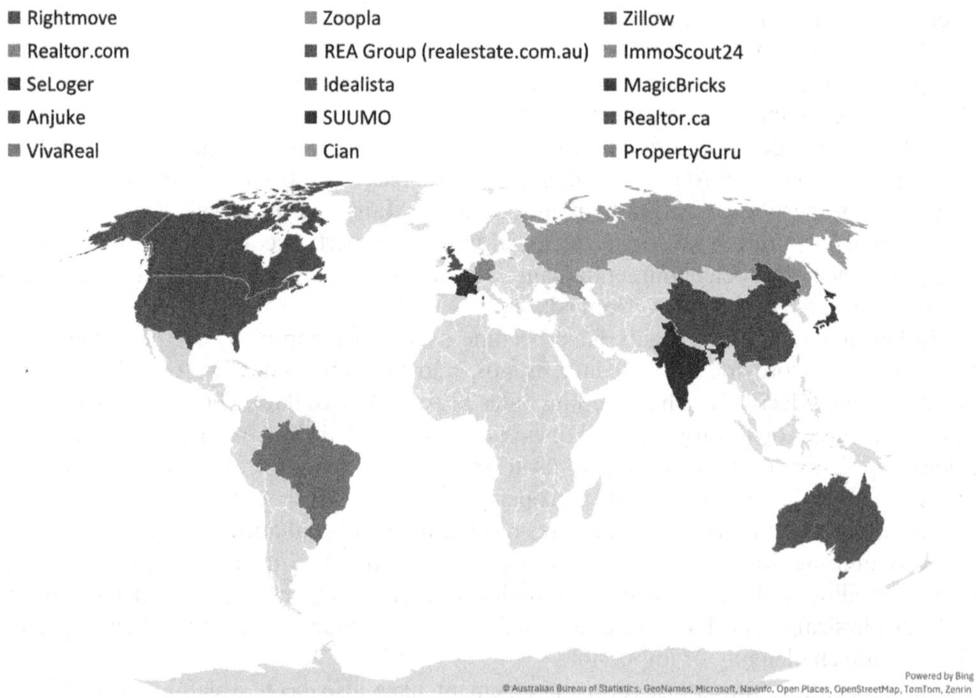

Figure 4.4 Online Real Estate Marketplaces Globally.
By author (2024).

Platforms like Zillow and Realtor.com dominate the United States market. Zillow is renowned for its "Zestimate" AVM. Realtor.com, as the official site of the National Association of Realtors, provides access to multiple listing service (MLS) listings, ensuring comprehensive coverage of available properties.

The REA Group, which operates realestate.com.au, has a significant influence on Australia's property market. Owned by News Corp, this platform has established itself as the dominant Australian property portal. In neighbouring New Zealand, Trade Me Property holds a significant market share.

European countries have developed their national champions. Germany's real estate searches are often conducted through ImmoScout24, the country's largest marketplace. SeLoger serves France's property market well, while Idealista has emerged as a major portal across Southern Europe, covering Spain, Italy, and Portugal.

The variety of platforms in Asia reflects the diversity of markets. India's property seekers often turn to MagicBricks, one of the country's largest property portals. Despite its unique characteristics, China's real estate market is served by platforms like Anjuke, which is part of the larger 58.com group. SUUMO, a leading service in the country, meets Japan's property information needs.

Realtor.ca, the official site of the Canadian Real Estate Association, represents Canada's real estate market online. This platform offers a national database of properties listed through the MLS system.

Brazil's VivaReal has established itself as a major real estate portal in South America, catering to one of the continent's largest property markets. Cian, a leading property classifieds platform, dominates Russia's online real estate landscape.

PropertyGuru heavily influences Southeast Asia's property market, particularly in Singapore. This platform has also expanded its operations to other regional countries, demonstrating the potential for cross-border expansion in some markets.

While sharing features like property listings and search functionalities, these platforms have each evolved to meet the specific needs of their local markets. The prevalence of strong national platforms rather than global players shows the localised nature of real estate markets and the importance of understanding regional nuances in property search and transactions.

4.2.1.4 Case Study: Rightmove in the UK

Rightmove, founded in 2000, is the UK's largest online real estate portal and property website, making up approximately 80% of all time spent on property portals. Top UK estate agencies (Countrywide, Connells, Halifax, Royal & Sun Alliance, and Woolwich) established a joint venture to create a common platform for property listings. This origin as an industry initiative is significant, as it positioned Rightmove as an enabler rather than a disruptor in the property market ecosystem.

- *Business Model and Revenue:* Rightmove operates on a subscription-based model, charging estate agents and letting agents list properties on its platform. The company also generates revenue through additional advertising services and data products. This model has proven highly profitable, making Rightmove one of the most profitable UK companies (Rae, 2024).

- *Key Features and Innovations:* Rightmove offers a comprehensive suite of features to facilitate property searches and provide valuable market insights. At its core is an extensive database of properties for sale and rent across the UK, complemented by an advanced search function that allows users to filter properties based on specific criteria. The platform goes beyond mere listings by providing historical sold prices and current market valuations, giving users a deeper understanding of property values. To help users make informed decisions, Rightmove offers local area guides and school performance data, providing context to the neighbourhoods where properties are located. The user experience is enhanced by a mobile app, enabling on-the-go property searches with a user-friendly interface. Rightmove provides market analytics tools for estate agents, allowing them to analyse market trends and property values, thus serving consumers and real estate professionals.
- *Evolution and Impact on the UK Property Market:* Rightmove's evolution aligns with the stages of Internet business strategy in real estate (Rowley, 2005), as illustrated in Figure 4.5. Rightmove established itself as a pioneering online property portal, consolidating listings from multiple agencies between 2000 and 2005. From 2005 to 2010, the platform soon expanded its reach, becoming the go-to site for property searches and promoting properties over the internet. Rightmove started focusing on customer service in 2010 by enhancing its offerings with more detailed property information, area guides, and tools for consumers and estate agents. The platform has become deeply integrated into the property search process, with its brand becoming synonymous with property search in the UK since 2015.

  Rightmove has significantly influenced UK housing market practices in several ways. First and foremost, it has revolutionised information accessibility in the property sector. Rightmove has effectively collapsed the traditionally separate stages of the housing search into a single, streamlined exercise (Rae, 2015). This consolidation has made property information far more accessible to consumers, allowing them to conduct comprehensive property searches from the comfort of their homes. Rightmove has played a crucial role in increasing market transparency. The platform has shed light on previously opaque aspects of the real estate market by providing comprehensive market data and analytics (Rowley, 2005). This increased transparency has empowered buyers and sellers with a wealth of information, leading to more informed decision-making processes. The platform's extensive property database and sophisticated analytics tools have enabled a shift towards more data-driven decision-making in the real estate sector. Both consumers and estate agents now have access to a vast array of data points and market insights, allowing for more nuanced and informed choices. This trend aligns with the broader digitalisation of real estate markets (Goodchild & Ferrari, 2024). Fields and Rogers

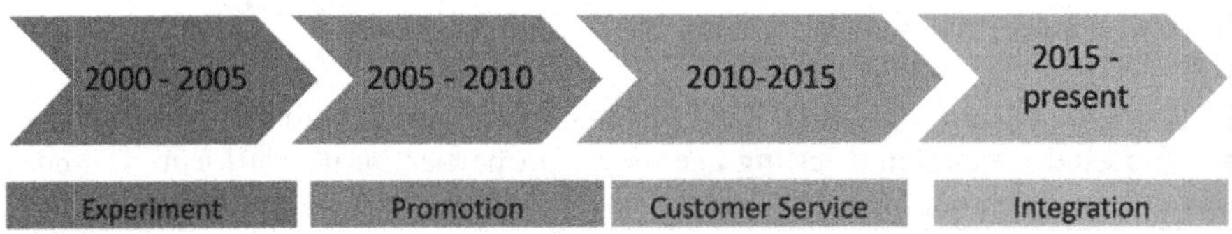

Figure 4.5 Rightmove's Evolution.
By author (2024).

(2021) suggest that platforms like Rightmove encourage a 'calculative mentality' among users, fundamentally altering how properties are valued and compared.

- *Challenges, Criticisms, and Future Outlook:* Rightmove's dominant position in the UK property portal market has led to criticism over its pricing structure. Many estate agents feel that Rightmove's fees are too high and that the company exploits its market position. This has led to attempts by agents to negotiate better terms or seek alternatives. The company faces pressure to continually innovate and add value to justify its fees to agents (Culot, 2023). The acquisition of OnTheMarket[6] by CoStar Group has intensified competition in the UK property portal market. CoStar's substantial resources and ambitious plans for OnTheMarket pose a significant challenge to Rightmove's dominance. Critics argue that Rightmove's dominant position has led to complacency in innovation (Dearsley, 2019). The rise of blockchain and other technologies in real estate transactions could disrupt the traditional portal model, requiring Rightmove to adapt its offerings. As the dominant player in the market, Rightmove may also face increased regulatory scrutiny, particularly if complaints about its market power persist.

Despite these challenges, Rightmove's position remains strong. Innovation is becoming increasingly important for Rightmove's future success. The company is expected to invest more in product development and marketing to maintain its competitive edge. This may lead to operating profit growth lagging behind revenue growth in the near term.

4.2.1.5 Impact on Real Estate Investment

Online real estate marketplace platforms have dramatically transformed the landscape of real estate investment, democratising access to opportunities that were once the preserve of large institutional investors. These platforms have significantly increased accessibility, allowing more individual investors to participate in the market. By providing comprehensive market data and analytics, they have also enhanced market transparency, helping to reduce the information asymmetry that has long characterised real estate transactions.

These platforms have brought about substantial efficiency gains. Tasks related to finding and evaluating investment properties that once took weeks can now be accomplished in hours, streamlining the entire investment process. Moreover, these platforms have expanded investors' geographical reach, facilitated cross-border real estate investments, and opened up new markets for diversification.

Data-driven decision-making has become a hallmark of modern real estate investment, largely thanks to the information available through these platforms. Investors can now access and analyse vast amounts of data to inform their investment strategies, leading to more sophisticated and potentially more successful investment decisions.

While real estate agents continue to be involved in the industry, online marketplaces have reduced the need for intermediaries in initial property discovery and information gathering. This partial disintermediation has changed the real estate market dynamics, although the complexity of property transactions ensures that human expertise remains valuable.

These platforms' increased transparency and accessibility have also contributed to more efficient price discovery in real estate markets. Connecting a larger pool of buyers and sellers can increase liquidity, particularly for investment properties. This increased liquidity could make real estate a more attractive and viable investment option for a wider range of investors.

Online real estate marketplaces represent a significant evolution in how real estate is bought, sold, and invested. They successfully combine the convenience and accessibility of e-commerce platforms with the unique characteristics of real estate transactions. The ongoing development of these platforms promises to streamline processes further, enhance transparency, and open new investment opportunities in the global real estate market.

4.2.2 Real Estate Crowdfunding Platforms

Real estate crowdfunding platforms are online platforms that enable many individuals to pool their money to invest in real estate projects or properties that would typically be beyond the reach of most individual investors. These platforms leverage technology to connect real estate developers or property owners with a broad base of potential investors, democratising access to real estate investments (Lowies et al., 2017). Montgomery et al. (2018) defined real estate crowdfunding as "a method of financing whereby real estate project developers or property owners raise capital for their projects or properties from a large number of investors through an internet-based platform."

These platforms typically operate under specific regulatory frameworks, such as the Jumpstart Our Business Startups (JOBS) Act in the United States, which allows for broader participation in private investments (Maarbani, 2015). They often provide detailed information about investment opportunities, including property details, financial projections, and risk assessments, to help investors make informed decisions.

4.2.2.1 Key Features of Real Estate Crowdfunding Platforms

Real estate crowdfunding platforms offer several key features that make them attractive to investors. One of the most significant advantages is the lower investment minimums, with some platforms allowing investments starting from as low as USD 10, dramatically increasing the accessibility of real estate investing to a broader range of individuals (Davis, 2024). These platforms also facilitate diversification by enabling investors to spread their capital across multiple properties or projects, potentially reducing risk. Transparency is another crucial feature, with platforms typically providing detailed information about each investment opportunity, including financials, market analysis, and risk factors, allowing investors to make more informed decisions. Furthermore, many platforms offer automated investment management, handling administrative tasks such as rent collection, distributions, and tax documentation. This streamlines the investment process for participants, reducing the time and effort required to manage real estate investments.

4.2.2.2 Types of Real Estate Crowdfunding

Real estate crowdfunding platforms typically offer three main types of investment opportunities.

1 *Equity Crowdfunding:* In this model, investors receive ownership shares in a property or a portfolio of properties.

   Equity crowdfunding allows investors to become partial owners of real estate projects, potentially benefiting from both property appreciation and rental income. Investors' returns are typically proportional to their ownership stake and are often distributed as a share of the property's cash flow or upon the sale of the property. This type of investment usually offers higher potential returns but comes with higher risk, as investors' returns are not guaranteed and depend on the property's performance. For example, a platform might offer shares in a multi-family apartment complex. Investors will receive a portion of the rental income and a share of the profits if the property is sold at a higher value.

2 *Debt Crowdfunding:* In this model, investors fund loans secured by real estate, earning returns through interest payments.

   Debt crowdfunding, also known as real estate crowdlending, involves investors collectively funding a loan for a real estate project. The property typically secures the loan, providing some protection for investors. Returns come in regular interest payments over a set period, ranging from 6 months to 5 years. While the potential returns are generally lower than equity investments, they are more predictable and carry less risk. An example might be a short-term loan for a property flip, where investors fund the purchase and renovation costs and receive monthly interest payments plus their principal back when the property is sold.

3 *Hybrid Models:* Some platforms offer investments that combine elements of both equity and debt.

   Hybrid models aim to balance the higher potential returns of equity investments with the more stable, predictable returns of debt investments. These might be preferred equity, mezzanine debt, or convertible notes. For instance, a platform might offer an investment that combines a fixed interest rate (akin to a debt investment) with a share of the profits upon selling the property (similar to an equity investment).

Each model has its risk-return profile, allowing investors to choose opportunities that best align with their investment goals and risk tolerance. Many platforms offer a mix of these investment types, enabling investors to diversify their real estate portfolios across various structures and property types.

4.2.2.3 Impact on Real Estate Investment

Real estate crowdfunding platforms have profoundly impacted the investment landscape, most notably through democratising access. These platforms have opened up real estate investment opportunities to a much wider range of investors, including those with limited capital who were previously excluded from this asset class. By allowing investments with relatively low minimums, crowdfunding has allowed retail investors to participate in real estate projects that were once the exclusive domain of high-net-worth individuals and institutional investors.

Another significant impact is the potential for increased liquidity in real estate investments. Some crowdfunding platforms offer secondary markets where investors can trade shares or investments. While not as liquid as publicly traded securities, these secondary markets offer potentially greater liquidity than traditional real estate investments, which often require long-term commitments.

Through these platforms, diversification has become much more accessible. Investors can easily spread their capital across multiple properties and geographic locations, potentially reducing risk and increasing the stability of their real estate portfolio. This level of diversification was previously challenging for individual investors to achieve in the real estate sector.

Crowdfunding platforms have also disintermediated the real estate investment process. By connecting investors directly with property developers or owners, these platforms often reduce the need for traditional intermediaries such as banks or investment brokers. This disintermediation can lower costs for investors and those seeking capital, improving the market's overall efficiency.

For real estate developers and property owners, crowdfunding platforms offer the potential for faster access to capital compared to traditional financing methods. The streamlined process and broader investor base can accelerate funding timelines, allowing projects to move forward more quickly.

In conclusion, digital platforms, including both online marketplaces and crowdfunding platforms, are significantly reshaping the real estate investment landscape. They offer increased accessibility, transparency, and efficiency while presenting new opportunities and challenges for investors and developers alike. The ability to invest with smaller amounts, diversify easily, and potentially access more liquid investments is changing how individuals approach real estate investing. For developers and property owners, these platforms provide new opportunities for raising capital and expanding their reach to a broader investor base.

4.3 Data Analytics for Informed Decision-Making

4.3.1 Big Data in Real Estate

The real estate industry has fundamentally transformed by adopting data analytics and big data technologies (Barkham et al., 2022). This transformation has revolutionised how stakeholders approach property valuation, investment decisions, market analysis, and operational efficiency (Baldominos et al., 2018). Big data and advanced analytical capabilities have created unprecedented opportunities to gain deeper market insights and make more informed decisions. This shift represents what Lee et al. (2023) described as a "digital renaissance" in real estate, fundamentally altering how professionals understand and interact with market dynamics, property performance, and investment opportunities.

4.3.1.1 Types of Data Analytics in Real Estate

As Grover et al. (2018) identified in their review of real estate analytics frameworks, the real estate data analytics field encompasses three distinct but interconnected approaches: descriptive analysis, predictive analysis, and prescriptive analysis. Singh et al. (2020) highlighted the importance of diagnostic analysis in big data analytics for property price prediction. Combining these approaches, which serve different purposes and provide unique insights into the market, we construct the "analytical hierarchy" of real estate decision-making (Figure 4.6).

Descriptive analytics forms the foundation of real estate data analysis, summarising historical data to depict past events and the current state. This approach employs various statistical methods, including time series analysis and pattern recognition, to understand historical price trends, occupancy rates, and revenue performance. Through business intelligence dashboards and visualisation tools, descriptive analytics helps stakeholders understand market dynamics.

Diagnostic analytics takes this understanding further by explaining why certain outcomes occurred. It examines relationships and dependencies within the data through

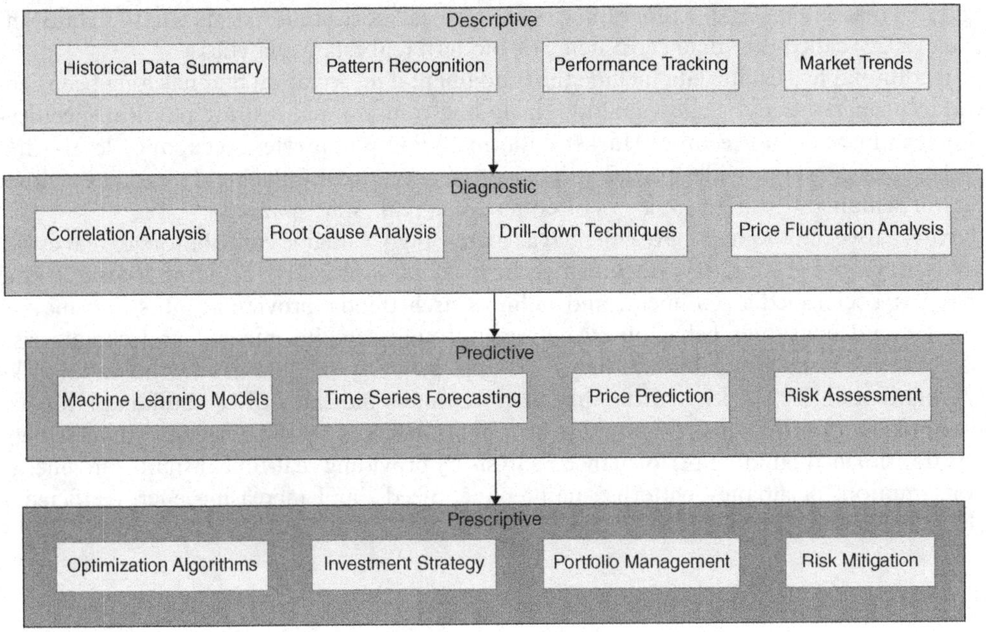

Figure 4.6 Types of Data Analytics in Real Estate.
By author (2024).

correlation analysis and root cause investigation. Diagnostic analytics can reveal causality in property value fluctuations and changes in vacancy rates, enabling more informed decision-making based on demonstrated relationships between various market factors.

Predictive analytics represents a more sophisticated approach, which Choi et al. (2018) describe as the cornerstone of modern real estate investment strategy. Their research shows how advanced statistical modelling and machine learning algorithms can forecast future outcomes based on historical data. Predictive analytics provides insights that help stakeholders prepare for and capitalise on future market conditions via regression analysis, time series forecasting, and, more recently, neural networks.

Prescriptive analytics uses advanced data analysis techniques to predict future market trends and recommend specific actions a real estate professional should take to optimise investment strategies, maximise return, and minimise risk based on current market conditions and property characteristics (Patten, 2020). Their research demonstrates how this approach combines optimisation algorithms, simulation modelling, and expert systems to suggest optimal strategies for investment, portfolio management, and risk mitigation. Hirvonen et al. (2024) further show how prescriptive analytics helps real estate professionals understand what might happen and determine the best course of action in response to predicted scenarios.

4.3.1.2 Data Sources in Real Estate Analytics

Comprehensive real estate market analysis requires integrating diverse data sources, which can be divided into three main groups: traditional real estate data, alternative data sources, and macroeconomic and demographic data. This categorisation reflects what D'Hauwens et al. (2022) called the "data ecosystem" of modern real estate analytics.

Traditional real estate data include the fundamental information that has long been central to property analysis and valuation, including transaction histories, physical specifications, and ownership records. Market data, including rental rates, occupancy levels, and construction costs, provides crucial insights into market performance and trends, as Geltner (2007) demonstrated in the analysis of commercial real estate markets.

Alternative data sources have emerged as increasingly valuable contributors to real estate analysis. Ghose et al. (2019) documented how digital footprints, including website traffic patterns, social media sentiment, and online search trends, provide insights into market interest and consumer behaviour. Analysed through satellite imagery and climate risk assessments, environmental data helps evaluate property conditions and potential risks (Zennaro et al., 2021). Integrating the smart readiness indicator (SRI), building information modelling (BIM), and cyber-physical systems in smart buildings' energy management has transformed building performance analysis by providing real-time insights into energy consumption, occupancy patterns, maintenance needs, and increasing energy efficiency (Hernández et al., 2024).

4.3.1.3 Technology Stack for Real Estate Analytics

Implementing effective real estate analytics requires a complete technology stack that Patten (2020) describe as a "multi-layered technological ecosystem." This framework increasingly incorporates a wide range of data management, analysis, and visualisation tools.

Business intelligence (BI) tools represent the "visualisation layer" of the real estate analytics technology stack. Platforms like Tableau,[7] Power BI,[8] and Looker[9] have become essential for transforming raw data into actionable insights. These tools' collaborative features facilitate knowledge sharing across organisations, ensuring that insights derived from data analysis can inform decision-making at all levels.

Statistical software and programming languages form the computational core of the analytics technology stack. With their extensive ecosystems of specialised libraries, languages such as R and Python provide the foundation for sophisticated data analysis. Integrating specialised packages enables analysts to perform complex data manipulations that reveal deeper insights into real estate market dynamics.

Machine learning frameworks have become increasingly vital to real estate analytics (Naeem et al., 2023; Menin Machado, 2018). Platforms such as TensorFlow[10] and PyTorch[11] support various applications, from property valuation to market forecasting. Natural language processing capabilities enable the analysis of unstructured data sources, adding another dimension to real estate analytics (Munawar et al., 2020; Sinyak et al., 2021; Cardenas et al., 2022).

4.3.1.4 Impact on Investment Decision-Making

The integration of advanced analytics, such as big data, AI, and machine learning, has fundamentally transformed real estate investment decision-making, a "paradigm shift" in investment strategy (Alsahan & AlZaidan, 2024; Martínez & Rodríguez, 2024). Enhanced market intelligence is the primary impact of advanced analytics on investment decision-making, leading to more accurate valuations and improved risk assessment (Battisti et al., 2020; Munawar et al., 2020). The ability to process vast amounts of market data enables a more sophisticated understanding of market dynamics.

4.3.1.5 Challenges and Future Outlook

Implementing advanced analytics in real estate presents several substantial challenges in the technical, organisational, and regulatory domains. Data quality and integration are primary challenges in implementing real estate analytics. Organisations consistently encounter data accuracy, completeness, and standardisation issues across different sources. Integrating data from disparate systems demands significant expertise and resources. Technical challenges persist in the infrastructure layer of analytics implementation. Organisations must address system scalability, processing speed, and storage capacity. The complexity of modern analytics platforms requires substantial investment in infrastructure and expertise.

Advanced analytics capabilities continue to evolve, particularly in AI and machine learning applications for real estate. These technologies enable increasingly sophisticated approaches to property analysis and market forecasting. Natural language processing and computer vision applications are opening new possibilities for analysis.

Big data analytics has become indispensable in real estate investment and decision-making, fundamentally transforming how organisations approach market analysis and investment decisions. The ability to effectively collect, analyse, and act upon data insights will become increasingly critical for success in the real estate industry. Organisations need to develop and enhance their analytics capabilities while addressing key data quality, privacy, and technical implementation challenges.

4.3.2 Virtual and Augmented Reality (VR/AR)

Virtual reality (VR) and augmented reality (AR) technologies have emerged as transformative forces in real estate investment and finance. These immersive technologies revolutionise how properties are marketed, evaluated, and transacted. According to Starr et al. (2021), VR/AR technologies stand out among PropTech innovations because they offer unparalleled visualisation and interaction capabilities.

4.3.2.1 Technology Overview and Applications

VR creates virtual worlds without spatial reference to the real world, while AR combines actual life with digital reality by superimposing computer-created virtual objects, scenes, and systems on actual scenes (Naeem et al., 2023). These technologies have a substantial market impact—by 2025, revenues generated from VR/AR software are expected to reach USD 2.6 billion, with the market reaching at least USD 80 billion in real estate (Deep et al., 2023).

Virtual property tours represent one of the most significant applications of VR technology in real estate. These tours enable potential buyers to virtually inspect properties from any location, providing immersive experiences through VR headsets. Xiong et al. (2024) demonstrate that this capability facilitates global property transactions while significantly reducing the time and resources traditionally required for property viewing.

The technology's application extends to enhanced property visualisation, where AR and VR support interactive design review and collaboration. Deep et al. (2023) note that these technologies create intuitive 3D communication settings that enable thorough analysis of building models before physical construction begins. This capability has proven valuable in new development projects and major renovation planning.

In terms of marketing and promotion, VR and AR technologies have transformed traditional property advertisements and search functions. Naeem et al. (2023) describe how these technologies provide platforms for virtual property presentation and architectural visualisation, facilitating virtual trade opportunities that were previously impossible. This has particularly benefited international property transactions and cross-border investments.

4.3.2.2 Impact on Investment Decision-Making

Research has demonstrated significant effects of VR/AR on real estate investment decisions and property marketing outcomes. In their empirical study, Xiong et al. (2024) found that properties with VR tours attract 72.3% more prospective buyers for physical inspections. Furthermore, their research revealed that a 1% increase in online followers leads to a 21% increase in physical property visits. Perhaps most significantly, they documented that VR tours reduce marketing time by 6.4% and narrow the bid–ask spread by 2%.

These technologies have fundamentally altered the decision-making process in real estate investment. Deep et al. (2023) identified four key constructs that influence investment decisions through VR/AR: confident value perception (CVP), innovative investment appeal (IIA), trusted property transactions (TPT), and effortless property engagement (EPE). These constructs provide a framework for understanding how virtual and augmented reality technologies influence investor behaviour and decision-making processes.

4.3.2.3 Digital Twin Analytics: Real Estate Data Analysis Through VR/AR

While VR/AR technologies are commonly associated with property visualisation and virtual tours, their integration with digital twin capabilities creates powerful new real estate data analysis tools. Digital twins utilise spatial data to provide the core analytical framework while incorporating equipment and engineering data to understand building systems. Combined with VR/AR interfaces, this creates an immersive analytical environment that transforms how stakeholders interact with property data.

The analytical capabilities of digital twins in VR/AR environments operate across multiple levels:

- At the component level, the digital twin monitors individual building elements—for instance, tracking HVAC system performance metrics or analysing lighting system energy consumption patterns. VR/AR interfaces allow users to virtually "walk through" the building and interact with its components, accessing real-time performance data and historical trends through intuitive visual overlays.
- At the asset level, comprehensive building performance analysis becomes possible by integrating multiple data streams, including building management system (BMS) data, IoT sensor feeds, and environmental metrics. VR/AR interfaces allow users to visualise building-wide patterns, identify inefficiencies, and simulate potential improvements. For example, facility managers can use VR to explore energy usage patterns throughout a building, with heat maps and data visualisations overlaid on the virtual model.

- At the portfolio level, VR/AR-enabled digital twins support comparative analysis across multiple properties. Investment managers can virtually navigate through their entire portfolio, analysing performance metrics, identifying trends, and making data-driven investment decisions. The technology enables side-by-side comparison of properties in virtual space, with relevant data and analytics readily accessible through the AR interface.

Predictive analytics capabilities are enhanced by integrating AI and machine learning with VR/AR visualisation. The digital twin can identify patterns, forecast potential issues, and suggest optimisations, presenting these insights through intuitive visual interfaces in virtual space. This helps stakeholders understand complex data relationships and make more informed property operations and investment decisions.

The value of this integrated approach is demonstrated through measurable outcomes in operational efficiency, maintenance optimisation, and decision support. For example, organisations implementing VR/AR-enabled digital twins have reported up to 35% reductions in operational costs through data-driven optimisation and 90% increases in maintenance efficiency through predictive scheduling (Ottinger et al., 2021). The immersive nature of VR/AR interfaces makes complex data more accessible and actionable, supporting better-informed property management and investment decisions.

VR/AR integration particularly enhances real-time monitoring and decision-support features. Instead of reviewing data on traditional dashboards, stakeholders can immerse themselves in the virtual property environment, interacting directly with real-time data feeds and analytical insights. Mobile VR/AR applications extend these capabilities beyond the office, allowing property managers and investors to access and analyse property data anywhere.

Combining digital twin analytics with VR/AR represents a significant evolution of real estate data analysis. These technologies move beyond traditional visualisation to create interactive and intelligent analytical environments. They are expected to be increasingly central in property management, investment analysis, and portfolio optimisation.

4.3.2.4 Implementation Challenges and Future Prospects

Despite the clear benefits, several challenges exist in implementing VR/AR technologies in real estate. Deep et al. (2023) highlight technical issues, including difficulties detecting real-time tracking sensors, user discomfort with extended use, and dissatisfaction with visual quality in some applications. Display screen limitations and integration challenges with existing systems also present ongoing concerns for implementation.

The adoption barriers extend beyond technical issues. Starr et al. (2021) identify several implementation challenges, including high initial investment costs and the need for specialised expertise. Integrating these technologies with existing business processes requires significant organisational change, while user training requirements can present additional hurdles for widespread adoption.

The future of VR/AR in real estate investment appears promising, with several significant developments on the horizon. Naeem et al. (2023) discuss the emergence of spatial augmented reality (SAR) and other technological improvements that promise enhanced visualisation capabilities, improved user comfort, and better integration with other PropTech solutions.

The market impact of these technologies continues to evolve. Xiong et al. (2024) project that VR/AR technologies will continue to reduce property marketing time, increase property viewing efficiency, improve market information transparency, and enhance decision-making processes. These improvements are expected to drive further adoption across the industry.

It is also noteworthy that VR/AR technologies are transforming the real estate industry. Starr et al. (2021) project fundamental changes in property marketing strategies, investment evaluation processes, client engagement methods, and transaction procedures. These changes suggest a future where virtual and augmented reality become integral to real estate investment and finance practices.

Virtual and augmented reality technologies are fundamentally changing real estate investment and finance by providing immersive, data-rich environments for property evaluation and decision-making. They are expected to further enhance market efficiency, reduce transaction costs, and improve investment outcomes. Recent empirical studies (Deep et al., 2023; Xiong et al., 2024) demonstrate that VR/AR technologies enhance the property viewing experience, contributing to more efficient market operations and better-informed investment decisions.

4.4 PropTech and Financial Modelling in Real Estate

4.4.1 *Artificial Intelligence (AI) in Real Estate*

Artificial intelligence (AI) represents a fundamental transformation in real estate investment and finance, introducing new paradigms for valuation, financial modelling, operations, and market analysis. Recent advances in AI technologies, particularly large language models (LLMs) and explainable AI frameworks, have created opportunities for more transparent and efficient real estate operations (Acharya et al., 2024; Haurum et al., 2024). The integration of these technologies addresses traditional challenges in real estate, including information asymmetry between buyers and sellers, bounded rationality in decision-making, and the need for more sophisticated analysis tools. This section examines the foundational AI technologies and their applications in real estate, laying the groundwork for more specific implementations, which will be explored in subsequent sections.

4.4.1.1 AI Framework in Real Estate

The implementation of AI in real estate operates across three primary layers (shown in Figure 4.7).

1. *Data Processing Layer:* This layer involves integrating structured and unstructured data from multiple sources, the automated processing and standardisation of real estate data, and creating relevant variables for analysis through feature engineering.
2. *Analytics Layer:* This layer focuses on pattern recognition to identify market trends and correlations, predictive modelling to forecast market movements and property performance, and natural language processing to analyse textual data and market sentiment.

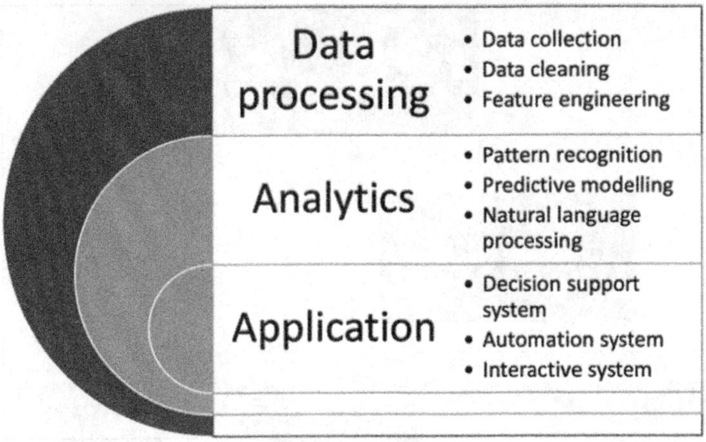

Figure 4.7 Layers of AI Implementation in Real Estate.
By author (2024).

3 *Application Layer:* This layer encompasses AI-powered decision support systems that provide recommendations for investment decisions, automation systems that streamline routine tasks, and interactive systems that enable users to access AI-derived insights.

4.4.1.2 *Core AI Technologies in Real Estate*

Among all the AI technologies that can be applied in real estate, three core technologies are most commonly used in real life.

1 *Machine Learning Frameworks:* Machine learning serves as the foundation for various real estate applications, ranging from market analysis to the development of investment strategies. Recent implementations using LangChain and advanced language models have demonstrated superior capabilities in processing real estate data and providing professional insights (Haurum et al., 2024). Tree-based models like LightGBM and XGBoost have shown particular effectiveness in real estate predictions, with the ability to handle complex non-linear relationships while maintaining interpretability (Trindade Neves et al., 2024). These frameworks are increasingly complemented by explainable AI techniques that ensure transparency and fairness in decision-making processes (Acharya et al., 2024).
2 *Natural Language Processing (NLP):* NLP technologies enable the analysis of unstructured data sources in real estate. This includes processing market reports and news articles to identify relevant trends, analysing legal documents, assessing social media sentiment, and extracting insights from property descriptions.
3 *Computer Vision:* Advanced image processing capabilities support various real estate-related tasks. This includes property feature recognition to identify key characteristics, condition assessment to evaluate a property's state, location analysis to assess factors like accessibility and amenities, and development potential evaluation to identify opportunities for improvement or expansion.

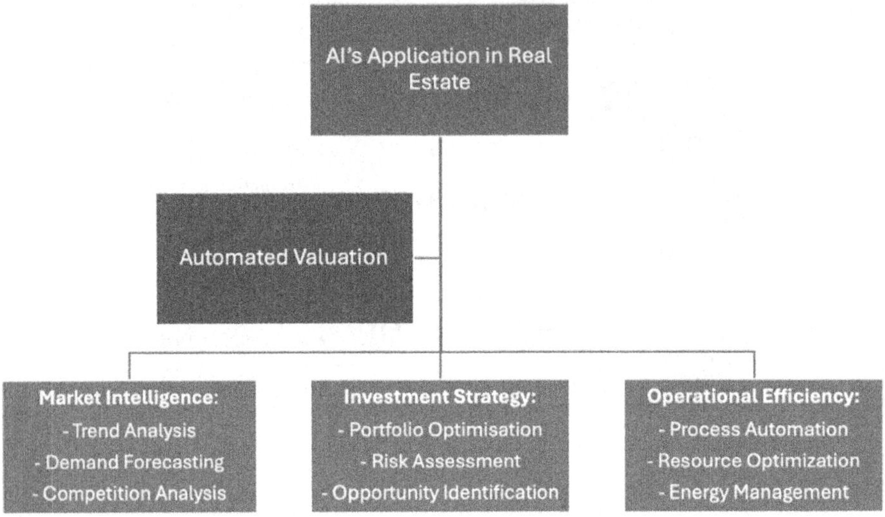

Figure 4.8 AI's Application in Real Estate
By author (2024).

4.4.1.3 AI Applications Beyond Valuation

AI's application in real estate extends beyond automated valuation (detailed in Section 4.4.2) and several other key industry areas, as shown in Figure 4.8.

In market intelligence, AI systems are very good at identifying emerging market patterns through sophisticated trend analysis. These systems also provide valuable demand forecasting capabilities, helping stakeholders anticipate future market movements.

Investment strategy is another area in which AI has made significant impacts. Through advanced algorithms, AI delivers data-driven portfolio optimisation recommendations, helping investors achieve optimal asset allocation. The technology also implements multi-factor risk assessment frameworks that provide deeper insights into potential investment risks. Furthermore, AI systems are particularly adept at opportunity identification, automatically detecting and flagging promising investment prospects that might otherwise go unnoticed.

Regarding operational efficiency, AI has revolutionised how real estate businesses function on a daily basis. Process automation has streamlined numerous administrative tasks, reducing manual workload and minimising human error. Resource optimisation has become more sophisticated through AI-driven analysis, ensuring more effective allocation of both human and material resources. The technology has also significantly contributed to energy management, enabling smarter building operations that reduce costs while improving sustainability.

4.4.1.4 Implementation Framework

Successful AI implementation in real estate requires three key dimensions (Figure 4.9).

First, strategic planning forms the foundation of any AI initiative. This begins with a comprehensive needs assessment to identify specific business requirements that AI can

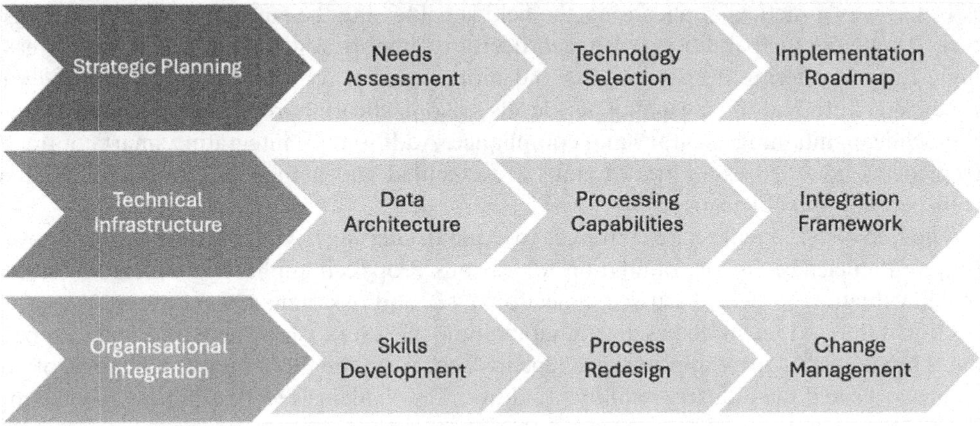

Figure 4.9 Dimensions of Implementing AI in Real Estate.
By author (2024).

address. Organisations must then carefully select appropriate AI technologies that align with these needs. A well-defined implementation roadmap that outlines a phased deployment plan ensures a systematic and manageable rollout.

The second critical dimension involves establishing the necessary technical infrastructure. This requires thoughtful design of data architecture systems that can effectively collect and store the vast amounts of information AI systems need to function. Organisations must also ensure they have sufficient processing capabilities by implementing the required computing infrastructure. Furthermore, a robust integration framework must be developed to connect AI systems seamlessly with existing business technologies and databases.

The final dimension focuses on organisational integration, essential for long-term success. This involves developing comprehensive skills development programs to build internal capabilities and ensure staff can effectively work with AI systems. Business processes must often be redesigned to accommodate new AI-enabled workflows and maximise efficiency gains. Additionally, organisations must implement robust change management initiatives to support cultural transformation and promote the adoption of AI technologies among employees. This holistic approach helps ensure that AI implementation succeeds technically and becomes deeply embedded in the organisation's operational fabric.

4.4.1.5 Future Directions

Recent research by Haurum et al. (2024) and Trindade Neves et al. (2024) indicates that the evolution of AI in real estate is advancing along several key trajectories, shaped by breakthroughs in language models, explainable AI, and automated systems. One significant development is the movement towards enhanced integration across platforms and systems. This encompasses the seamless connection between different AI applications, allowing them to work in concert rather than in isolation. Furthermore, this integration extends to the broader PropTech ecosystem, creating interconnected networks of technology solutions. The industry is increasingly moving towards unified platforms that can manage diverse real estate operations cohesively, streamlining workflows and improving efficiency.

In terms of practical applications, the field is witnessing the emergence of more sophisticated tools and capabilities. Advanced decision support systems are becoming increasingly refined, offering deeper insights and more nuanced recommendations. Transaction processing automation is reaching new levels of sophistication, reducing manual intervention while maintaining accuracy and compliance. Additionally, integrating smart contracts transforms how real estate transactions are executed and managed, introducing greater transparency and efficiency.

The role of AI in real estate continues to expand, creating new opportunities for innovation and efficiency. As the foundation for various PropTech applications, including automated valuation models (discussed in Section 4.4.2) and risk management tools (covered in Section 4.4.3), AI technologies are reshaping how real estate investment and finance operate. The success of these applications depends heavily on the proper implementation of AI frameworks and the industry's ability to address key challenges while adapting to evolving market needs.

4.4.2 Automated Valuation Models (AVMs)

AVMs represent a significant advancement in PropTech, transforming traditional property valuation methods through sophisticated algorithms and data analytics. These models utilise computational techniques to estimate property values, incorporating multiple data sources and advanced statistical methods to provide rapid, objective valuations (Glumac & Des Rosiers, 2021). Recent studies have demonstrated AVMs' increasing sophistication and accuracy, with modern AI-driven systems achieving accuracy rates above 90% when tested on large transaction datasets (Sing et al., 2021). The evolution of AVMs from simple regression models to sophisticated AI-driven systems reflects the broader technological transformation in real estate valuation practices. However, their effectiveness depends heavily on the quality of underlying data and the proper implementation of performance metrics (Isakson et al., 2020).

Since their inception in the 1980s, AVMs have progressed through several distinct phases. Initially, these models relied primarily on basic regression analysis and limited datasets. Contemporary AVMs have evolved to incorporate AI, machine learning, and vast property and market data. This evolution has been driven by the need to overcome the limitations of traditional valuation methods, particularly in accuracy, consistency, and speed (Demirci, 2021).

Modern AVMs integrate diverse data sources through a sophisticated technical architecture. At the foundation lies a comprehensive data integration layer that combines property characteristics, transaction histories, market comparables, economic indicators, and geographic information systems (GIS) data. The quality and reliability of this data integration process are important for AVM performance, with recent studies showing that data quality directly impacts valuation accuracy (Isakson et al., 2020). Advanced techniques such as fuzzy logic and rough set theory can help generate reliable valuations for markets with insufficient source data (Renigier-Biłozor et al., 2019).

The analytical engine of contemporary AVMs employs multiple valuation methodologies to generate accurate property valuations. Machine learning algorithms, including neural networks, random forests, and gradient boosting techniques, enhance prediction accuracy by identifying subtle patterns in property data. Recent implementations using boosted tree ensembles have shown promise, demonstrating the ability to explain over 91%

of price variance in residential markets (Sing et al., 2021). These advanced methods significantly improve traditional mass appraisal techniques (Demirci, 2021).

Quality control systems form an essential component of AVM architecture. These systems incorporate confidence scoring mechanisms, outlier detection algorithms, and market condition adjustments to ensure valuation reliability. The development of standardised performance metrics has become increasingly important, with research showing that proper implementation of metrics like forecast standard deviation (FSD) is crucial for assessing AVM reliability (Isakson et al., 2020). This is particularly important in markets with limited transaction data, where conventional validation methods may be insufficient (Renigier-Bilozor et al., 2019).

In investment decision-making, AVMs serve multiple crucial functions. They enable rapid assessment of large property portfolios, facilitate regular revaluation for mark-to-market purposes, and help identify value drivers and trends. Automating these processes has significantly reduced the time and cost associated with portfolio valuation while maintaining or improving accuracy levels (Sing et al., 2021). The integration of advanced valuation methods has made it possible to value properties even in markets with limited comparable sales data (Renigier-Bilozor et al., 2019).

The effectiveness of modern AVMs is well illustrated by their implementation in Singapore's housing market. Sing et al. (2021) developed and tested an AI-based AVM using boosted tree ensembles on over 300,000 housing transactions across Singapore's unique two-tier housing system. The model achieved remarkable accuracy, explaining 91.33% of price variance in the public housing market and 94.28% in the private housing market, with mean absolute percentage errors of 8.55% and 5.34%, respectively. Even in out-of-sample testing using 2018 data, the model maintained error rates between 5% and 9%, demonstrating its robust predictive capabilities. This success was attributed to sophisticated data integration, including property-specific attributes, spatial characteristics, market timing factors, and one-year-lagged average housing values at the building level.

The Singapore case demonstrates how AVMs can effectively handle complex market structures and large-scale valuation tasks while maintaining high accuracy. However, AVMs still face several significant challenges. Technical limitations include data quality and availability issues, model assumptions and biases, and difficulty capturing unique property characteristics. The transition from traditional valuation models to automated systems requires careful consideration of these limitations and the development of appropriate validation frameworks (Glumac & Des Rosiers, 2021). While mass valuation methods offer numerous advantages, they must be implemented with an understanding of their limitations and appropriate contexts for use (Demirci, 2021).

The impact of AVMs on valuation practices has been substantial. They have augmented traditional valuation methods, enhanced efficiency in mass valuations, and improved consistency while reducing subjective bias. Integrating automated systems has led to the development of new taxonomies and frameworks for understanding real estate valuation in the digital age (Glumac & Des Rosiers, 2021). This transformation represents a fundamental shift in how property valuations are conducted and validated (Demirci, 2021).

Looking towards the future, several trends are emerging in AVM development. Technical advancements promise enhanced AI capabilities, improved data integration, and real-time valuation updates. The ongoing evolution of performance metrics and validation frameworks will be crucial for maintaining market confidence and regulatory compliance (Isakson et al., 2020). The development of more sophisticated approaches for addressing data limitations and market uncertainties will further enhance the applicability of AVMs across various market contexts (Renigier-Bilozor et al., 2019).

4.4.3 Risk Management Tools

Risk management in real estate has undergone significant transformation due to the integration of PropTech solutions. Modern risk management tools leverage AI, big data analytics, and advanced modelling techniques to provide more comprehensive and accurate risk assessment capabilities. These technological innovations enable real estate professionals to identify, assess, and mitigate risks more effectively across various dimensions of real estate investment and operations.

4.4.3.1 Technological Framework for Risk Management

The contemporary approach to real estate risk management is built upon a technological framework that includes multiple components:

1 *Data Integration Systems:* These systems aggregate data from various sources, including market transactions, economic indicators, environmental data, and property-specific information. Using diverse data sources enables a more comprehensive understanding of real estate investment risk factors.
2 *Analytics Engines:* Advanced analytics engines process the integrated data using machine learning algorithms, statistical models, and predictive analytics to identify patterns, correlations, and potential risk factors. These engines can process structured and unstructured data, providing insights that might not be apparent through traditional analysis methods.
3 *Risk Assessment Platforms:* These platforms provide user interfaces and visualisation tools that enable stakeholders to interact with risk analysis results, conduct scenario planning, and make informed decisions based on the processed data and analytics.

4.4.3.2 Key Risk Categories and Associated Tools

Modern risk management tools address several key categories of real estate risk:

Market Risk Assessment Tools: These tools assess market conditions, trends, and potential volatility to evaluate investment risks. They use machine learning algorithms to analyse historical data, current market indicators, and predictive models to forecast market movements. Using advanced sentiment analysis, they monitor news feeds, social media, and additional sources to measure market sentiment and detect potential shifts before they appear in conventional indicators.

Financial Risk Analysis Systems: These systems evaluate financial risks associated with real estate investments. They include:

- Cash flow analysis tools that model potential variations in income and expenses
- Debt service coverage ratio (DSCR) monitoring systems
- Leverage risk assessment platforms
- Interest rate sensitivity analysis tools

Environmental Risk Management Platforms: Environmental and climate risk tools have evolved significantly, becoming essential real estate risk management components. These platforms integrate multiple data sources and advanced analytics to provide comprehensive climate risk assessment capabilities that address the real estate sector's immediate and long-term challenges.

Climate Risk Assessment Capabilities: The platforms employ sophisticated approaches to assess physical and transition risks. For physical risks, these systems analyse acute events such as extreme weather incidents alongside chronic risks, including sea-level rise and changing temperature patterns. The platforms also evaluate transition risks, considering the impact of regulatory changes, market shifts, and technological developments related to climate change adaptation and mitigation.

These platforms incorporate several analytical layers that work together. The climate scenario analysis component examines property vulnerability under different climate scenarios, typically aligned with IPCC projections and incorporating various warming scenarios such as 1.5°C, 2°C, and 4°C pathways. This analysis gives investors crucial insights into potential long-term impacts on property values and operational viability. Through advanced GIS integration, the hazard exposure mapping functionality enables a detailed assessment of exposure to various climate hazards, including flooding, hurricanes, wildfires, and heat stress. These tools synthesise historical data with forward-looking climate models to project future risk exposure with increasing accuracy.

These platforms also conduct detailed asset-level vulnerability assessments at the individual property level. This involves evaluating specific property characteristics that might affect climate resilience, such as building materials, elevation, HVAC systems, and flood protection measures. Through this analysis, organisations can identify specific adaptation needs and develop targeted mitigation strategies.

Operational Risk Management Systems: These systems monitor and analyse risks associated with property operations, including:

- Tenant default risk assessment
- Maintenance risk prediction
- Occupancy rate analysis
- Operating expense monitoring: Advanced AI algorithms can predict potential operational issues before they occur, enabling proactive risk management.

4.4.3.3 *Integration with Other PropTech Solutions*

Risk management tools increasingly integrate with other PropTech solutions to create comprehensive risk management ecosystems. This integration enables more effective risk monitoring and mitigation:

Integration with Property Management Systems: Risk management tools connect with property management platforms to access real-time operational data, enabling continuous risk monitoring and early warning systems for potential issues.

Connection with Financial Planning Tools: Integration with financial planning and analysis tools enables real-time updating of risk assessments based on financial performance and market conditions.

Linkage with Smart Building Systems: Connection to smart building infrastructure provides real-time data on building performance, occupancy patterns, and maintenance needs, enhancing operational risk management capabilities.

4.4.3.4 Advanced Features and Capabilities

Modern risk management tools have several advanced features that enhance their effectiveness. Scenario analysis and stress testing capabilities allow users to model multiple potential future scenarios, assess the impact of various risk factors, test portfolio resilience under different market conditions, and evaluate the effectiveness of different risk mitigation strategies.

Real-time monitoring and alert systems provide continuous risk assessment updates through automated alert systems for risk threshold violations, early warning indicators for emerging risks, and customisable monitoring parameters based on risk tolerance levels. Portfolio-level risk analytics enable comprehensive portfolio risk assessment, correlation analysis between different properties, geographic concentration risk evaluation, and sector exposure analysis.

4.5 Concluding Remarks

PropTech has emerged as a transformative force in real estate investment and finance, fundamentally altering how properties are valued, traded, and managed. The technological innovations discussed in this chapter demonstrate the industry's evolution from traditional, relationship-based practices to data-driven, digitally enabled operations.

Digital platforms, particularly online marketplaces and crowdfunding platforms, have democratised real estate investment by reducing barriers to entry and increasing market accessibility. These platforms have enhanced market transparency and efficiency while creating new opportunities for both individual and institutional investors. The success of platforms like Rightmove illustrates how digital transformation can revolutionise market operations while maintaining the essential role of human expertise.

The integration of advanced data analytics and AI has significantly improved real estate decision-making processes. Big data analytics, combined with virtual and augmented reality technologies, enable more sophisticated market analysis and property evaluation. These tools provide stakeholders with unprecedented insights into market trends, property performance, and investment opportunities.

Automated valuation models (AVMs) represent a significant advancement, demonstrating how AI can enhance traditional valuation practices. The high accuracy rates achieved by modern AVMs, as evidenced by implementations in markets like Singapore, suggest that these tools will play an increasingly important role in property valuation while complementing rather than replacing human expertise.

PropTech solutions that enable more comprehensive and accurate risk assessment have revolutionised risk management in real estate. Integrating climate risk analysis, financial modelling, and operational monitoring provides stakeholders with robust tools for identifying and mitigating various forms of risk. These advances are particularly crucial as the industry faces evolving challenges related to climate change and market volatility.

However, implementing PropTech solutions requires careful consideration of several factors. Organisations must develop appropriate technical infrastructure, ensure regulatory compliance, and manage the cultural transformation necessary for effective technology adoption. The challenge lies in implementing individual technologies and creating integrated systems that enhance overall operational efficiency while maintaining security and reliability.

PropTech is likely to continue evolving, driven by advances in artificial intelligence, blockchain technology, and data analytics. Future developments may include enhanced integration across platforms, more sophisticated predictive capabilities, and improved automation of complex processes. However, the industry must balance technological innovation with practical considerations of data quality, privacy concerns, and the need for human oversight.

The future of real estate investment and finance appears increasingly technology-driven, yet the fundamental principles of property value and market dynamics remain crucial. Success in this evolving landscape will depend on effectively combining technological capabilities with traditional real estate expertise, ensuring that innovation enhances rather than replaces human judgement in investment and financial decision-making.

Notes

1 https://www.redfin.com/.
2 https://www.zillow.com/.
3 https://fundrise.com/.
4 https://www.housecanary.com/.
5 https://propy.com/home/.
6 https://www.onthemarket.com/.
7 https://www.tableau.com/en-gb.
8 https://app.powerbi.com.
9 https://cloud.google.com/looker/?hl=en#analyze-governed-data-deliver-business-insights-and-build-ai-powered-applications.
10 https://www.tensorflow.org/.
11 https://pytorch.org/.

References

Acharya, D. B., Divya, B., & Kuppan, K. 2024. Explainable and fair AI: Balancing performance in financial and real estate machine learning models. *IEEE Access*, 12, 154022–154034.

Aghion, P., Bergeaud, A., & Van Reenen, J. 2023. The impact of regulation on innovation. *American Economic Review*, 113, 2894–2936.

Alsahan, I. M., & Alzaidan, Z. I. 2024. Unleashing the power of artificial intelligence in real estate valuation: Opportunities and challenges ahead. *Journal of Knowledge Learning and Science Technology*, 3, 1–10.

Baldominos, A., Blanco, I., Moreno, A. J., Iturrarte, R., Bernárdez, Ó., & Afonso, C. 2018. Identifying real estate opportunities using machine learning. *Applied Sciences*, 8, 2321.

Barkham, R., Bokhari, S., & Saiz, A. 2022. Urban big data: City management and real estate markets. In P. M. Pardalose, S. T. Rassia, & A. Tsokas (eds.), *Artificial intelligence, machine learning, and optimization tools for smart cities: Designing for sustainability*, 177–209. Springer.

Battisti, E., Shams, S. R., Sakka, G., & Miglietta, N. 2020. Big data and risk management in business processes: Implications for corporate real estate. *Business Process Management Journal*, 26, 1141–1155.

Cardenas, E., Shorten, C., Khoshgoftaar, T. M., & Furht, B. 2022. A comparison of house price classification with structured and unstructured text data. *The International FLAIRS Conference Proceedings*, Hutchinson Island, Jensen Beach, Florida.

Choi, T. M., Wallace, S. W., & Wang, Y. 2018. Big data analytics in operations management. *Production and Operations Management*, 27, 1868–1883.

Culot, C. 2023. "I feel like I'm working to pay Rightmove", says disgruntled agent. *Property Industry Eye*. https://propertyindustryeye.com/i-feel-like-im-working-to-pay-rightmove-says-disgruntled-agent/

D'Hauwens, R., Walravens, N., & Ballon, P. 2022. Data ecosystem business models: Value and control in data ecosystems. *Journal of Business Models*, 10, 1–30.

Davis, C. 2024. 4 best real estate crowdfunding investment platforms of October 2024. https://www.nerdwallet.com/best/investing/real-estate-crowdfunding-platforms

Dearsley, J. 2019. PropTech today: Rightmove fails to understand first rule of innovation. *Estate Agent Today*. https://www.estateagenttoday.co.uk/breaking-news/2019/07/proptech-today-rightmove-fails-to-understand-first-rule-of-innovation/

Deep, S., Vishnoi, S., Malhotra, R., Mathur, S., Yawale, H., Kumar, A., & Singla, A. 2023. Influence of augmented reality and virtual reality on real estate investment decisions: Understand consumer perspective in Indian AEC industry. *Engineering, Construction and Architectural Management*, 32(2), 1122–1140.

Demirci, O. 2021. *Automated valuation models (AVMs): Machine learning, namely mass (advanced) valuation methods and algorithms*. Property Elite.

Fields, D., & Rogers, D. 2021. Towards a critical housing studies research agenda on platform real estate. *Housing, Theory and Society*, 38, 72–94.

Geltner, D. 2007. *Commercial real estate analysis and investments*. Thompson Higher Education Mason.

Ghose, A., Ipeirotis, P. G., & Li, B. 2019. Modeling consumer footprints on search engines: An interplay with social media. *Management Science*, 65, 1363–1385.

Glumac, B., & Des Rosiers, F. 2021. Towards a taxonomy for real estate and land automated valuation systems. *Journal of Property Investment & Finance*, 39, 450–463.

Goodchild, B., & Ferrari, E. 2024. Intermediaries and mediators: An actor-network understanding of online property platforms. *Housing Studies*, 39, 102–123.

Grover, V., Chiang, R. H., Liang, T.-P., & Zhang, D. 2018. Creating strategic business value from big data analytics: A research framework. *Journal of Management Information Systems*, 35, 388–423.

Haurum, K. R., Ma, R., & Long, W. 2024. Real estate with AI: An agent based on LangChain. *Procedia Computer Science*, 242, 1082–1088.

Hernández, J. L., De Miguel, I., Vélez, F., & Vasallo, A. 2024. Challenges and opportunities in European smart buildings energy management: A critical review. *Renewable and Sustainable Energy Reviews*, 199, 114472.

Hirvonen, M., Kauppi, K., & Liesiö, J. 2024. Identifying enablers for the successful deployment of prescriptive analytics–a multiple case study. *European Business Review*, 37 (2), 245–277.

Isakson, H. R., Ecker, M. D., & Kennedy, L. 2020. Principles for calculating AVM performance metrics. *The Appraisal Journal*, 88, 14–29.

Lee, C. H., Chan, C., He, Y., Zhao, C., Wang, H., & Zhou, G. Y. 2023. Framework of digital renaissance with human-in-the-loop. *Integration of Energy, Information, Transportation and Humanity: Renaissance from Digitization*, 169.

Lowies, B., Viljoen, C., & McGreal, S. 2017. Investor perspectives on property crowdfunding: Evidence from Australia. *Journal of Financial Management of Property and Construction*, 22, 303–321.

Maarbani, S. 2015. *Real estate crowdfunding: The digital evolution of real estate funding and investment*. PricewaterhouseCoopers.

Martínez, J., & Rodríguez, M. 2024. Investment analysis in the era of big data: Tools and techniques. *International Journal of Accounting, Finance, and Economic Studies*, 2.

Menin Machado, G. 2018. Big data analytics for real estate asset management. PhD thesis. https://www.politesi.polimi.it/handle/10589/148475

Montgomery, N., Squires, G., & Syed, I. 2018. Disruptive potential of real estate crowdfunding in the real estate project finance industry: A literature review. *Property Management*, 36, 597–619.

Munawar, H. S., Qayyum, S., Ullah, F., & Sepasgozar, S. 2020. Big data and its applications in smart real estate and the disaster management life cycle: A systematic analysis. *Big Data and Cognitive Computing*, 4, 4.

Naeem, N., Rana, I. A., & Nasir, A. R. 2023. Digital real estate: A review of the technologies and tools transforming the industry and society. *Smart Construction and Sustainable Cities*, 1, 15.

NAR. 2024. *2024 NAR home buyers and sellers generational trends*. National Association of REALTORS, Research Group.

Ottinger, N. B., Jordan Stein, E., Crandon, M. G., & Jain, A. 2021. *Digital twin: The age of Aquarius in construction and real estate*. Ernst & Young Global Limited.

Patten, J. M. 2020. Data: Unlocking hidden insight and value across dynamic real estate portfolios. *Corporate Real Estate Journal*, 9, 317–331.

Rae, A. 2015. Online housing search and the geography of submarkets. *Housing Studies*, 30, 453–472.

Rae, A. 2024. The housing market, technology and new platforms. In Gibb, K., Leishman, C., Marsh, A., Meen, G., Viforl, R. O., & Watkins, C. (eds.), *The Routledge handbook of housing economics*. Routledge.

Renigier-Biłozor, M., Janowski, A., & D'Amato, M. 2019. Automated valuation model based on fuzzy and rough set theory for real estate market with insufficient source data. *Land Use Policy*, 87, 104021.

Rowley, J. 2005. The evolution of internet business strategy: The case of UK estate agency. *Property Management*, 23, 217–226.

Sing, T. F., Yang, J. J., & Yu, S. M. 2021. Boosted tree ensembles for artificial intelligence based automated valuation models (AI-AVM). *The Journal of Real Estate Finance and Economics*, 65, 1–26.

Singh, A., Sharma, A., & Dubey, G. 2020. Big data analytics predicting real estate prices. *International Journal of System Assurance Engineering and Management*, 11, 208–219.

Sinyak, N., Tajinder, S., Kumari, J., & Kozlovskiy, V. 2021. Predicting real estate market trends and value using pre-processing and sentiment text mining analysis. *Real Estate: Economics, Management*, 35–43. https://doi.org/10.22337/2073-8412-2021-1-35-43

Starr, C. W., Saginor, J., & Worzala, E. 2021. The rise of PropTech: Emerging industrial technologies and their impact on real estate. *Journal of Property Investment & Finance*, 39, 157–169.

Trindade Neves, F., Aparicio, M., & De Castro Neto, M. 2024. The impacts of open data and explainable AI on real estate price predictions in smart cities. *Applied Sciences*, 14, 2209.

Xiong, C., Cheung, K. S., Levy, D. S., & Allen, M. 2024. The effect of virtual reality on the marketing of residential property. *Housing Studies*, 39, 671–694.

Zennaro, F., Furlan, E., Simeoni, C., Torresan, S., Aslan, S., Critto, A., & Marcomini, A. 2021. Exploring machine learning potential for climate change risk assessment. *Earth-Science Reviews*, 220, 103752.

Chapter 5

PropTech Strategies in Sustainable Real Estate

5.1 Integration of PropTech in Sustainable Real Estate Practices

The integration of PropTech into sustainable real estate practices represents a significant shift in how the industry approaches environmental, social, and governance (ESG) challenges. This convergence of technology and sustainability is reshaping the built environment, offering innovative solutions to long-standing issues and creating new opportunities for value creation.

5.1.1 PropTech as a Strategy

PropTech has emerged as a powerful tool for pursuing sustainable real estate practices. The industry is using PropTech to find new ways to improve its environmental performance, enhance operational efficiency, and improve the well-being of building occupants.

5.1.1.1 Energy Efficiency Solutions

Energy efficiency is one of the most critical aspects of sustainable real estate, and PropTech plays an important role in this area. Smart energy management systems, powered by artificial intelligence (AI) and machine learning algorithms, are revolutionising how buildings consume and conserve energy.

For instance, predictive energy management systems can analyse historical data, weather forecasts, and occupancy patterns to optimise HVAC operations. These systems can automatically adjust temperature settings, reducing energy waste while maintaining comfort levels. Some advanced systems can learn from occupant behaviour and preferences over time, further refining their energy-saving strategies.

Smart lighting systems are another area where PropTech is making significant strides. These systems use a combination of sensors, AI, and Internet of Things (IoT) connectivity to adjust lighting levels based on occupancy, time of day, and natural light availability. Some systems can even adjust the colour temperature of lighting to support occupants' circadian rhythms, potentially improving well-being and productivity.

PropTech also enables more efficient integration of renewable energy sources into buildings. Smart grid technologies and energy storage solutions allow buildings to consume, produce, and store energy. This bidirectional flow of energy, facilitated by PropTech, is paving the way for net-zero energy buildings and even energy-positive structures that produce more energy than they consume (Ghaffarianhoseini et al., 2018).

DOI: 10.1201/9781003504436-5

5.1.1.2 *Smart Building Technologies*

Today's smart buildings are complex ecosystems of interconnected technologies that work together to optimise performance, enhance user experience, and reduce environmental impact.

The building management system (BMS) is the heart of smart building technology. It acts as the building's central nervous system. Modern BMS platforms leverage IoT sensors, cloud computing, and AI to monitor and control various building systems in real time. These include everything from HVAC and lighting to security and elevator systems.

One key advantage of smart building technologies is their ability to generate and analyse vast amounts of data. This data-driven approach allows more precise control of building systems, predictive maintenance, and continuous performance optimisation. For example, smart buildings can automatically adjust ventilation rates by analysing occupancy patterns and space utilisation data, potentially reducing energy consumption while improving indoor air quality.

Smart building technologies also enhance the user experience in ways that support sustainability goals. For instance, mobile apps that allow occupants to control their immediate environment (like lighting or temperature) improve comfort and lead to energy savings by ensuring that resources are used only when and where needed.

5.1.1.3 *Waste Reduction Technologies*

PropTech is making significant contributions to waste reduction in the real estate sector during construction and building operations. During construction, building information modelling (BIM) technology helps reduce material waste by enabling more precise planning and coordination. BIM allows for the virtual construction of a building before physical construction, identifying potential conflicts and inefficiencies that could lead to waste.

In operational buildings, smart waste management systems use IoT sensors and AI to optimise waste collection and recycling processes. For example, sensors in waste bins can monitor fill levels and composition, allowing for more efficient collection routes and better sorting of recyclables. Some systems can even use machine learning to predict waste generation patterns, further optimising collection schedules.

5.1.1.4 *Digital Twin Technology*

Digital twin technology represents one of the most exciting developments in PropTech for sustainable real estate. A digital twin is a virtual replica of a physical building that can be used to monitor, simulate, and optimise building performance in real time.

Digital twins integrate data from various sources, including IoT sensors, BMS, and external data feeds (like weather forecasts), to create a comprehensive, real-time model of the building. This technology allows for unprecedented levels of analysis and optimisation. For example, facility managers can utilise digital twins to simulate various operational scenarios, testing the impact of different strategies on energy consumption, occupant comfort, and other key performance indicators.

5.1.1.5 *Blockchain for Sustainable Real Estate*

While still in its early stages of adoption in real estate, blockchain technology offers promising applications for enhancing sustainability. Blockchain can provide transparent, immutable records of a building's sustainability credentials, from energy performance to material sourcing. This can help combat greenwashing and provide more reliable data for green building certifications.

Blockchain is also being explored as a means to facilitate peer-to-peer energy trading in microgrids, which could potentially enhance the viability of local renewable energy generation. Moreover, blockchain-based smart contracts might automate and enforce sustainability-linked agreements, such as green leases or performance-based maintenance contracts.

5.1.2 *Fundamental Principles of Sustainable PropTech*

Several fundamental principles (Figure 5.1) guide the integration of sustainability into PropTech strategies. These principles ensure that technological innovations are not pursued for their own sake but are aligned with broader sustainability goals and contribute to creating more environmentally and socially responsible built environments.

5.1.2.1 Lifecycle Thinking

One of the core principles guiding sustainable PropTech is lifecycle thinking. This approach considers the environmental, social, and economic impacts of a building or technology throughout its entire lifecycle, from raw material extraction through manufacturing, use, and end-of-life disposal or recycling. Sustainable PropTech solutions are designed for durability and longevity, with the ability to be updated and adapted over time rather than replaced entirely. This reduces waste and resource consumption associated with frequent technology turnover. The energy and resources consumed by PropTech solutions are carefully considered alongside the efficiencies they create, as sustainable PropTech solutions aim for a net positive impact throughout their lifecycle. End-of-life considerations are also crucial, as sustainable PropTech strategies must account for what will happen to hardware components when they are no longer useful. This might include designing for easy disassembly and recycling or implementing take-back programs. Additionally, there is growing recognition of the need to consider the embodied carbon of PropTech solutions – the emissions associated with their production and installation – alongside the operational efficiencies they provide.

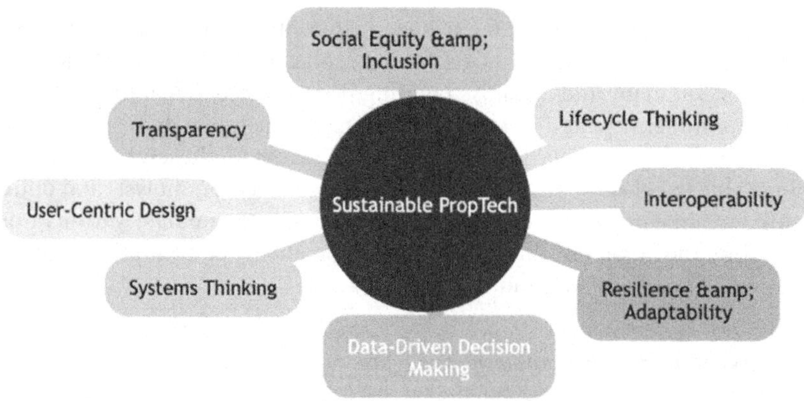

Figure 5.1 Fundamental Principles for Sustainable PropTech.
By author (2024).

5.1.2.2 Systems Thinking

Systems thinking is another crucial principle in sustainable PropTech. This approach recognises that buildings are complex systems embedded within larger urban and ecological systems. Sustainable PropTech strategies, therefore, aim to optimise performance at multiple scales. At the individual building level, PropTech solutions work together as an integrated system rather than as isolated components—for example, smart HVAC systems might communicate with occupancy sensors and weather forecast data to optimise energy use. Sustainable PropTech often extends beyond individual buildings to optimise performance at a district or neighbourhood scale, including shared energy systems, district-wide waste management solutions, or coordinated transport systems. At the broadest scale, sustainable PropTech considers how buildings interact with urban systems, potentially involving integration with smart city initiatives, participation in demand response programs for the electric grid, or contribution to urban heat island mitigation strategies.

5.1.2.3 Data-Driven Decision Making

The ability to collect, analyse, and act on data is a cornerstone of sustainable PropTech strategies. This principle is based on the understanding that you can't manage what you don't measure. Several key aspects define this data-driven approach. Sustainable PropTech solutions typically involve real-time monitoring of various building parameters, including energy use and indoor air quality, enabling the immediate identification and resolution of issues. Advanced PropTech solutions leverage historical data and machine learning algorithms to predict future performance and potential issues, enabling proactive management. The data collected through PropTech solutions facilitates internal benchmarking, where performance is compared over time, and external benchmarking, where performance is measured against similar buildings or industry standards. This comprehensive data-driven approach enables continuous improvement, with performance data informing ongoing adjustments and upgrades to optimise building operations.

5.1.2.4 User-Centric Design

While technology is at the heart of PropTech, sustainable strategies recognise that buildings ultimately exist to serve human needs. The principle of user-centric design ensures that PropTech solutions enhance rather than detract from the user experience. Sustainable PropTech aims to optimise environmental conditions for occupant comfort and well-being, recognising that this often aligns with energy efficiency goals. Additionally, PropTech solutions should be intuitive and easy to use for building managers and occupants, increasing the likelihood that sustainable features will be used effectively. User-centric design also acknowledges that needs may change over time and, therefore, builds flexibility to adapt to evolving requirements. Furthermore, as buildings become more connected and data-driven, ensuring the privacy and security of user data has become a crucial aspect of sustainable PropTech strategies.

5.1.2.5 Interoperability and Open Standards

The principle of interoperability recognises that sustainable buildings require multiple systems to work together seamlessly. PropTech supports open communication protocols and data standards that enable effective system interaction, scalability for future technology integration without wholesale infrastructure replacement, and vendor independence that avoids restrictive proprietary systems. These elements create a foundation for sustainable buildings.

5.1.2.6 *Resilience and Adaptability*

Sustainable PropTech strategies recognise the need for buildings to be resilient to changing conditions and adaptable to future needs. This principle includes three key aspects: climate resilience, where PropTech solutions help buildings adapt to and mitigate climate change impacts; technological resilience, ensuring systems can function effectively even if components fail or external systems are disrupted; and futureproofing, which builds capacity to adapt to advancing technology and evolving sustainability requirements without requiring complete system replacement.

5.1.2.7 *Transparency and Accountability*

Transparency is crucial in sustainable PropTech strategies to combat greenwashing and drive continuous improvement. This principle centres on defining measurable sustainability metrics that can be tracked and reported, employing independent third-party verification of sustainability claims and performance, and providing clear, accessible information about building performance and sustainability initiatives to all stakeholders, including occupants, investors, and the broader community.

5.1.2.8 *Social Equity and Inclusion*

Ultimately, truly sustainable PropTech strategies take into account issues of social equity and inclusion. This encompasses ensuring PropTech solutions are accessible to all users, including those with disabilities; considering how PropTech can be implemented in affordable housing rather than just high-end developments; and assessing and optimising the impact of PropTech strategies on surrounding communities, not just building occupants.

These principles provide a framework for integrating sustainability into PropTech strategies. By adhering to these principles, the real estate industry can ensure that technological innovations create more sustainable, resilient, and socially responsible built environments.

5.1.3 *Alignment with Sustainable Development Goals*

PropTech strategies are increasingly designed to align with broader environmental sustainability objectives (Figure 5.2). This alignment ensures that technological innovations in real estate contribute meaningfully to addressing global environmental challenges.

5.1.3.1 *Climate Change Mitigation*

Climate change mitigation is an environmental challenge where PropTech strategies significantly contribute to reducing real estate's carbon footprint. This is achieved through multiple approaches: smart energy management systems and AI-driven solutions can reduce building energy consumption by 10–15% or more (Kontokosta & Tull, 2017); integration of

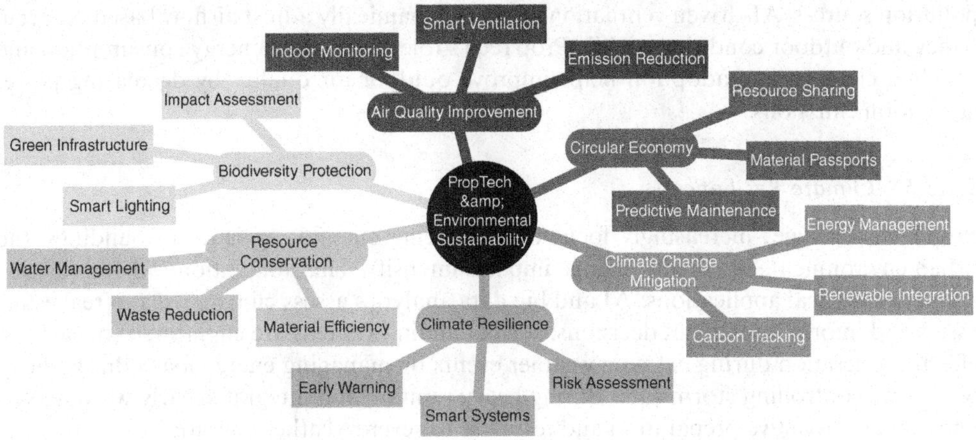

Figure 5.2 How PropTech Aligns with Sustainable Development Goals.
By author (2024).

renewable energy is facilitated through smart grid technologies and blockchain-based energy trading platforms; and carbon emissions tracking and reporting are enhanced through PropTech solutions, particularly blockchain-based systems that provide transparent, tamper-proof carbon accounting for regulatory compliance and carbon market participation.

5.1.3.2 Resource Conservation

PropTech strategies significantly advance resource conservation through multiple channels. Smart water management systems utilising sensors and AI can reduce water consumption by up to 30% (Peña-García et al., 2015). At the same time, BIM technology and digital twin simulations enhance material efficiency through precise planning and IoT-enabled predictive maintenance. Additionally, smart waste management systems incorporating IoT sensors and AI optimise collection processes, demonstrating up to 50% cost reductions while improving recycling performance.

5.1.3.3 Biodiversity Protection

PropTech plays a valuable role in biodiversity protection through several technological applications. IoT sensors and drone technology enable the monitoring of green infrastructure, such as urban forests and green roofs, while AI and machine learning assist in assessing the biodiversity impacts of real estate developments. Smart lighting systems can also be optimised to minimise light pollution that would otherwise disrupt wildlife behaviour and ecosystems.

5.1.3.4 Air Quality Improvement

PropTech plays a significant role in air quality management through comprehensive monitoring and control systems. IoT sensors enable continuous tracking of indoor air quality parameters such as CO_2, VOCs, and particulate matter to optimise ventilation and identify pollution sources. AI-driven ventilation systems dynamically adjust airflow based on occupancy and outdoor conditions, while PropTech's role in reducing energy consumption and enabling clean energy adoption helps improve outdoor air quality by decreasing power generation emissions.

5.1.3.5 Climate Resilience

PropTech strategies increasingly focus on enhancing climate resilience in buildings and urban environments as climate change impacts intensify. This adaptation is driven by several technological applications: AI and big data analytics assess climate risks to real estate assets and inform investment decisions; smart building systems are engineered to maintain effective operation during extreme weather events by managing energy loads during heatwaves and controlling stormwater during heavy rainfall; and integrated early warning systems enable proactive preparation and response to severe weather conditions.

5.1.3.6 Circular Economy

PropTech is vitally advancing the circular economy within real estate through several key innovations. Digital material passports, powered by blockchain technology, enable comprehensive tracking of building materials to facilitate future reuse and recycling initiatives. Implementing predictive maintenance systems extends the operational lifespan of building components, adhering to the circular principle of maximising material use duration. Additionally, PropTech platforms are revolutionising space utilisation by enabling resource sharing, which improves efficiency and reduces the demand for new construction projects.

While PropTech presents compelling opportunities for sustainable real estate, we must recognise that technology is not a complete solution. PropTech's success in advancing sustainability depends heavily on thoughtful implementation and careful consideration of its environmental impact, from data centre energy consumption to hardware materials. The future holds promise for even greater alignment between PropTech and environmental objectives. As quantum computing, AI, and renewable energy technologies mature, new possibilities for sustainable real estate practices will emerge. Yet, achieving PropTech's full potential requires sustained collaboration across disciplines, including technologists, real estate professionals, environmental scientists, and policymakers.

5.2 Data-Driven Sustainability Decision-Making

5.2.1 Evolution of Data Analytics in Real Estate

The transformation of real estate data analytics represents a significant shift in how the industry approaches information management and decision-making, particularly in the context of sustainability. This evolution has been driven by technological advancement and increasing emphasis on environmental responsibility in the built environment (Figure 5.3).

Before the 2000s, real estate analytics relied primarily on manual data collection and basic financial metrics. Property managers and owners tracked fundamental indicators such as occupancy rates and rental income through paper-based systems, which could not monitor or analyse environmental performance.

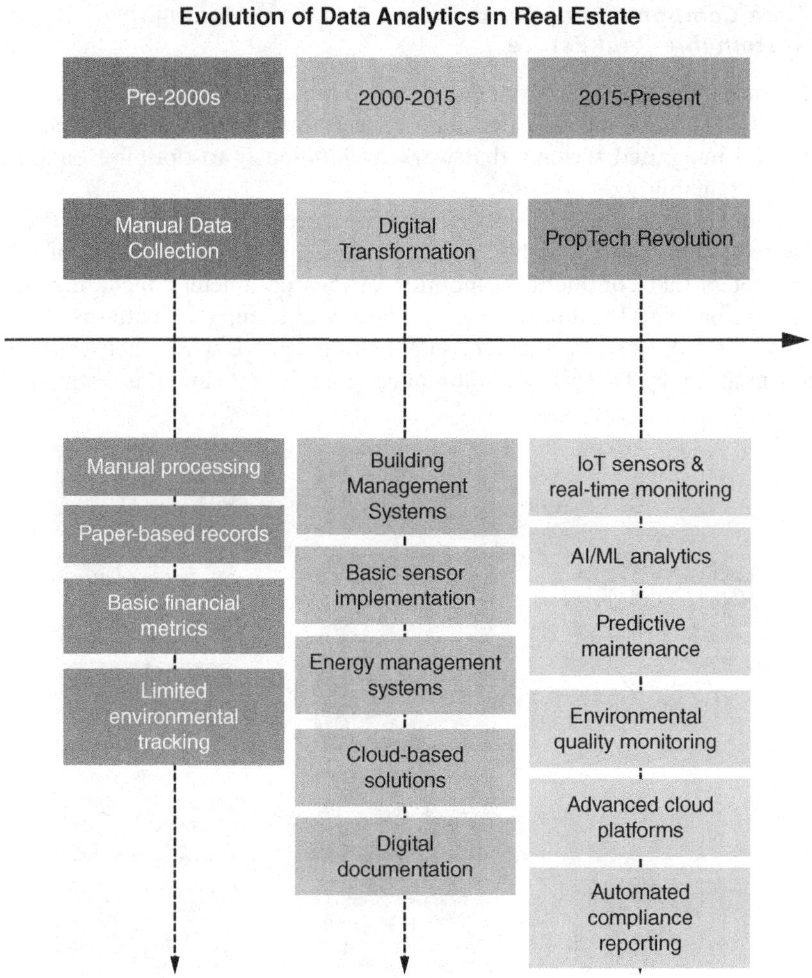

Figure 5.3 Evolution of Data Analytics in Real Estate.
By author (2024).

The period from 2000 to 2015 marked a significant transition as digital technologies began transforming the real estate sector. This era saw the introduction of computerised property management systems, digital data storage solutions, and basic energy monitoring tools that began to provide insights into building performance beyond traditional financial metrics.

From 2015 onwards, the integration of IoT devices, cloud computing, and advanced analytics platforms has revolutionised data-driven decision-making in real estate. Modern buildings now employ extensive sensor networks that continuously monitor energy consumption, environmental conditions, water usage, and occupancy patterns, facilitating sophisticated analysis and optimisation of building performance.

5.2.2 *Core Components of Data-Driven Decision-Making in Sustainable Real Estate*

The core components of data-driven decision-making in sustainable real estate (as shown in Figure 5.4) include a sophisticated data collection infrastructure, advanced analytics platforms, and integrated systems that work harmoniously to optimise building performance and sustainability outcomes.

At the foundation lies the data collection infrastructure, which serves as the building's nervous system. Modern sustainable buildings utilise an extensive network of IoT sensors and smart meters that continuously monitor various parameters, including energy consumption, environmental conditions, water usage, and occupancy patterns. According to Kontokosta and Tull (2017)'s research, such comprehensive sensor networks can reduce energy consumption by 10–15% without requiring significant capital investments.

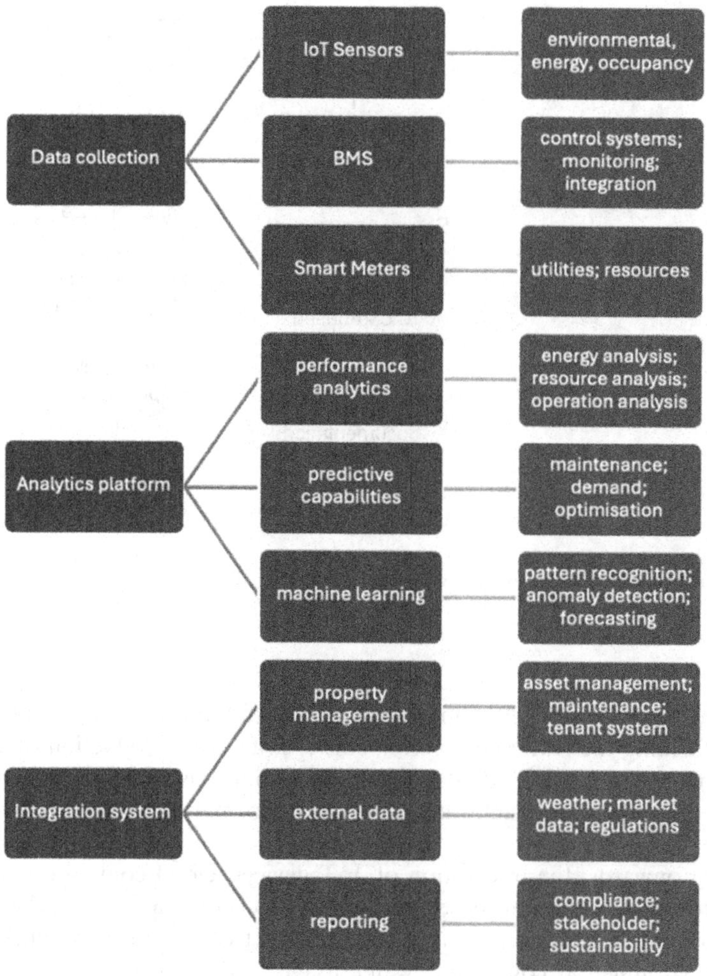

Figure 5.4 Data-Driven Decision-Making.

By author (2024).

Building management systems (BMS) act as the central nervous system, integrating data from multiple sources and providing centralised control over building operations. Modern BMS platforms offer real-time monitoring capabilities, automated response mechanisms for system optimisation, historical data storage for trend analysis and benchmarking, and interfaces with external systems to provide a comprehensive view of building performance.

Analytics platforms transform the collected data into actionable insights through various analytical tools and techniques. Performance analytics examine energy consumption patterns, resource utilisation, and operational efficiency. These platforms can identify anomalies in system performance, track sustainability metrics, and benchmark performance against similar properties.

5.3 Blockchain Applications in Sustainable Real Estate

Building on the foundational blockchain concepts established in Chapters 2 and 4, this section explores specific applications in the context of sustainable real estate.

5.3.1 Smart Contracts for Sustainable Real Estate

Smart contracts are self-executing contracts with terms directly written into code, as originally conceptualised by Szabo (1994), who proposed that smart contracts could be used for synthetic assets and help reduce transaction costs. In sustainable real estate, smart contracts represent a novel approach to handling transactions at lower costs while offering automation, transparency, and security (Hewa et al., 2021). These digital protocols can automate various processes while ensuring compliance with sustainability standards and environmental regulations, addressing several critical challenges in sustainable real estate practice.

5.3.2 Current Challenges and Smart Contract Solutions

5.3.2.1 Green Bond Mechanisms

Traditional green financing in real estate faces significant hurdles related to the verification and accountability of sustainability performance. Research indicates that sustainable features often come with higher upfront costs, which can deter developers and investors when relevant financial benefits cannot be adequately evidenced (Devine et al., 2023). Smart contracts can revolutionise green financing by providing reliable, real-time sustainability performance data that enables lenders to offer preferential rates for properties with verifiably strong sustainability profiles (Shankar, 2022).

Smart contracts can automate the release of funds for sustainable building projects only when verified sustainability milestones are achieved, as demonstrated in blockchain-enabled green bond structures. These contracts can be programmed to comply with green bond principles and climate bond standards, ensuring funds are exclusively used for environmental upgrades (Climate Bonds Initiative, 2024).

The automated verification process works by integrating with IoT devices and third-party auditing systems. When predetermined sustainability targets are met—such as achieving specific energy efficiency improvements or renewable energy generation milestones—the smart contract automatically recognises the achievement and releases the next tranche of funding. This mechanism significantly reduces the administrative burden and potential for disputes while ensuring that green finance is genuinely tied to environmental performance outcomes.

5.3.2.2 Performance-Based Payments

The difficulty of quantifying the long-term benefits of sustainable investments has historically made it challenging to justify the initial expenses of green building projects (von Paumgartten, 2003; Fuerst, 2015). Smart contracts address this challenge by enabling transparent tracking of financial benefits through automated monitoring and payment systems.

Rental income distribution can be tied to energy efficiency improvements through smart contracts that automatically monitor building performance using IoT sensors and smart meters. Ahn and Cho (2017) and Png et al. (2019) demonstrate that IoT-based smart control systems can reduce energy consumption by 17–30% in buildings, providing measurable data for performance-based payment calculations.

Smart contracts can implement sustainability-linked returns where automatic bonus payments are triggered when reduction targets are met. For example, a 5% bonus distribution can be programmed when annual energy consumption reduction targets are achieved, as verified by real-time data from building management systems (Mistrangelo & Tagliabue, 2022). The smart contract maintains a running tally of cumulative savings and calculates return on investment for specific sustainability initiatives, providing transparent reporting to both lenders and borrowers.

Property owners, tenants, or regulators can input specific sustainability targets into the smart contract, which then tracks progress using data from IoT devices and other verified sources. When targets are met, predefined rewards are automatically distributed, which may include direct payments, credits applied to rent or mortgage payments, automatic updates to green building certifications, or the issuance of carbon credits for exceeding emissions reduction targets.

5.3.2.3 Carbon Credit Management

Building energy savings can generate tokenised carbon credits that are automatically distributed to stakeholders through smart contract mechanisms. This application addresses the challenge of efficiently managing and trading carbon credits generated by sustainable real estate practices (Tian et al., 2022).

Smart contracts enable the efficient tokenisation and trading of carbon credits by automating the verification process through integration with building management systems and IoT sensors. When a building achieves verified energy savings or renewable energy generation, the smart contract automatically calculates the corresponding carbon offset and issues tokenised carbon credits to designated stakeholders. This creates an additional revenue stream tied directly to environmental performance, incentivising continued sustainability improvements.

The blockchain-based carbon credit system guarantees transparency and prevents double-counting problems that have historically affected carbon markets. Each carbon credit is uniquely identified and tracked on the blockchain, creating an immutable record of its generation, ownership, and retirement. This enhanced transparency and traceability boosts confidence in carbon markets and facilitates more efficient price discovery.

5.3.2.4 *Maintenance Automation*

Smart contracts can trigger maintenance requests and payments when IoT sensors detect performance deviations, enabling predictive maintenance approaches that significantly reduce downtime and repair costs (Himeur et al., 2023). This application addresses the traditional challenges of reactive maintenance in building management.

By integrating with comprehensive IoT sensor networks, smart contracts can continuously monitor equipment performance across building systems, including HVAC, lighting, and water management. When sensors detect anomalies or performance degradation that exceeds predetermined thresholds, the smart contract automatically generates maintenance requests and can initiate service provider scheduling and payment processes.

Research indicates that predictive maintenance systems can extend equipment lifespan by 1–3 years while reducing energy consumption by 15–45% (Tagliabue et al., 2021). Smart contracts enhance this approach by automating the entire maintenance workflow: from detection and diagnosis to service provider engagement and payment processing. All maintenance activities are recorded on the blockchain, creating a transparent and auditable trail of property upkeep that can be valuable for regulatory compliance, insurance purposes, and asset valuation.

The automated payment mechanism ensures that service providers are compensated promptly upon completion of verified work, thereby reducing disputes and enhancing service quality. Smart contracts can also implement performance-based payment structures where service providers receive bonuses for achieving specific outcomes, such as restoring equipment to optimal efficiency levels or completing work within accelerated timeframes.

5.3.2.5 *Mapping the Sustainability Benefits*

Smart contracts provide innovative solutions for incorporating and enforcing sustainability measures in the real estate sector. From the perspective of the sustainable development goals (SDGs), which are the core of the United Nations 2030 Agenda for Sustainable Development (UN, 2015), we mapped how smart contract technology can contribute to achieving these global objectives in the context of real estate.

As shown in Figure 5.5, smart contracts contribute to health and well-being (SDG 3) by enforcing and verifying compliance with healthy building standards, thus contributing to occupant well-being. In terms of economic development and innovation (SDGs 8 and 9), smart contracts offer substantial benefits. They reduce transaction costs by automating many processes, including document verification, fund transfers, and property ownership changes. This increased efficiency contributes to economic growth and more streamlined economic activities in the real estate sector.

Regarding energy efficiency (SDG 7), smart contracts can automate the monitoring and reporting of building energy performance. This automation incentivises the use of clean energy in real estate and promotes more efficient energy use in buildings. Smart contracts also have the potential to reduce inequality (SDG 10). By reducing costs and barriers to entry, smart contracts could make real estate investments more accessible to a broader range of people. The increased transparency in transactions could lead to fairer pricing, potentially reducing inequalities in the property market.

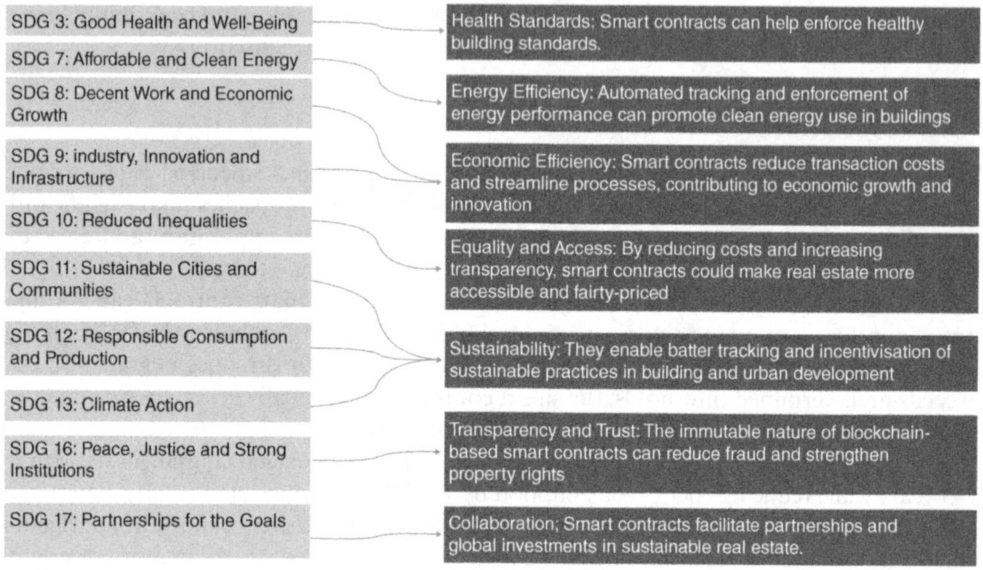

Figure 5.5 Mapping Benefits of Smart Contracts in Real Estate to the UN Sustainable Development Goals (SDGs).

By author (2024).

Smart contracts also promote sustainability in urban development and real estate (SDGs 11, 12, and 13). They facilitate more efficient property registration and transfer, supporting better urban management and leading to more equitable and sustainable urban development. In terms of responsible consumption, smart contracts can automate and enforce energy performance contracts, promoting more responsible energy use in buildings. They can also track and verify the sustainability credentials of materials used in real estate development. For climate action, smart contracts enable efficient tokenisation and trading of carbon credits generated by sustainable real estate practices. They can provide automated reward systems for meeting or exceeding sustainability targets in construction and building management.

The impact of smart contracts on institutional strength and trust (SDG 16) is significant. Blockchain-based smart contracts, with their immutability and transparency, can substantially reduce fraud in real estate transactions. Automated and tamper-proof property registries can strengthen property rights and reduce disputes.

Lastly, smart contracts facilitate partnerships and collaboration (SDG 17) by enabling more efficient collaboration between various stakeholders, including developers, investors, regulators, and occupants. Smart contracts can foster international partnerships and investments in sustainable real estate projects by reducing barriers and increasing transparency.

5.3.3 Blockchain Technology and Financial Inclusion in Real Estate

Blockchain technology presents unique opportunities to address long-standing issues of financial inclusion and accessibility in real estate, particularly through fractional ownership that lowers investment hurdles and creates new financing channels for affordable housing development.

Real estate has traditionally been viewed as an investment vehicle for the wealthy, with high entry barriers and complex processes often excluding a significant portion of the population from participating in property ownership and its associated benefits. According to the World Bank, approximately 1.7 billion adults remain unbanked globally, limiting their access to formal financial services, including real estate investments (Bhardwaj et al., 2018).

Blockchain technology, with its decentralised and transparent nature, has the potential to revolutionise how we approach real estate transactions, ownership, and financing. The Fidelity International report highlights that blockchain is "laying the groundwork for asset tokenisation and the empowerment of underbanked communities, particularly in Asia" (Fidelity International, 2022).

5.3.3.1 Fractional Ownership: Democratizing Real Estate Investment

Fractional ownership refers to the division of any asset into smaller, more affordable shares. In real estate, this concept allows multiple investors to own a portion of a property, sharing both the costs and benefits of ownership.

Blockchain technology revolutionises fractional ownership by providing a secure, transparent, and efficient system for managing shared ownership through:

- Easy division of property into smaller, tradable units
- Transparent and immutable record-keeping of ownership
- Efficient transfer of ownership shares
- Automated distribution of rental income or other benefits

5.3.3.2 Case Studies of Blockchain-Based Real Estate Platforms

To better understand how blockchain technology is being applied to enhance financial inclusion in real estate, it's valuable to examine concrete examples from the market. In this section, we will explore several pioneering platforms that are leveraging blockchain to democratise real estate investment. These case studies showcase different approaches to fractional ownership, tokenisation, and peer-to-peer transactions in various global markets. Following the individual case studies, we present a comparative analysis that highlights the similarities, differences, and unique features of each platform. This analysis provides insights into how different implementations of blockchain technology address specific challenges in real estate financial inclusion, from lowering entry barriers to increasing liquidity and transparency. By examining these real-world applications, we can gain a clearer picture of the current state and future potential of blockchain in transforming real estate investment accessibility.

5.3.3.2.1 REALT

RealT[1] is a platform that allows users to invest in tokenised shares of rental properties in the United States. It uses Ethereum blockchain for tokenisation and focuses on residential rental properties in Detroit, Michigan. The minimum investment can be as low as USD 50.

It offers fractional ownership through RealTokens and daily rent distributions to token holders.

RealT purchases properties and creates a unique LLC for each of them. The LLC is then tokenised, with each token representing partial ownership. Investors can buy these tokens, becoming fractional owners of the property. Rental income is distributed to token holders proportional to their ownership.

RealT offers a significantly lower entry requirement for real estate investment and enables global capital to enter the U.S. property market. It provides liquidity to traditionally illiquid real estate assets and financial inclusion to global real estate investors.

5.3.3.2.2 PROPERTYSHARE

PropertyShare[2] is an Indian platform offering fractional ownership in commercial real estate. It focuses on Grade A commercial properties in major Indian cities and offers a minimum investment of approximately USD 10,000. Similar to RealT, it also provides regular rental income and potential capital appreciation to the investor.

PropertyShare identifies high-quality commercial properties. The properties are structured into a special purpose vehicle (SPV), which is then tokenised and offered to investors. Investors can buy and sell their shares on the PropertyShare platform.

PropertyShare made prime commercial real estate accessible to individual investors, improved transparency in the Indian real estate market, and provided a new investment avenue for the growing Indian middle class.

5.3.3.2.3 ATLANT

ATLANT[3] is a global platform working on blockchain-based solutions for fractional real estate ownership and peer-to-peer rentals. It uses blockchain for both fractional ownership and a rental marketplace with the aim of tokenising both commercial and residential properties globally. Meanwhile, it also offers a decentralised exchange for property token trading.

Property owners can list their properties for tokenisation on the ATLANT platform. The properties are vetted and then tokenised. Investors can buy tokens representing fractional ownership of these properties. The platform also facilitates P2P rentals, using smart contracts to manage agreements and payments. In this way, ATLANT is working towards creating a global, decentralised real estate market.

These platforms demonstrate how blockchain technology is being applied to democratise access to real estate investments, increase market liquidity, and improve transparency in property transactions. While each platform has its unique focus and features, they all leverage blockchain to address traditional barriers to real estate investment, thus contributing to greater financial inclusion in the sector.

5.3.3.2.4 COMPARATIVE ANALYSIS OF BLOCKCHAIN-BASED REAL ESTATE PLATFORMS

RealT stands out with its very low minimum investment, making it the most accessible for small investors. PropertyShare focuses on high-end commercial properties, potentially attracting more affluent investors. ATLANT aims for a more global and diverse property portfolio (as shown in Table 5.1).

Table 5.1 Target Market and Property Types

Platform	*Geographic Focus*	*Property Types*	*Minimum Investment*
RealT	USA (Detroit)	Residential rentals	USD 50
PropertyShare	India	Commercial (Grade A)	~USD 10,000
ATLANT	Global	Residential and commercial	Not specified

Table 5.2 Blockchain Implementation

Platform	*Blockchain Used*	*Token Type*	*Smart Contract Features*
RealT	Ethereum	ERC-20 (RealTokens)	Automated rent distribution
PropertyShare	Not specified	Proprietary tokens	Ownership transfer
ATLANT	Ethereum	Property tokens, utility tokens	P2P rentals, exchanges

Ethereum is the most common blockchain platform, likely due to its smart contract capabilities. ATLANT's approach is the most comprehensive, covering ownership, rentals, and token exchange. Table 5.2 compares their implementation of blockchain.

In terms of liquidity (in Table 5.3), RealT and ATLANT offer the highest liquidity with their secondary markets, while PropertyShare is the most restricted in terms of global accessibility.

Given the nature of real estate, income distribution is the most important for investors. RealT offers the most frequent income distribution, potentially attracting investors seeking regular cash flow. All platforms leverage blockchain for more efficient income distribution compared to traditional real estate investments (see Table 5.4).

All platforms offer fractional ownership, which is the key factor in financial inclusion. RealT and ATLANT facilitate cross-border investments, opening up global real estate markets. PropertyShare stands out for its focus on investor education, which is crucial for financial inclusion in emerging markets. The comparison is shown in Table 5.5.

Each platform leverages blockchain technology to increase accessibility, transparency, and liquidity in real estate investments. However, they take different approaches to address specific market needs and investor preferences. The variety of approaches demonstrates the flexibility of blockchain technology in addressing different aspects of financial inclusion in real estate.

- *Accessibility:* RealT leads in terms of lowering the entry barrier with its USD 50 minimum investment.
- *Diversity:* ATLANT offers the most comprehensive platform, covering various property types and investment models.
- *Market Focus:* PropertyShare's concentration on the Indian market demonstrates how blockchain can address specific regional needs.

Table 5.3 Liquidity and Trading Facility

Platform	*Secondary Market*	*Trading Frequency*	*Global Accessibility*
RealT	Yes	Daily	High
PropertyShare	Limited	Not specified	Limited to Indian investors
ATLANT	Yes (DEX)	Continuous	High

Table 5.4 Income Distribution

Platform	*Income Type*	*Distribution Frequency*	*Automated*
RealT	Rental income	Daily	Yes
PropertyShare	Rental income	Monthly/quarterly	Not specified
LABS Group	Usage fees, appreciation	As accrued	Yes
ATLANT	Rental income	As accrued	Yes
BitProperty	Rental income, Interest	Not specified	Not specified

Table 5.5 Financial Inclusion Aspects

Platform	*Fractional Ownership*	*Cross-Border Investment*	*Investor Education*
RealT	High	Yes	Moderate
PropertyShare	Moderate	Limited	High
ATLANT	High	Yes	Low

This analysis shows how different platforms are using blockchain technology to address various aspects of financial inclusion in real estate, from lowering minimum investments to enabling global property investments.

5.3.3.3 New Financing Channels for Affordable Housing Development

Housing affordability has been a huge issue in many countries around the world. Challenges in financing new developments worsen the crisis. Traditionally, the funding for affordable housing largely relied on government funds and subsidies. Traditional funding sources often fail due to high capital costs, difficulty assessing and managing risk, and limited private financing methods (Rizvi, 2008; Williams & Whitehead, 2015; Pomeroy, 2018; Lawson et al., 2024).

Blockchain technology enables new fundraising models for affordable housing projects, such as tokenised crowdfunding and blockchain-based peer-to-peer lending platforms.

5.3.3.3.1 TOKENISED CROWDFUNDING

With the help of blockchain technology, developers can issue tokens representing shares in the affordable housing project, allowing many small investors to participate. All transactions and ownership records are stored on the blockchain, providing a clear and unalterable history. Smart contracts can automatically manage the distribution of rental income or other returns to token holders.

The key difference between traditional crowdfunding and tokenised crowdfunding lies in how the investment is structured and recorded, with tokenised crowdfunding utilising blockchain technology. Table 5.6 lists the differences.

Muneeza et al. (2018) evaluated the value of crowdfunding for financial inclusion by exploring the Malaysian and Shariah-compliant crowdfunding, and they found that the blockchain technology has significant impacts on the growth of crowdfunding as well as financial inclusion. It suggests that tokenised crowdfunding, especially in real estate, is an emerging area that offers potential benefits like increased liquidity and global accessibility but also faces regulatory challenges. Blockchain technology distinguishes it from traditional crowdfunding by offering a more transparent, potentially more liquid, and globally accessible investment structure.

Table 5.6 Comparison Between Traditional Crowdfunding and Tokenised Crowdfunding

Aspect	*Traditional Crowdfunding*	*Tokenized Crowdfunding*
Technology	Centralized platforms	Blockchain-based
Investment structure	Direct investments or donations	Digital tokens representing ownership
Record keeping	Centralized ledgers or platforms	Distributed ledger (blockchain)
Liquidity	Often limited, no secondary market	Potentially higher via token trading
Geographical reach	Often limited by jurisdiction	Potentially global (subject to regulations)
Regulatory Framework	More established, varies by country	Often in regulatory grey area, evolving
Investor access	May be limited to accredited investors	Potentially open to wider range of investors
Fractional ownership	Less common, more complex	Easily facilitated through token division
Transparency	Varies by platform	High, due to blockchain's immutable record
Cost structure	Platform fees, often higher	Can be lower due to disintermediation
Investment size	Often larger minimum investments	Can allow for very small investments
Secondary trading	Usually not available	Possible through token exchanges
Investor rights	Defined by traditional legal contracts	Encoded in smart contracts on blockchain
Cross-border transactions	Can be complex and costly	Potentially simpler and less costly

An Argentinian platform, Xcapit, has been studied by Mhlanga (2023) and Chapiro (2021) as an example that uses blockchain to provide an alternative investment platform for individuals without bank accounts or financial expertise. It developed a blockchain platform to manage the distributed and community-based renewable energy generation in Córdoba, Argentina, in a decentralised and autonomous way. This system ensures transparency and efficiency. Although there is no use case in real estate yet, the platform demonstrates how blockchain can democratise access to real estate investments through crowdfunding mechanisms. Several similar platforms, including RealT, Blocksquare,[4] and Smartlands,[5] are leveraging blockchain for tokenised crowdfunding in real estate.

5.3.3.3.2 BLOCKCHAIN-BASED PEER-TO-PEER (P2P) LENDING PLATFORMS

Blockchain-based platforms can connect borrowers directly with lenders, potentially offering more favourable terms than traditional financial institutions. Blockchain-based P2P lending platforms offer several advantages over traditional lending models. By directly connecting borrowers and lenders, blockchain removes the need for traditional financial intermediaries, potentially reducing costs and increasing efficiency. Automated, self-executing contracts can manage loan terms, payments, and collateral, reducing administrative overhead and increasing transparency. Similar to fractional ownership, blockchain allows multiple lenders to contribute to a single loan, lowering the entry barrier for individual lenders. Blockchain-based platforms can connect lenders and borrowers globally, increasing the pool of available capital and investment opportunities.

Blockchain-based P2P lending in real estate offers investors access to a new asset class with potentially higher yields. It enables investors to diversify investments across multiple properties or loans with lower minimum investment amounts compared to traditional real estate investments.

For borrowers, blockchain-based P2P lending platforms potentially provide lower interest rates due to reduced overhead costs, faster loan approval and funding processes, and access to alternative financing options. This is especially beneficial for those underserved by traditional banks.

Similar to crowdfunding platforms, the key difference between traditional P2P and blockchain-based P2P lending is how the debt is structured and recorded. Table 5.7 highlights the differences:

5.4 Collaboration and Partnerships for Sustainable PropTech

5.4.1 Industry Collaboration: Enhancing Sustainability Through PropTech

Collaborative efforts between technology firms and industry stakeholders increasingly drive the intersection of PropTech and sustainability in real estate. Collaboration is an ongoing process where knowledge sharing, active participation, outcomes, trust development, and cumulative experiences interplay and repeatedly influence each other (Martinez-Moyano, 2006). KPMG (2017) highlights that collaboration and partnerships will be key for real estate companies to adopt PropTech, as many need more internal capabilities to develop solutions themselves. In their survey, 49% expect to collaborate with an existing or new supplier to develop their technological innovation capability, 28% expect to

Table 5.7 Comparison Between Traditional P2P Lending and Blockchain-Based P2P Lending in Real Estate

Aspect	*Traditional P2P Lending*	*Blockchain-Based P2P Lending*
Platform	Centralized online platforms	Decentralized blockchain technology
Intermediation	Often involves third-party intermediaries	Potential for disintermediation through smart contracts
Security	Traditional security measures and financial regulations	Cryptographic security inherent in blockchain
Transparency	Limited; information controlled by platform	Higher; transactions recorded on public ledger
Geographical reach	Often limited to specific jurisdictions	Potentially global, subject to regulations
Liquidity	Limited secondary market for loans	Potentially higher through tokenisation and secondary token markets
Cost structure	Platform fees, potentially higher	Can be lower due to reduced intermediation
Regulatory framework	More established, varies by country	Often in regulatory grey area, evolving
Asset representation	Traditional loan agreements	Digital tokens representing loan or property fraction
Contract execution	Manual, often paper-based	Automated through smart contracts
Data management	Centralised databases	Distributed ledger (blockchain)
Identity verification	Traditional KYC processes	Potential for blockchain-based digital identities
Investment size	Often larger minimum investments	Can allow for very small investments (fractional)
Cross-border transactions	Can be complex and costly	Potentially simpler and less costly

collaborate or partner with an existing supplier to develop their technological innovation capability, and 17% expect to collaborate with a new supplier. In a more recent survey by Jones-Lang-LaSalle (2022), it was found that 45% of respondents consider the facilitation of collaborative working to be one of the primary purposes of office space. This rises to 55% among corporations with over 10,000 employees. And 69% have already introduced or will soon introduce technology to boost in-office collaboration. The finding not only implies the potential for co-working spaces but also indicates that the industry is well aware of the importance of collaboration. Collaborations in PropTech have various forms, such as partnerships, mergers and acquisitions (M&A), joint research and development, and technology adoption. The synergy is important in addressing complex sustainability challenges and accelerating the adoption of sustainable practices. Figure 5.6 presents the types of collaborations in PropTech that can help enhance sustainability in the real estate industry.

In recent years, the industry has seen significant growth in investments and M&A activity. In 2021, PropTech investments reached a record high of over USD 24 billion globally (Houlihan-Lokey, 2024). The United States has been a leader in PropTech M&A deals, with 162 transactions in 2021, although this number decreased to 99 in 2022

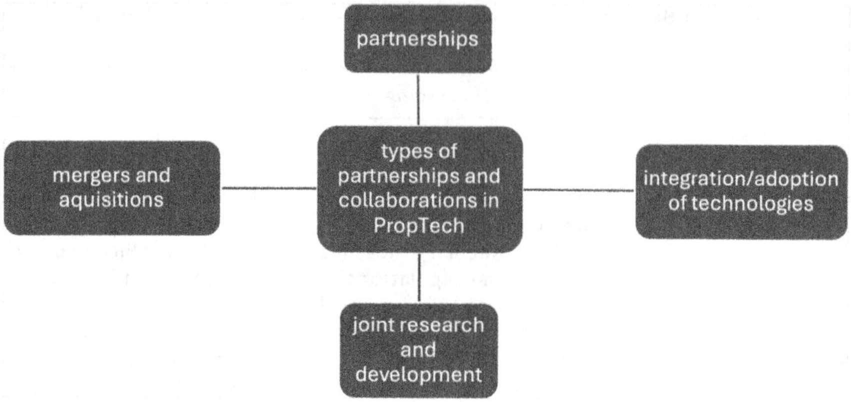

Figure 5.6 The Types of Collaborations in PropTech.
By author (2024).

(Houlihan-Lokey, 2024). Worldwide, PropTech M&A activity has shown an upwards trend, with 24 deals closed in the fourth quarter of 2019, compared to just nine in the second quarter of 2017 (Andrew et al., 2020). The details of M&A trends in PropTech will be discussed in Chapter 7.

Collaborations led by universities and research institutions are driving the development of new technologies in the real estate industry.

The MIT Real Estate Innovation Lab, financially supported by Jones Lang LaSalle (JLL), Real Capital Analytics (RCA), and Ernst & Young (EY), collaborates with various PropTech firms to provide a platform for knowledge exchange. They have developed several AI-driven tools for sustainable urban planning. Bokhari et al. (2021) from MIT's Real Estate Innovation Lab have illuminated the role of accelerator programs in fostering PropTech start-up growth. Their analysis reveals that well-structured accelerator programs can boost a start-up's lifetime cumulative funding by an average of 7%. This collaborative effort between academic researchers and the PropTech sector illustrates how university-led studies can offer valuable insights for enhancing start-up development and driving innovation in real estate technology. The research shows the potential for academic–industry partnerships to shape more effective strategies for nurturing emerging PropTech ventures.

Another example is the joint R&D partnership between the Urban Land Institute (ULI) Greenprint Center for Building Performance and Measurabl, announced in March 2022 (ULI & Measurabl, 2022). This alliance combines ULI Greenprint's role as a global membership consortium of leading real estate stakeholders with Measurabl's expertise as the most widely adopted ESG software solution for the commercial real estate industry. The partnership enables ULI Greenprint members, who collectively manage over USD 1.3 trillion in assets, to leverage Measurabl's comprehensive ESG data platform covering 12 billion square feet of commercial real estate across 90 countries. This collaboration facilitates measuring, managing, reporting, and improving environmental performance in the real estate industry. ULI Greenprint uses data from the Measurabl platform for its annual "State of Green" publication, which serves as a proxy for the industry's progress on decarbonisation. The partnership also supports ULI Greenprint's ambitious goals of reducing

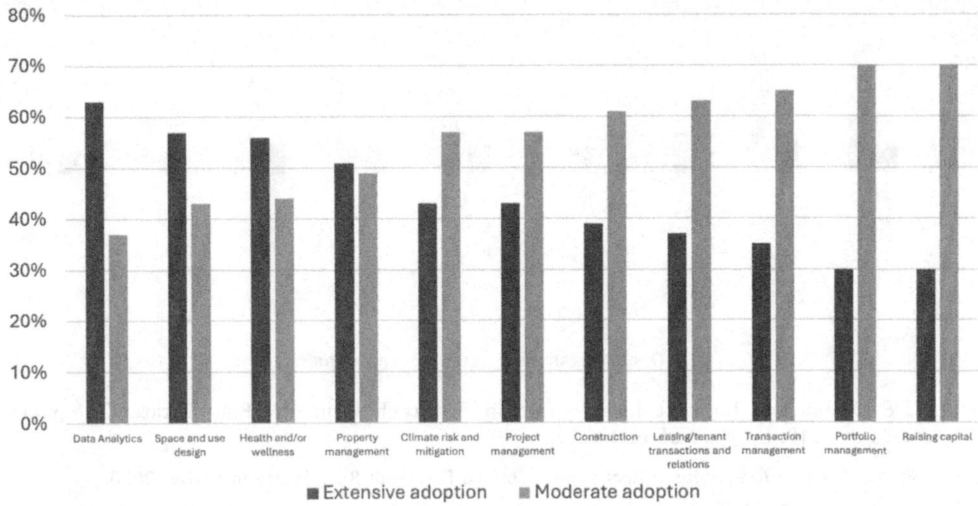

Figure 5.7 Recent Integration/Adoption of Proptech Adoption by Companies in the United States in 2022, by Business Area.

Source: Proptech: Changing the Way Real Estate Is Done, Urban Land Institute, 2021.

greenhouse gas emissions by 50% by 2030 and achieving net-zero carbon operations by 2050. Through this collaboration, research institutions and PropTech firms are working together to drive meaningful improvements in ESG performance and sustainability in the real estate sector.

Partnerships between PropTech start-ups and established real estate firms are leading to the integration and adoption of innovative technologies into existing real estate business areas. According to ULI (2021), in the United States, real estate companies are extensively adopting PropTech in various business areas, with data analytics (63%), space and use design (57%), and health and wellness technology (56%) leading the way. Figure 5.7 exhibits the recent adoption of new technology across various business areas in the real estate industry in the United States.

Globally, real estate companies are investing both directly and indirectly in PropTech firms, particularly in smart building systems, online lending marketplaces, and AI (Andrew et al., 2020). Figure 5.8 shows the investments in PropTech firms by real estate companies globally in 2019.

Many of these adoptions have helped to enhance sustainability. For example, Spacebring partnered with Kisi, a PropTech start-up, to integrate automated access control systems into its coworking spaces. This collaboration improved energy efficiency by automatically managing lighting and HVAC based on real-time occupancy data (Kisi, 2024).

The landscape of collaborations in this field is characterised by various partnerships spanning various sectors and organisational types. These collaborative efforts are not limited to a single model but manifest in a broad spectrum of arrangements, each tailored to leverage the partners' unique strengths. Here are some main areas where the partnerships boost sustainability in the real estate industry (Figure 5.9).

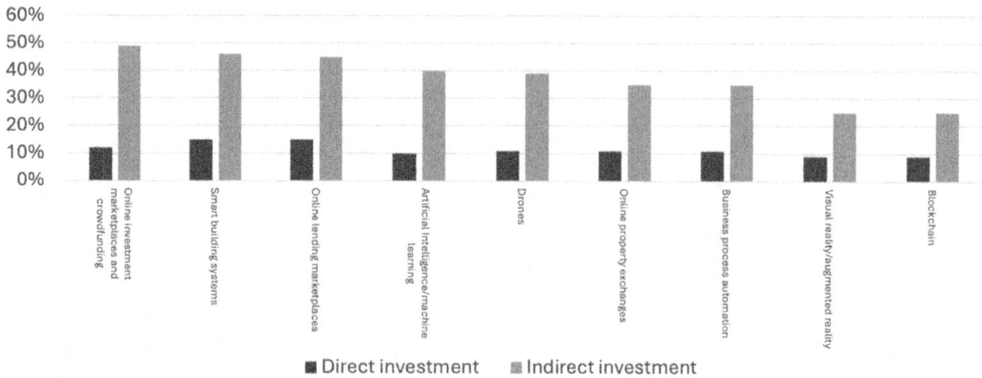

Figure 5.8 Direct and Indirect Investments in Proptech Firms by Real Estate Companies Worldwide in 2019, by Type of Technology.

Source: PropTech 2020: The Future of Real Estate, Oxford Future of Real Estate Initiative, 2020.

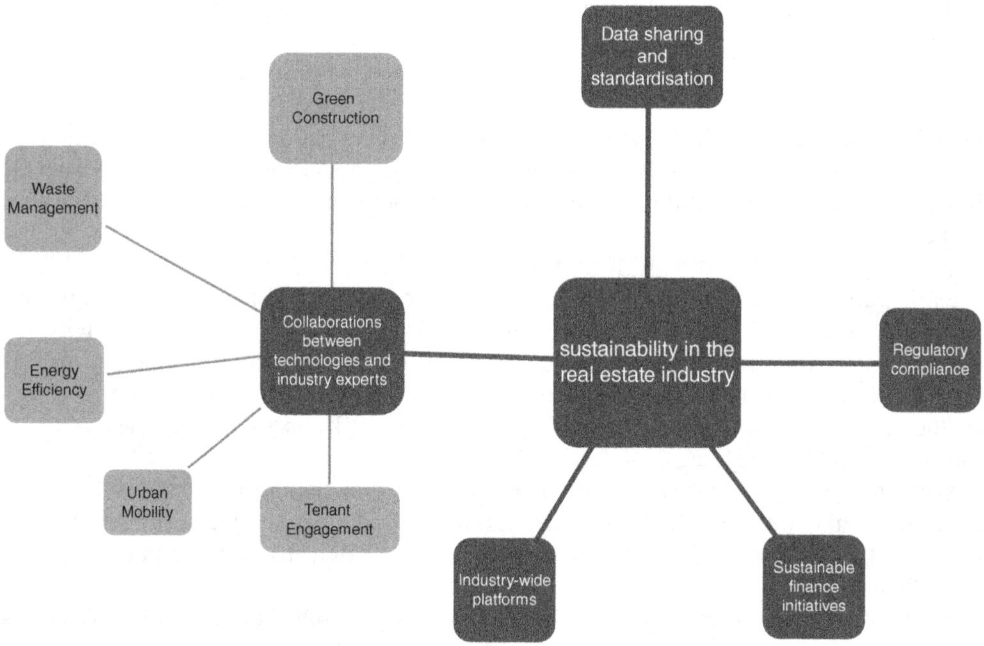

Figure 5.9 Major Areas Where PropTech Partnerships Boost Sustainability.

By author (2024).

1 *Data Sharing and Standardisation*: PropTech firms work with real estate companies, government agencies, and industry bodies to establish common data standards for sustainability metrics. The Global Real Estate Sustainability Benchmark (GRESB) reported that by 2023, over 2,000 participants collectively managing more than USD 7.2 trillion in assets were using their standardised framework for sustainability reporting (GRESB, 2024). For example, in June 2024, Measurabl partnered with Susteco to create a

centralised data-driven ecosystem for the real estate industry, integrating ESG data and software within Susteco's smart building ecosystem. This integration enables more efficient monitoring and management of sustainability metrics, streamlining ESG reporting and compliance processes (Measurabl, 2024).

2 *Research and Development Initiatives*: Academic institutions partner with PropTech companies to advance sustainability research and innovation. For instance, the joint R&D partnership between the Urban Land Institute (ULI) Greenprint Center and Measurabl combines ULI Greenprint's role as a global consortium with Measurabl's ESG software expertise, enabling over USD 1.3 trillion in assets to leverage comprehensive ESG data platforms.
3 *Government and Energy Efficiency Programs*: Public–private partnerships drive sustainability adoption across the industry. The great example is that the ENERGY STAR program has 291 active service and product providers (mostly PropTech firms) benchmarking 399,891 buildings and gaining 13,765 ENERGY STAR certifications from June 2023 to June 2024 (ENERGYnSTAR, 2024).
4 *Sustainable Finance Initiatives*: PropTech firms partner with real estate investors and fund managers to develop new solutions and strategies for green financing and responsible investment. According to Climate Bonds Initiative (2024), the use of proceeds of green bonds in the building sector, often facilitated by PropTech platforms, grew from USD 7.7 billion in 2014 to USD 107.7 billion in 2023 (as shown in Figure 5.10).
5 *Industry-wide Platforms*: Creating shared platforms for sustainability data and best practices fosters knowledge exchange and accelerates the adoption of sustainable practices across the sector. WiredScore (2024) certified over 2,000 buildings across 27 countries by 2024, while SmartScore (2024) saw a 300% increase in certifications between 2022 and 2024.

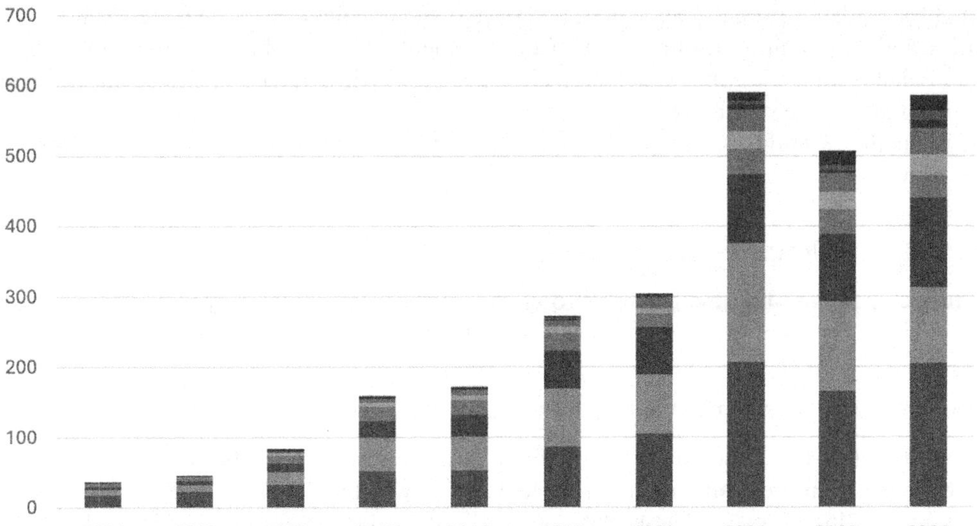

Figure 5.10 Use of Proceeds: Global Green Bonds (Billion USD).

Source: Climate Bonds Initiative (2024).

These collaborative efforts are not only enhancing sustainability in real estate but also driving innovation and creating new business opportunities in the PropTech sector.

5.4.2 *Connected Communities: PropTech for Collaborative Sustainable Real Estate Initiatives*

PropTech has emerged as a driving force in transforming how we interact with and manage real estate. As sustainability becomes an increasingly critical concern in urban development and property management, PropTech's role in fostering community engagement for sustainable real estate initiatives has gained significant attention. From mobile applications that facilitate resident feedback to blockchain systems ensuring transparency in green building certifications, PropTech is reshaping the landscape of community engagement in real estate. This transformation can be understood through the lens of socio-technical systems theory (Trist & Bamforth, 1951), which suggests that the successful implementation of new technologies depends on the interaction between social and technical factors. The socio-technical systems theory also explains how PropTech tools for community engagement can lead to more successful and sustainable outcomes by aligning technological solutions with social needs and behaviours.

Sustainable real estate includes designing, constructing, and operating buildings and communities that minimise environmental impact while maximising social and economic benefits. This approach considers energy efficiency, water conservation, waste reduction, and the overall well-being of occupants and surrounding communities. PropTech is transforming the real estate industry by introducing innovative solutions that streamline processes, enhance user experiences, and provide data-driven insights. PropTech's potential in promoting sustainability lies in its ability to bridge the gap between technological innovation and community engagement. By providing accessible tools and platforms, PropTech enables residents, property managers, and developers to collaborate more effectively in creating and maintaining sustainable living environments. This collaborative approach echoes the collaborative planning theory (Healey, 1997), which emphasises the importance of inclusive dialogue in urban planning.

There are several PropTech tools for community engagement, described in the following sections.

5.4.2.1 Mobile Applications and Platforms

Mobile apps have become powerful tools for community engagement in sustainable real estate initiatives. These apps can facilitate community engagement (Figure 5.11).

5.4.2.1.1 CASE STUDY: COMMONPLACE (HTTPS://WWW.COMMONPLACE.IS/)

Commonplace is a community engagement platform used on over 2,000 projects across the UK, engaging more than 6 million people. It offers web platforms designed to foster local community involvement. These platforms are utilised by various entities, including local governments, community groups, private developers, and other organisations. Their primary functions are disseminating information about planned projects and collecting public feedback.

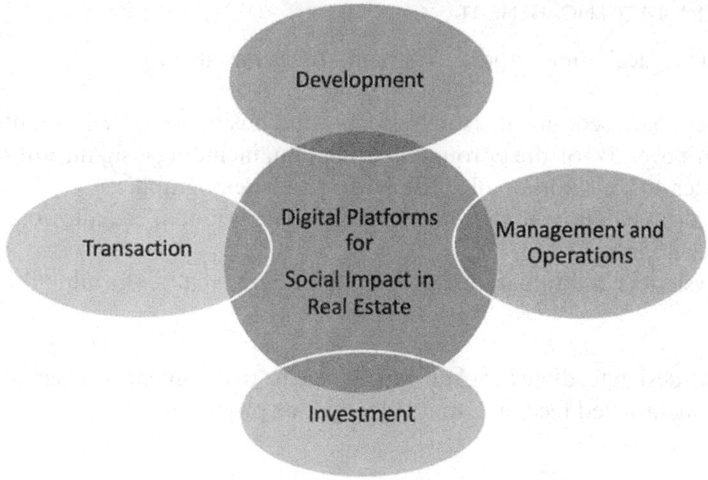

Figure 5.11 What Mobile Apps or Online Platforms Can Facilitate Community Engagement. By author (2024).

The data gathered through Commonplace are unique due to its large volume of self-selected respondents. This characteristic stems from Commonplace's core purpose as an engagement tool to encourage broad community participation. While the platform provides valuable insights, policymakers should recognise potential demographic gaps in the data. If certain groups are under-represented, decision-makers should seek additional data sources to ensure a comprehensive understanding. Commonplace's dashboard feature presents real-time data, enabling users to identify and address any representational imbalances quickly.

Among numerous successful community engagement cases, the "Mini Holland" project in Waltham Forest (UK) might be the best example of how PropTech, community engagement, and sustainability connect (Gulliford, 2020).

5.4.2.1.2 BACKGROUND

In 2014, Waltham Forest secured £20 million in investment to develop Mini-Holland, a program to improve cycle routes and public spaces across the borough. This initiative was part of the Mayor of London's Healthy Streets Approach, targeting outer London boroughs with more car-dependent residents than inner London.

5.4.2.1.3 THE PROPTECH SOLUTION

Waltham Forest partnered with Commonplace, a community engagement platform, to facilitate public participation in the planning and implementation of the Mini-Holland project. This digital solution represents a prime example of PropTech enhancing urban planning and development processes.

5.4.2.1.4 COMMUNITY ENGAGEMENT

Using Commonplace's digital tools, Waltham Forest was able to:

- Gather over 15,000 comments from residents and businesses over 18 months.
- Engage with over 5% of the borough's population, including a significant proportion of people under 40—a traditionally difficult-to-reach demographic.
- Provide real-time data and analytics to critical decision-makers, allowing for informed strategy development.
- Create newsletters and infographics from the gathered data, informing the public about changes.

The platform's design feedback tool allowed residents to comment on specific geographical areas, providing targeted feedback that helped refine proposals.

5.4.2.1.5 SUSTAINABILITY OUTCOMES

The community engagement facilitated by the PropTech solution led to several sustainable outcomes:

- *Infrastructure Improvements*: The project created 22 km of segregated cycle lanes, 700 new trees, 300 new bike hangars, 100 junction changes, 40 modal filters to prevent local rat runs, and 15 new pocket parks.
- *Increased Active Travel*: Research showed that people in the borough were walking and cycling for an average of 41 minutes more per week compared to those in similar areas.
- *Health Benefits*: A Kings College report calculated that the scheme increased the life expectancy of children living in the borough by six weeks.
- *Reduced Car Dependency*: The project aimed to make the borough as walking and cycle-friendly as Dutch cities, where over 50% of journeys are considered active travel in some areas.

This case study demonstrates how PropTech can be a powerful enabler of community engagement in urban development projects, leading to more sustainable outcomes. The Waltham Forest Mini Holland project showcases how PropTech can bridge the gap between community engagement and sustainable urban development. By utilising digital platforms to involve residents in planning, cities can create more inclusive, sustainable environments that meet their communities' needs.

5.4.2.2 Internet of Things (IoT) and Smart Home Devices

The IoT has enabled the development of novel participatory sensing technologies that allow citizens to collect and share data about their environment. Smart meters with in-home displays have been shown to lead to considerable and persistent household energy reductions, with pilot studies demonstrating average savings of up to 6% for electricity and 7% for gas (Koski et al., 2015). These devices act as a crucial stepping stone in sparking active consumer interest and engagement with energy information. IoT devices can make energy use more tangible and actionable for residents by providing real-time feedback on energy consumption and costs. Smart home devices could enable community-wide initiatives, such as coordinated demand response programs or neighbourhood energy-saving challenges, fostering a sense of shared responsibility and collective action towards sustainability goals.

Another example of IoT devices engaging citizens is the deployment of Smart Citizen sensor kits (SCKs) in Amsterdam. Waag Society (2014), a cultural institution in Amsterdam, organised a three-month deployment of SCKs with 73 users. They took several actions to foster community engagement: They (1) organised an "install party" to help users set up their sensors, (2) held meetups and workshops for users to interact, (3) enabled peer-to-peer technical assistance between users, and (4) brought in experts to discuss air quality monitoring. This orchestrated approach led to higher levels of participation compared to the Barcelona community, which just crowdfunded the devices without organised activities. The Amsterdam community developed a sense of meaningful participation, including users feeling responsible for the data they produced. They could envision how the data could lead to collective action (e.g., mapping noise levels to revise legislation), and discussions emerged about how citizens can use technology to participate in civic life. The orchestrated deployment in Amsterdam showed how IoT devices and community-building activities can contribute to sustained engagement and dialogue around civic issues. This case demonstrates that simply distributing IoT devices is helpful but not enough; orchestrated community activities are crucial for fostering meaningful engagement with the technology and data.

Implementing PropTech in community engagement for sustainability takes work. Figure 5.12 shows the good practices in the process:

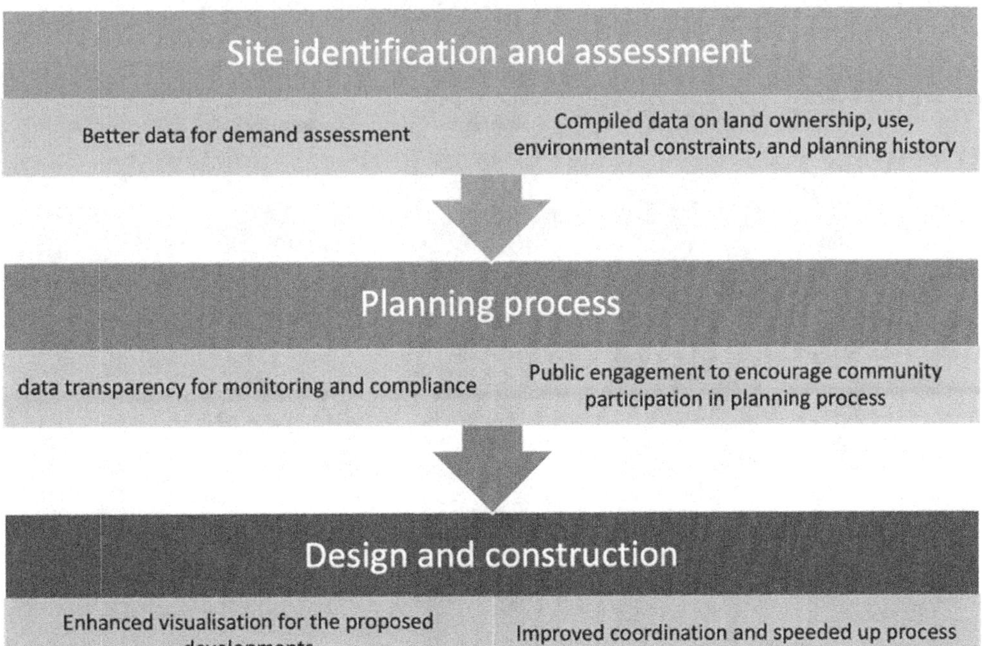

Figure 5.12 Best Practices in Implementing PropTech in Community Engagement.

By author (2024).

The landscape of community engagement is rapidly evolving, driven by emerging trends that present challenges and opportunities. As communities increasingly demand transparent, real-time information on sustainability metrics and show growing interest in participatory design and decision-making processes, there's a rising expectation for seamless technology integration in daily community life. However, this shift challenges ensuring equal access and familiarity with PropTech tools across diverse community members. To address this, implementing training programs and providing accessible devices can foster inclusive participation.

The case studies and examples discussed in this chapter demonstrate the transformative potential of technology in fostering community involvement. From mobile apps that gamify sustainable behaviours to blockchain-powered energy trading systems, PropTech is creating new avenues for residents to participate in and benefit from sustainable real estate practices.

5.4.3 Digital Platforms for Social Impact in Real Estate

The real estate industry, including development, investment, management, and transactions, has significant potential to create a positive social impact. Digital platforms are increasingly important in realising this potential across various aspects of the real estate sector. Figure 5.13 shows how these digital platforms facilitate social responsibility in different areas of real estate.

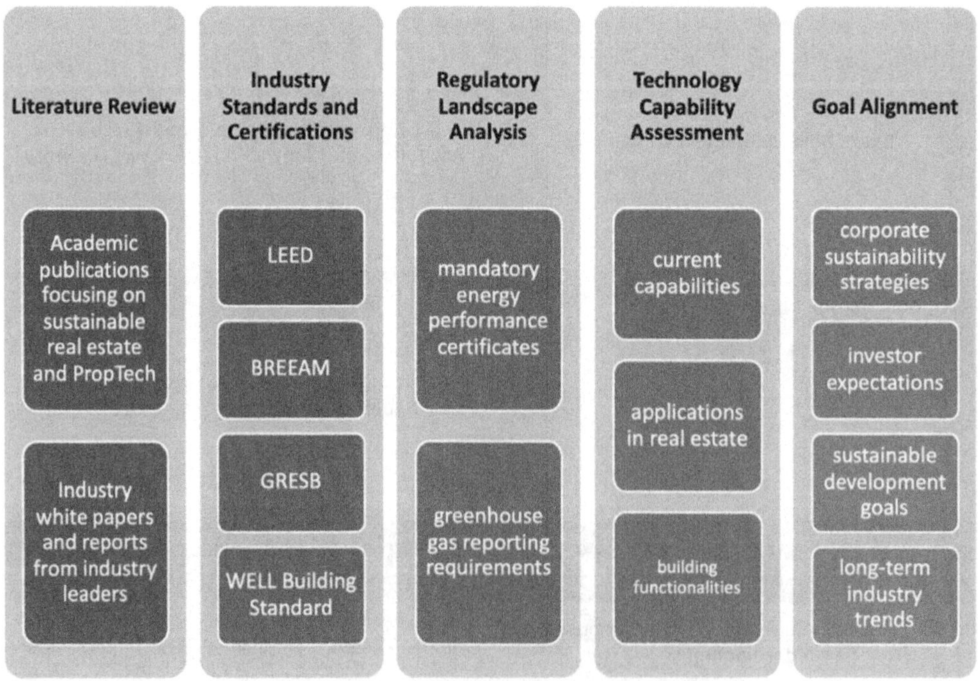

Figure 5.13 Digital Platforms for Social Impact in Various Real Estate Sectors.

By author (2024).

5.4.3.1 Real Estate Development

In development, digital platforms continue to enhance community engagement, sustainability, and inclusive design and improve data transparency, monitoring, and assessment. From planning and site identification to design and construction, digital platforms have shown increasing importance. Figure 5.14 demonstrates how digital platforms have increased the social impact of each stage of real estate development.

Digital platforms have enhanced the social impact of real estate development across all stages, from site selection to final construction. In the initial phase, tools like LandInsight (https://land.tech/) streamline site identification and assessment by consolidating crucial data while providing better insights for needs assessment. During the planning stage, these platforms enhance data accessibility and transparency, levelling the playing field among stakeholders and improving compliance tracking. They also facilitate public engagement through participatory planning tools like coUrbanize (https://www.courbanize.com/) and Commonplace (https://www.commonplace.is/), fostering community input throughout development. In design and construction, advanced visualisation technologies such as VR and 3D modelling tools like VU.CITY (https://www.vu.city/) allows all parties to better understand a project's potential impact. These digital platform solutions improve coordination among various stakeholders and accelerate housing delivery by digitising and streamlining processes (Burgess & Quinio, 2018), ultimately leading to more informed, inclusive, and efficient real estate development practices.

As showcased in Burgess and Quinio's (2018) report, Singapore's Construction and Real Estate Network (CORENET) program is an example of advanced digital innovation in planning and construction. Singapore is described as a leader in the BIM industry, having reached BIM Level 3 in 2015, while the UK was still aiming for Level 2 by 2016. The

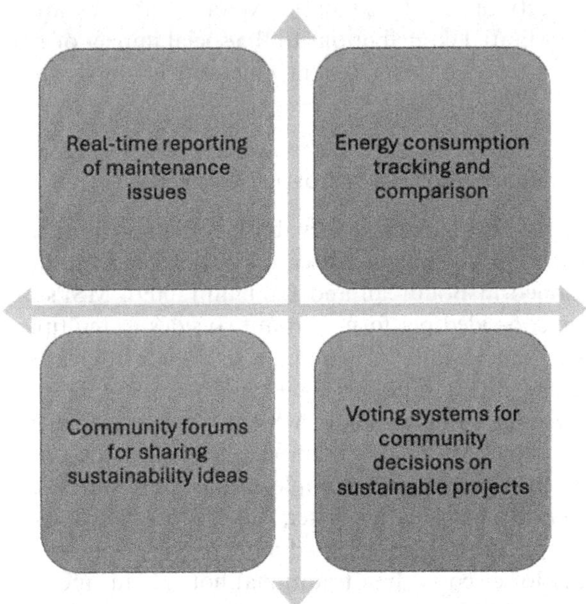

Figure 5.14 Digital Platforms' Use in Each Stage of Real Estate Development.
By author (2024).

CORENET program was launched in 2002 by the Building and Construction Authority (BCA) of Singapore. It involves an e-submission system that includes 16 regulatory authorities and eight government ministries responsible for the construction and real estate industry. This system allows applicants to submit planning applications online, eliminating the need for hard copies of plans, paper planning records, or on-site visits. The e-submission system adds transparency to the process and allows users to track and monitor their applications online. In 2016, Singapore launched the e-Plan Check system, making it mandatory for new build projects over 5,000 square meters. The e-Plan Check system, fully funded by the government, allows architects and engineers to check their planning applications against plan objectives. It uses automated models to check BIM e-submission compliance. The system aims to shift from traditional 2D design drafts to digital modelling, where the BIM gathers project information throughout the building lifecycle. This new database can be progressively completed at every stage of the project, allowing building professionals and regulatory officers to check regulatory compliance and conduct audit checks through the same platform. The CORENET case is an example of how a comprehensive, government-led approach to digitising planning and construction processes can lead to significant advancements in efficiency and transparency in the real estate development sector.

5.4.3.2 Real Estate Management and Operations

Digital platforms are transforming how real estate assets are managed and operated, with a focus on social impact. This transformation extends beyond mere efficiency gains, significantly impacting the social fabric of communities and the way people experience built environments. As technology advances, these platforms are becoming increasingly sophisticated, offering innovative solutions that address long-standing challenges in the real estate sector while simultaneously creating new opportunities for social engagement, sustainability, and improved quality of life. This section explores three key areas where digital platforms are making substantial contributions to the social impact of real estate: the sharing economy and multi-sided platforms, tenant engagement apps, and IoT-enabled smart building platforms.

5.4.3.2.1 SHARING ECONOMY AND MULTI-SIDED PLATFORMS

The sharing economy creates numerous new business opportunities, and the digital multi-sided platform (MSP), which has increased the social impact in the real estate sector, is one of them. As defined in Boudreau and Lakhani (2009), MSPs are a straightforward generalisation of the two-sided platform—from two sides to multiple sides. Hagiu (2014) further suggests that the technologies, products, or services that can create value mainly through enabling direct interactions between multiple customers are all MSPs. Petroni (2024) argues that systems and data allow MSPs to bring together people with the same purpose to co-create value. MSPs are commonly believed to be important in facilitating interactions or value creation between multiple groups. MSPs allow customers to compare and find real estate services much easier, faster, and cheaper (because intermediary costs are minimised). For example, Airbnb has enabled more people to access short-term accommodations at potentially lower costs than traditional hotels. This increases affordability and options for travellers. Similarly, platforms connecting renters and landlords directly could potentially reduce costs and increase housing access (PwC, 2021). MSPs give customers access to trusted and verified real estate service providers (PwC, 2021). Review and rating systems on MSPs can increase transparency and trust in real estate transactions. This can empower consumers and incentivise better service from providers. PwC (2021) notes that MSPs provide greater customer reach for real estate service providers. These platforms can reduce barriers for small property owners or service providers to enter the market, potentially increasing diversity in the real estate ecosystem. Petroni's (2024) introduction to MSPs suggests that MSPs create enormous value by reducing search and/or transaction costs for participants. Sharing economy platforms allow for more efficient use of existing real estate. For example, Airbnb allows homeowners to rent out spare rooms or entire homes when not in use, making better use of underutilised space. This can have environmental benefits by reducing the need for new construction.

Moreover, MSPs can make recommendations to users based on data and machine learning (Petroni, 2024). MSPs can use their data and matching algorithms to better connect individuals with properties and services that meet their specific needs, potentially improving overall satisfaction and quality of life. Hagiu (2014) discussed how MSPs bring together different user groups. Some real estate MSPs, like co-living platforms, aim to build community and social connections among residents. This addresses social isolation issues that can occur in dense urban environments. Meanwhile, MSPs support direct interactions between different sides (Hagiu, 2014), allowing for more decentralised provision of real estate services, potentially shifting power away from large intermediaries towards individual providers and consumers.

However, it's important to note that these platforms also come with challenges and potential negative impacts that need to be managed. For example, if residential properties are converted to short-term rentals, they may contribute to housing shortages or increased rents in some areas. Hagiu (2014) discusses various challenges MSPs face, including balancing the interests of different sides and potential conflicts. Concerns exist about labour protections and rights for service providers on these platforms. Discrimination and bias issues in platform algorithms and user behaviour should be addressed. Regulatory challenges arise around taxation, zoning laws, and consumer protections. To maximise the positive social impact of MSPs in real estate, careful design of platform rules, pricing structures, and governance models is critical (Hagiu, 2014). Additionally, collaboration between platform companies, policymakers, and communities will be important to address challenges and ensure these innovations contribute positively to social outcomes in the real estate sector.

5.4.3.2.2 TENANT ENGAGEMENT APPS

Tenant engagement is crucial in generating social impact within the social housing sector. Research indicates that meaningful tenant participation can improve housing satisfaction, more effective landlords, and cost savings for housing organisations (Blackford & LeBrasseur, 1992; Bliss & Lambert, 2016; Hickman & Preece, 2019). However, the implementation of tenant engagement initiatives often needs help. While housing providers view tenant participation as important, tensions exist between empowering tenants and maintaining organisational control over decision-making processes (Hickman & Preece, 2019). The digitalisation of tenant engagement through social media and other platforms presents both opportunities and risks—it can potentially broaden participation but also risks weakening tenant influence if critical voices are suppressed to protect organisational reputations (Condie & Ayres, 2024). Some innovative approaches are emerging, such as co-design processes for new circular economy housing projects, but resource constraints and technical complexities can limit tenant involvement (Sacranie & Çetin, 2022). Previous studies suggest tenant engagement can generate positive social impacts, but realising this potential requires housing providers to prioritise meaningful tenant influence over marketing or reputational concerns.

In practice, digital platforms and apps have improved tenant engagement in both residential and commercial real estate sectors, enhancing the industry's social impact. Tenant engagement apps are emerging as powerful tools to enhance social impact in real estate by fostering community, improving communication, and optimising building operations. These applications offer various features that benefit tenants and property managers. CREtech (2018) shows that apps like Comfy and HqO allow tenants to control aspects of their environment, such as temperature and lighting, while providing analytics to improve building performance. Others, like Office App and The HIVE, facilitate maintenance requests, room bookings, and community messaging, enhancing the overall tenant experience. Some platforms, such as Equiem, offer integrated solutions that deliver customised online and offline experiences, helping landlords attract and retain tenants while improving return on investment (CREtech, 2018). Zettl (2023) found that the adoption of these technologies is expected to grow, with 77% of commercial real estate leaders anticipating increased investments in workplace technology by 2025. These apps streamline operations and create more engaging and efficient environments, potentially addressing challenges like low office occupancy in the post-pandemic era.

These digital platforms and apps are transforming traditional office spaces into vibrant, community-oriented environments that better meet the needs of modern workers and tenants by providing valuable amenities and fostering engagement.

5.4.3.2.3 IOT-ENABLED SMART BUILDING PLATFORMS

IoT-enabled platforms can significantly optimise energy use, improve air quality, and enhance occupant comfort in buildings and cities through several mechanisms. IoT sensors and systems allow for real-time monitoring and control of building systems like HVAC, lighting, and appliances. This enables more precise management of energy consumption based on actual occupancy and usage patterns. The integration platform of diverse IoT sensors, data analytics, and automated controls allows for a more holistic, data-driven approach to optimising building performance across energy, air quality, and comfort parameters (Jia et al., 2019). This represents a significant advancement over traditional building management approaches.

5.4.3.3 Real Estate Investment

Digital platforms are supporting socially responsible real estate investing, creating new opportunities for investors to make a positive impact. The impact real estate investment platforms bridge investors and real estate projects that focus on social or environmental benefits. They make it easier for socially conscious investors to find and support projects aligned with their values. For example, the crowdfunding platform SmallChange allows people to invest in real estate projects that change cities and neighbourhoods for the better, and the proprietary Change Index was created to track that change. The Change Index uses data points such as an area's walk- and bike-ability, public transit access, proximity to green space, and availability of commercial and cultural amenities, along with the team's makeup and benefits to the community, to measure the possibilities of a project. The index is people-centric, and the projects on the platforms focus on socially impactful themes, including affordable housing/business incubators, renovation/brownfield, activating the street, reduced parking, minimal site impact, and net zero and alternative energy sources. So far, 57% of SmallChange funded projects are with minority or women developers, and 65% are affordable housing projects. SmallChange projects have been accessible to many underserved populations and cover 93% of urban locations, with 78% being walkable locations (SmallChange, 2024).

As online platforms have become central to housing markets in the UK and elsewhere, Goodchild and Ferrari (2024) discuss how platforms may influence tenure preferences, mobility, and concepts of the sharing economy in housing. Their study examines how platforms mediate information and potentially alter market practices, such as making housing searches a more leisurely activity or promoting a calculative mentality. Housing investment portals dedicated to facilitating investment in affordable housing projects are making a significant difference in addressing housing accessibility issues. The impact of these digital platforms on affordable housing is particularly noteworthy. For instance, Rhino, a security deposit alternative platform, reported saving renters over USD 500 million in upfront costs as of 2024, significantly reducing barriers to housing access (Rhino, 2024). Similarly, the crowdfunding platform SmallChange (mentioned previously) has channelled over USD 100 million into affordable housing projects since its launch.

These examples demonstrate how digital platforms facilitate socially responsible real estate investment and drive tangible, positive outcomes in the housing (especially the affordable housing) sector. These digital platforms will be increasingly important in addressing social and environmental challenges through real estate investment.

5.5 Measuring the Impact of PropTech Strategies for Sustainability

5.5.1 Metrics for Success

5.5.1.1 Methodology for Identifying Key Metrics in Sustainable PropTech Strategies

We use a systematic approach (shown in Figure 5.15) to identify the key metrics that measure the success and impact of sustainable PropTech strategies and ensure that the chosen metrics are comprehensive, relevant, and aligned with industry standards and best practices.

The method begins with thoroughly examining existing academic literature and industry reports. This review helps identify metrics validated through research and practical application. We have also included the established frameworks and certifications, which significantly shape the metrics landscape. These frameworks provide a foundation for identifying widely recognised and comparable metrics across different properties and portfolios. Then, we analyse environmental regulations and reporting requirements across different regions to inform the selection of metrics. Aligning metrics with regulatory requirements ensures compliance and facilitates standardised reporting.

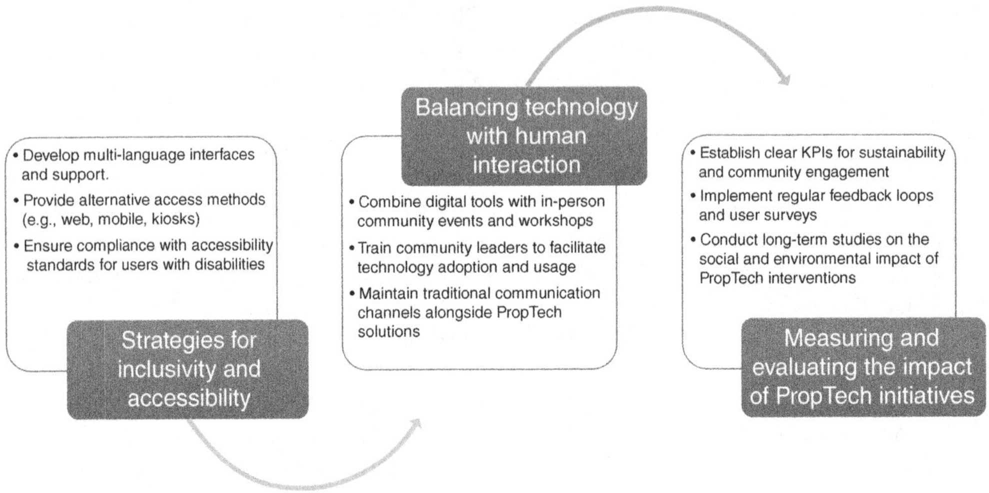

Figure 5.15 Methodology to Identify the Key Metrics That Measure the Success and Impact of Sustainable PropTech Strategies.

By author (2024).

The evolving landscape of PropTech solutions is influenced by embedded metrics that can be effectively measured and monitored. Understanding technological capabilities helps select metrics that can be accurately and efficiently measured using available PropTech solutions. Selected metrics must align with broader sustainability goals and business objectives. This alignment ensures that the chosen metrics drive actions contributing to overarching sustainability and business goals. In this way, we can identify a robust set of metrics that provide meaningful insights into the performance and impact of sustainable PropTech initiatives. The following sections will investigate these specific metrics, categorised by their respective focus areas within sustainable real estate practices.

5.5.1.2 *Specific Metrics for Sustainable PropTech Strategies*

Following the methodology outlined previously, we have identified key metrics across various focus areas within sustainable real estate practices. These metrics provide a comprehensive framework for evaluating the success and impact of sustainable PropTech initiatives aligned with established industry standards and academic research (GRESB, 2024; Doan et al., 2017).

1 *Energy Efficiency Metrics:* Energy efficiency is a cornerstone of sustainable building practices, directly impacting both environmental footprint and operational costs (UN, 2023). The measurable key performance indicators (KPIs) in this category are:

- *Energy Use Intensity (EUI):* The annual energy consumption per unit of gross floor area, measured by kWh/m^2/year or kBTU/ft^2/year. EUI allows for comparison across buildings of different sizes and types, providing a standardised energy efficiency measure (ENERGY STAR, 2024).
- *Peak Demand Reduction:* The decrease in maximum power demand during high-usage periods, measured by kilowatt reduction or percentage reduction. Lowering peak

demand can significantly reduce energy costs and strain on the power grid (Newsham et al., 2009).

2 *Water Management Metrics:* Water conservation is increasingly critical, especially in water-stressed regions (World Green Building Council, n.d.), and the KPIs are:

- *Water Use Intensity (WUI):* The annual water consumption per unit of gross floor area; its measurement is liters/m^2/year or gallons/ft^2/year. WUI helps identify opportunities for water conservation and compares efficiency across different properties (BREEAM, n.d.).
- *Water Recycling Rate:* The percentage of water recycled or reused on-site. Higher recycling rates indicate better water management practices and reduced reliance on municipal water supplies (LEED, n.d.).

3 *Indoor Environmental Quality (IEQ) Metrics:* IEQ metrics are essential for occupant health, comfort, and productivity, as highlighted by the International WELL Building Institute (WELL, n.d.), especially so after the COVID-19 pandemic.

- *Air Quality Index (AQI): A* composite measure of indoor air quality based on levels of key pollutants. It is typically measured on a scale of 0–500, with lower numbers indicating better air quality. Good air quality is essential for occupant health and can impact productivity and well-being (Potrč Obrecht et al., 2019).
- *Thermal Comfort Score:* The percentage of occupants satisfied or on a scale (e.g., −3 to +3). Thermal comfort directly affects occupant satisfaction and productivity (Janda, 2011).

4 *Waste Management Metrics:* Effective waste management is another important metric in promoting sustainable development and circular economy principles (Zuo & Zhao, 2014). Waste diversion rate and generation intensity can be used to indicate the performance in this category.

- *Waste Diversion Rate:* The percentage of waste diverted from landfills through recycling, composting, or other means. Higher diversion rates indicate more effective waste management practices and reduced environmental impact (GRESB, 2023).
- *Waste Generation Intensity:* The amount of waste generated per occupant or per unit area, measured by kg/person/year or kg/m^2/year. This indicator helps identify opportunities for waste reduction and track progress over time (Kamaruzzaman et al., 2016).

5 *Renewable Energy Metrics:* The adoption of renewable energy is critical to reducing carbon emissions and enhancing energy security (World Green Building Council, n.d.). The most widely used indicator is renewable energy percentage, which shows the proportion of total energy consumption sourced from renewables (both on-site and off-site). This metric shows progress towards clean energy goals and reduced reliance on fossil fuels (GRESB, 2023).

6 *Carbon Emissions Metrics:* As stated by the United Nations Environment Programme, carbon emissions metrics are essential for tracking progress towards climate change mitigation goals (UN, 2023). Carbon emission intensity is usually used to measure carbon emissions performance. Reducing carbon emissions can be included in the metrics to evaluate the impact of a company's strategy or practice.

- *Carbon Emission Intensity:* It can be calculated in different ways in various contexts. For example, for buildings, it is the amount of carbon dioxide equivalent (CO_2e) emissions per unit of floor area. For corporations, it is the amount of CO_2e emissions per annual revenue (LSEG, 2024). For portfolios, it is the amount of CO_2e emissions per portfolio investment value (Nagy et al., 2023). This metric allows for a standardised comparison of carbon performance across different buildings and portfolios (Scofield, 2013).
- *Carbon Emissions Reduction:* The changes in carbon emissions. It refers to the year-over-year reduction in total greenhouse gas emissions. Tracking emissions reductions helps monitor progress towards carbon neutrality goals (World Green Building Council, n.d.).

7 *Financial Performance Metrics:* Financial metrics help justify investments in sustainable PropTech and demonstrate their economic benefits (Newsham et al., 2009). The most direct indicator can measure the return on sustainability investment (ROSI). It is calculated as the financial return generated from investments in sustainable technologies and practices. ROSI helps make the business case for sustainability initiatives and PropTech investments (Zuo & Zhao, 2014).
8 *Occupant Engagement Metrics:* Due to the special nature of real estate (space), occupant involvement is very important in sustainability practice. Occupant engagement is critical for the success of many sustainable PropTech initiatives (Janda, 2011). Two indicators are used in measure the impacts on occupant engagement in sustainability.

- *Sustainability Awareness Score*
- *Technology Adoption Rate*

Real estate professionals can gain a comprehensive understanding of the performance and impact of their sustainable PropTech initiatives by tracking these metrics across various focus areas. Regular monitoring and analysis of these metrics enable data-driven decision-making, continuous improvement, and demonstration of tangible benefits to stakeholders (GRESB, 2023).

5.5.2 Case Studies on Successful Implementation

This section presents a case study demonstrating the positive outcomes of implementing PropTech in sustainable real estate projects.

5.5.2.1 Case Study: Hank AI—Revolutionising HVAC Management Across Real Estate Portfolios

5.5.2.1.1 BACKGROUND

Hank[6] is an AI-powered HVAC optimisation platform designed to reduce energy consumption, maximise tenant comfort, and exceed indoor air quality standards in commercial real estate. This case study examines Hank's implementation across three real estate companies, demonstrating its adaptability and effectiveness in various property types and management scenarios. The data is collected from Hank's official website.

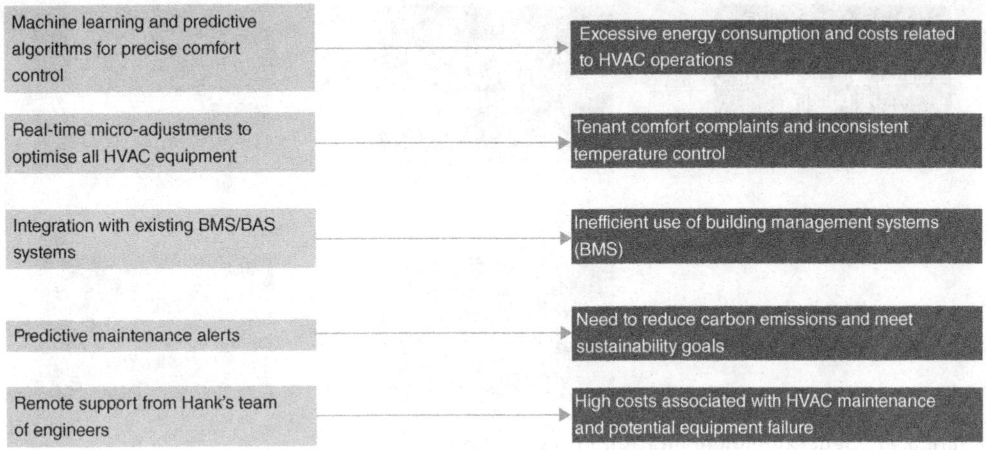

Figure 5.16 Challenges Versus PropTech Solutions—Hank Case.

5.5.2.1.2 CHALLENGES VERSUS PROPTECH SOLUTIONS

Commercial real estate companies often require assistance in managing their properties, despite unique circumstances. These challenges typically include high energy consumption and costs associated with HVAC operations, frequent tenant complaints about comfort and inconsistent temperature control, inefficient utilisation of BMSs, pressure to reduce carbon emissions to meet sustainability goals, and significant expenses related to HVAC maintenance and potential equipment failures. To address these issues, Hank's AI-powered HVAC optimisation platform offers a comprehensive solution. The platform employs machine learning and predictive algorithms to ensure precise comfort control, makes real-time micro-adjustments to optimise all HVAC equipment, seamlessly integrates with existing BMS/BAS systems, provides predictive maintenance alerts to prevent unexpected breakdowns, and offers remote support from Hank's team of experienced engineers. This technology-driven approach enables property managers to address multiple challenges simultaneously, enhancing energy efficiency, tenant satisfaction, and overall building performance. Figure 5.16 shows the general challenges versus PropTech solutions provided by Hank.

5.5.2.1.3 IMPLEMENTATION PROCESS

The general implementation process for each company is shown in Figure 5.17.

The three examples with Hank AI HVAC optimisation are:

1 *45 Church Street*[7]*:* A 125,000 sq ft commercial office building in Birmingham, UK, owned by Royal London Asset Management[8];
2 *Cerritos Center Court:* A 175,000 sq ft Class A office building in Cerritos, California, US, owned by Nome Capital Partners[9];
3 *Metro Center*[10]*:* A 258,806 sq ft Class-A office campus in Sacramento, California, United States, owned by GPR Ventures.[11]

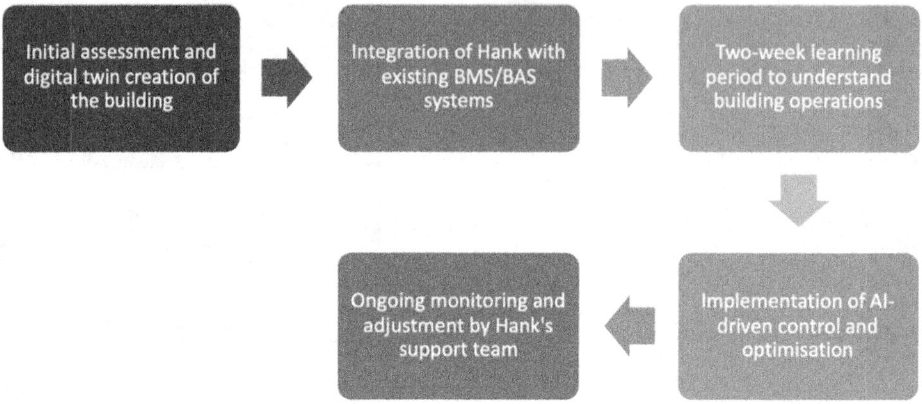

Figure 5.17 General Implementation Process—Hank Case.
By author (2024).

Table 5.8 details the three cases, highlighting the specific challenges and optimisation results in each case.

5.5.2.1.4 LESSONS LEARNED

- AI-driven HVAC optimisation can deliver significant energy savings and ROI across various property types and sizes;
- Improved energy efficiency and sustainability can be achieved without compromising tenant comfort—in fact, tenant satisfaction often improves;
- Predictive maintenance features can extend equipment life and reduce operational costs;
- The combination of AI technology and human expertise (remote support engineers) provides a powerful solution for building management;
- PropTech can help real estate companies meet sustainability goals and reduce carbon emissions without large capital investments.

5.5.2.1.5 KEY SUCCESS FACTORS

- Seamless integration with existing building management systems;
- Short learning period to understand building operations before full implementation;
- Ability to make real-time micro-adjustments based on various factors (occupancy, weather, etc.);
- Focus on both energy efficiency and tenant comfort;
- Continuous monitoring and optimisation by both AI and human experts.

5.5.2.2 Implications for the Industry

These case studies highlight the increasing significance of PropTech in the real estate industry, particularly in achieving sustainability goals and improving building performance. AI

Table 5.8. Three Cases of Hank AI HVAC Optimisation

Case Study	*Property Details*	*Specific Challenges*	*Results (After the HVAC Optimisation)*
1	125,000 sq ft commercial office building in Birmingham, UK	- HVAC system running on pre-set BMS schedules - Limited BMS control, leading to overrides	708% ROI 59% HVAC energy savings 21% total energy savings 500 metric tons of carbon emissions prevented annually £148,000 guaranteed annual savings 7% less deviation from tenant setpoint
2	175,000 sq ft Class A office building in Cerritos, California, US	- Inefficient use of air handling units - Complex HVAC setup with custom-built air handler - Anchor tenant dissatisfaction with temperature control - High energy consumption due to reactive HVAC management	1–2 year equipment life extension 45% HVAC energy savings 20% total energy savings 60% less deviation from tenant setpoint 2x increase in fresh airflow 3–5 years increase in equipment life
3	258,806 sq ft Class-A office campus occupied by major tenants, in Sacramento, California, US	- Over 80 weekly tenant complaints due to poor temperature control - Recommendation for costly new HVAC controls system (USD 400,000+) - High energy costs despite relatively new BMS	30% total energy savings 80% less deviation from tenant setpoint 50% savings on HVAC service calls 400% ROI Significant reduction in tenant complaints Remote resolution of tenant requests within 15 minutes

and machine learning technologies are proving to be powerful tools for optimising building systems, especially HVAC, with potential benefits for both new developments and existing buildings. The data-driven nature of these solutions enables more targeted and effective sustainability initiatives while integrating multiple PropTech solutions, which can yield synergistic benefits. However, successful implementation requires focusing on both technological and human factors, such as user engagement and behaviour change. This evolving landscape highlights the need for professionals who can bridge the gap between traditional real estate practices and innovative PropTech solutions. Additionally, regulatory frameworks and incentives are crucial in driving PropTech adoption for sustainability.

Importantly, these case studies demonstrate that PropTech solutions can deliver significant financial benefits alongside environmental improvements, creating a compelling business case for their adoption in the real estate sector.

5.6 Concluding Remarks

This chapter examines PropTech strategies in sustainable real estate and reveals several key conclusions. First, integrating technology in real estate has moved beyond mere efficiency improvements to become a crucial driver of sustainability transformation. Multiple case studies present evidence that PropTech solutions can deliver significant environmental benefits while maintaining or enhancing financial performance.

The success of PropTech implementations in sustainable real estate depends on several critical factors: complete integration, data-driven decision-making, stakeholder engagement, and measurable impact.

The evidence presented in this chapter indicates that PropTech will continue to play an increasingly crucial role in achieving sustainability goals within the real estate sector. The successful cases and frameworks discussed offer a valuable roadmap for industry practitioners aiming to leverage technology for environmental impact.

Notes

1 https://realt.co/.
2 https://www.propertyshare.in/.
3 https://atlant.io/.
4 https://blocksquare.io/.
5 https://smartlands.app/.
6 https://www.jllt.com/hank/.
7 https://www.45-churchstreet.co.uk/.
8 https://www.royallondon.com/.
9 https://nomecapitalpartners.com/index.html.
10 https://gprventures.com/portfolio/metro-center/.
11 https://gprventures.com/.

References

Ahn, J., & Cho, S. 2017. Anti-logic or common sense that can hinder machine's energy performance: Energy and comfort control models based on artificial intelligence responding to abnormal indoor environments. *Applied Energy*, 204, 117–130.

Andrew, B., Andrew, S., & Braesemann, F. 2020. *PropTech 2020: The future of real estate*. Oxford Future of Real Estate Initiative, University of Oxford.

Bhardwaj, M., Hedrick-Wong, Y., & Thomas, H. 2018. Financial inclusion for Asia's unbanked. *World Bank Blogs*. https://blogs.worldbank.org/en/allaboutfinance/financial-inclusion-asias-unbanked

Blackford, K. A., & Lebrasseur, R. 1992. First test of a tenant-centered instrument for measuring tenant participation in subsidized housing projects. *Journal of Planning Education and Research*, 11, 85–95.

Bliss, N., & Lambert, B. 2016. *Investing in involvement: An accessible framework for identifying the benefits of tenant involvement*. Tenants Leading Change.

Bokhari, S., Chegut, A., Frenchman, D., & Tausendschoen, I. 2021. *Is innovation really in a place? Accelerator program impacts on firm performance*. MIT Center for Real Estate.

Boudreau, K. J., & Lakhani, K. R. 2009. How to manage outside innovation. *MIT Sloan Management Review Sloanselect Collection*, 57(5), 50–57.
BREEAM. n.d. An introduction to BREEAM. https://breeam.com/standards/
Burgess, G., & Quinio, V. 2018. *Housing digital built Britain network: How could better use of data and digital technologies improve housing delivery through the UK planning system?* University of Cambridge.
Chapiro, C. 2021. Working toward financial inclusion with blockchain. *Stanford Social Innovation Review*. https://doi.org/10.48558/DZXJ-0Z18
Climate Bonds Initiative. 2024. *The climate bonds interactive data platform*. Climate Bonds Initiative.
Condie, J., & Ayres, L. 2024. Tenant participation and emerging social media practices in the social housing sector. *Housing Studies*, 40(3), 1–21.
CREtech. 2018. 8 apps to boost the tenant experience. https://www.cretech.com/cretech-blog/7-apps-to-boost-the-tenant-experience/
Devine, A., Kok, N., & Wang, C. 2023. Sustainability disclosure and financial performance: The case of private and public real estate. *Journal of Portfolio Management*, 49(10), 119–133.
Doan, D. T., Ghaffarianhoseini, A., Naismith, N., Zhang, T., Ghaffarianhoseini, A., & Tookey, J. 2017. A critical comparison of green building rating systems. *Building and Environment*, 123, 243–260.
ENERGY STAR. 2024. Most active service & product providers. https://www.energystar.gov/buildings/save-energy-commercial-buildings/expert-help/most-active
Fidelity International. 2022. How blockchain will revolutionise the financial system. https://www.fidelityinternational.com/editorial/article/how-blockchain-will-revolutionise-the-financial-system-1977b9-en5/
Fuerst, F. 2015, June 16. The financial rewards of sustainability: A global performance study of real estate investment trusts. SSRN: 2619434. http://dx.doi.org/10.2139/ssrn.2619434
Ghaffarianhoseini, A., Alwaer, H., Ghaffarianhoseini, A., Clements-Croome, D., Berardi, U., Raahemifar, K., & Tookey, J. 2018. Intelligent or smart cities and buildings: A critical exposition and a way forward. *Intelligent Buildings International*, 10, 122–129.
Goodchild, B., & Ferrari, E. 2024. Intermediaries and mediators: An actor-network understanding of online property platforms. *Housing Studies*, 39, 102–123.
GRESB. 2023. Real estate standard and reference guide. https://www.gresb.com/wp-content/uploads/resources-2023-real-estate-standard-and-reference-guide.pdf
GRESB. 2024. 2023 real estate assessment results. https://www.gresb.com/nl-en/2023-real-estate-assessment-results/
Gulliford, F. 2020. Creating "Mini Holland" in Waltham Forest. *Commonplace*. https://www.commonplace.is/customer-stories/creating-mini-holland-in-waltham-forest
Hagiu, A. 2014. Strategic decisions for multisided platforms. *MIT Sloan Management Review*, 55, 71.
Healey, P. 1997. Strategies, processes and plans. In Healey, P. (ed.), *Collaborative planning: Shaping places in fragmented societies*. Macmillan Education UK.
Hewa, T., Ylianttilla, M., & Liyanage, M. 2021. Survey on blockchain based smart contracts: Applications, opportunities and challenges. *Journal of Network and Computer Applications*, 177, 102857.
Hickman, P., & Preece, J. 2019. *Understanding social housing landlords' approaches to tenant participation*. UK Collaborative Center for Housing Evidence Glasgow.
Himeur, Y., Elnour, M., Fadli, F., Meskin, N., Petri, I., Rezgui, Y., Bensaali, F., & Amira, A. 2023. AI-big data analytics for building automation and management systems: A survey, actual challenges and future perspectives. *Artificial Intelligence Review*, 56, 4929–5021.
Houlihan-Lokey. 2024. *2023 PropTech year in review*. https://cdn.hl.com/pdf/2024/2023-proptech-year-in-review.pdf
Janda, K. B. 2011. Buildings don't use energy: People do. *Architectural Science Review*, 54, 15–22.
Jia, M., Komeily, A., Wang, Y., & Srinivasan, R. S. 2019. Adopting Internet of Things for the development of smart buildings: A review of enabling technologies and applications. *Automation in Construction*, 101, 111–126.
Jones-Lang-Lasalle. 2022. *The future of work survey 2022: Five critical challenges ahead for real estate*. JLL.

Kamaruzzaman, S. N., Lou, E. C. W., Zainon, N., Mohamed Zaid, N. S., & Wong, P. F. 2016. Environmental assessment schemes for non-domestic building refurbishment in the Malaysian context. *Ecological Indicators*, 69, 548–558.

Kisi. 2024. *Partner Spotlight: Kisi x Spacebring, connecting people and coworking spaces*. Kisi Incorporated.

Kontokosta, C. E., & Tull, C. 2017. A data-driven predictive model of city-scale energy use in buildings. *Applied Energy*, 197, 303–317.

Koski, P., Laitila, P., Martins, F., Edelenbos, E., Persson, A., & Ekander, F. 2015. *Smart meters and consumer engagement*. https://www.ca-eed.eu/ia-document/executive-summary-3-5-6-5-smart-meters-and-consumer-engagement/

KPMG. 2017. *Bridging the gap: How the real estate sector can engage with PropTech to bring the built and digital environments together*. Author.

Lawson, J., Troy, L., & Van Den Nouwelant, R. 2024. Social housing as infrastructure and the role of mission driven financing. *Housing Studies*, 39, 398–418.

LEED. n.d. *LEED rating system*. U.S. Green Building Council. https://www.usgbc.org/leed

LSEG. 2024. *Workspace data & analytics*. London Stock Exchange Group.

Martinez-Moyano, I. 2006. Exploring the dynamics of collaboration in interorganizational settings. *Creating a Culture of Collaboration: The International Association of Facilitators Handbook*, 4, 69.

Measurabl. 2024. *Measurabl and Susteco announce partnership accelerating data-driven business opportunities for real estate*. Measurabl Incorporated.

Mhlanga, D. 2023. Blockchain technology for digital financial inclusion in the industry 4.0, towards sustainable development? *Frontiers in Blockchain*, 6.

Mistrangelo, P., Tagliabue, L. C., & Tezel, A. 2022. Property tokenization digital framework for inclusive and sustainable asset markets development. *Proceedings of the 2022 European Conference on Computing in Construction*.

Muneeza, A., Arshad, N. A., & Arifin, A. T. 2018. The application of blockchain technology in crowdfunding: Towards financial inclusion via technology. *International Journal of Management and Applied Research*, 5, 82–98.

Nagy, Z., Giese, G., Wang, X., Mirabelli, A., Keywork, N., Turetsky, A., Griffiths, B., Bhansali, V., Chang, L., & Holdom, J. 2023. A framework for attributing changes in portfolio carbon footprint. *The Journal of Portfolio Management*, 49, 163–184.

Newsham, G. R., Mancini, S., & Birt, B. J. 2009. Do LEED-certified buildings save energy? Yes, but…. *Energy and Buildings*, 41, 897–905.

Peña-García, A., Hurtado, A., & Aguilar-Luzón, M. 2015. Impact of public lighting on pedestrians' perception of safety and well-being. *Safety Science*, 78, 142–148.

Petroni, M. 2024. *Introduction to multi-sided platforms*. Causeit, Inc. https://www.digitalfluency.guide/multi-sided-platforms/multi-sided-platforms

Png, E., Srinivasan, S., Bekiroglu, K., Chaoyang, J., Su, R., & Poolla, K. 2019. An Internet of Things upgrade for smart and scalable heating, ventilation and air-conditioning control in commercial buildings. *Applied Energy*, 239, 408–424.

Pomeroy, S. 2018. Challenges and opportunities in financing affordable housing in Canada. https://www.focus-consult.com/wp-content/uploads/Financing-affordable-housing-Mar-2017.pdf

Potrč Obrecht, T., Kunič, R., Jordan, S., & Dovjak, M. 2019. Comparison of health and well-being aspects in building certification schemes. *Sustainability*, 11, 2616.

PwC. 2021. Digitising real estate services: Use of digital multi-sided platforms (MSPs) in real estate sector. https://www.pwc.com/m1/en/blogs/pdf/use-of-digital-multi-sided-platforms-msp-real-estate-sector.pdf

Rhino. 2024. How Rhino's security deposit insurance works. https://www.sayrhino.com/products/security-deposit-insurance

Rizvi, Z. M. 2008. Housing and housing finance: A social challenge and economic opportunity. *Housing Finance International*, 23, 32–37.

Sacranie, H., & Çetin, S. 2022. Towards a socially inclusive circular economy: A study of tenant engagement in European social housing organisations. In V. Pál (ed.), *Social and cultural aspects of the circular economy*. Routledge.

Scofield, J. H. 2013. Efficacy of LEED-certification in reducing energy consumption and greenhouse gas emission for large New York City office buildings. *Energy and Buildings*, 67, 517–524.

Shankar, R. 2022. Potential of blockchain based tokenized securities for green real estate bonds. In R. Gangwar, A. Agrawalla, & S. Sreekumar (eds.), *Infrastructure Development—Theory, Practice and Policy*. Routledge.

SmallChange. 2024. How do we measure change? https://learn.smallchange.co/change-index

SmartScore. 2024. SmartScore certification. https://wiredscore.com/certify-a-building/smartscore/

Szabo, N. 1994. Smart contracts. https://www.fon.hum.uva.nl/rob/Courses/InformationInSpeech/CDROM/Literature/LOTwinterschool2006/szabo.best.vwh.net/smart.contracts.html

Tagliabue, L. C., Cecconi, F. R., Rinaldi, S., & Ciribini, A. L. C. 2021. Data driven indoor air quality prediction in educational facilities based on IoT network. *Energy and Buildings*, 236, 110782.

Tian, Y., Minchin, R., Chung, K., Woo, J., & Adriaens, P. 2022. Towards inclusive and sustainable infrastructure development through blockchain-enabled asset tokenization: An exploratory case study. In *IOP Conference Series: Materials Science and Engineering*. IOP Publishing, 012040.

Trist, E. L., & Bamforth, K. W. 1951. Some social and psychological consequences of the longwall method of coal-getting: An examination of the psychological situation and defences of a work group in relation to the social structure and technological content of the work system. *Human Relations*, 4, 3–38.

ULI. 2021. *Proptech: Changing the way real estate is done*. https://knowledge.uli.org/en/reports/research-reports/2021/proptech-changing-the-way-real-estate-is-done

ULI & Measurabl. 2022. *ULI Greenprint Center and measurable partner to help real estate owners maximize ESG performance*. Urban Land Institute (ULI) Greenprint Center for Building Performance.

UN. 2015. *Sustainable development goals*. United Nations Department of Global Communications. https://sdgs.un.org/goals

UNEP. 2023. *2022 global status report for buildings and construction: Towards a zero-emission, efficient and resilient buildings and construction sector*. United Nations Environment Programme.

Von Paumgartten, P. 2003. The business case for high performance green buildings: Sustainability and its financial impact. *Journal of Facilities Management*, 2, 26–34.

Waag Society. 2014. *Final report Smart Citizen Kit Amsterdam: Measuring is knowing?* https://waag.org/sites/waag/files/media/publicaties/eindrapportage-sck-asd.pdf

WELL. n.d. *WELL building standard v2*. International WELL Building Institute. https://v2.wellcertified.com/en/wellv2/overview

Williams, P., & Whitehead, C. M. 2015. Financing affordable social housing in the UK; Building on success? *Housing Finance International*, 14–19. International Union for Housing Finance.

WiredScore. 2024. WiredScore certification. https://wiredscore.com/certify-a-building/wiredscore/

World Green Building Council. n.d. The net zero carbon buildings commitment. https://worldgbc.org/thecommitment/

Zettl, M. 2023. Transforming office buildings: The game-changing possibilities of tenant experience apps. *Forbes*.

Zuo, J., & Zhao, Z.-Y. 2014. Green building research—current status and future agenda: A review. *Renewable and Sustainable Energy Reviews*, 30, 271–281.

Chapter 6

PropTech and Sustainability

6.1 Empirical Research Methodology

This chapter presents an empirical analysis of PropTech's impact on sustainability in the real estate industry. We employed a mixed-method research methodology that combines qualitative and quantitative approaches. This section outlines our data collection strategies, interview processes, sampling techniques, quantitative analysis, and case study methods.

6.1.1 *Data Collection*

Our empirical research relied on a wide range of data from multiple sources.

6.1.1.1 *Primary Data Collection*

To gather first-hand information and insights, we used semi-structured interviews to collect primary data. The participants are the key PropTech and sustainable real estate stakeholders. The interviews provided qualitative data on PropTech's challenges, opportunities, and real-world impacts on real estate sustainability.

6.1.1.2 *Secondary Data Analysis*

Our research utilised secondary data sources to analyse the PropTech and sustainability landscape quantitatively. We analysed reports from leading real estate firms, technology companies, and consultancies specialising in PropTech and sustainability, as well as listed companies' (real estate investment trusts, REITs) annual reports and sustainability reports, which provided valuable insights into market trends, technological advancements, and industry benchmarks. Government publications, including policy documents, regulatory guidelines, and reports on sustainable real estate practices and PropTech adoption, were examined to understand the regulatory landscape, including national and international sustainability standards, building codes, and energy efficiency regulations. We also collected and analysed various statistical datasets to support our findings with quantitative evidence. Furthermore, we compiled a database of real-world case studies that showcase PropTech implementations in sustainable real estate, offering concrete examples of the practical applications and impacts of these technologies in the field. This secondary data collection approach allowed us to gather information from various sources, ensuring a

DOI: 10.1201/9781003504436-6

robust and comprehensive understanding of the real estate PropTech and sustainability landscape.

6.1.2 *Interviews*

We conducted 15 in-depth, semi-structured interviews to allow for standardised questions and open-ended discussions.

6.1.2.1 *Participant Selection*

Interviewees were chosen based on their expertise and experience in PropTech, sustainability, or both. Participants included:

- Real estate developers/asset managers
- Sustainability consultants
- PropTech start-up founders or employees
- Institutional real estate investors
- Green building certification professionals (e.g., LEED, BREEAM)
- Government officials involved in building regulations or sustainability initiatives
- Real Estate Industry Association (i.e., RICS) representatives
- Environmental non-governmental association (NGO) representatives focused on the built environment

6.1.2.2 *Interview Themes*

The interviews aim at collecting key stakeholders' insights on how PropTech affects sustainability in the real estate industry, so the key themes explored in the interviews are set as follows:

- The role of PropTech in achieving sustainability goals
- Challenges in implementing PropTech solutions
- Measurable impacts on energy efficiency, resource management, and carbon footprint
- Future trends and potential innovations.

6.1.3 *Sampling Techniques*

To ensure a representative and diverse set of case studies and examples, we employed the following sampling techniques:

6.1.3.1 *Stratified Random Sampling*

We divided the population of PropTech implementations into categories based on:

- Geographic location (urban, suburban, rural)
- Property type (residential, commercial, industrial)

6.1.3.2 *Purposive Sampling*

For in-depth case studies, we deliberately selected examples that represented:

- Innovative PropTech solutions
- Measurable sustainability impacts
- Diverse challenges and success stories

6.1.3.3 *Snowball Sampling*

We used industry networks and recommendations from initial participants to identify additional relevant case studies and interviewees.

6.1.4 *Quantitative Data Analysis*

Our quantitative analysis employed rigorous statistical methods to derive meaningful insights from the collected data.

6.1.4.1 *Data Sources*

- Academic literature
- Annual and sustainability reports of listed companies (REITs)
- Industry reports, such as JLL Global Real Estate Technology Survey and KPMG Global PropTech Survey
- Government publications
- Statistical datasets, such as Unissu Global PropTech Database and CREtech database
- Real-world case studies.

6.1.4.2 *Analysis Methods*

We used descriptive statistics to summarise and describe the main features of the dataset.

6.1.4.3 *Software Tools*

We utilised statistical software package SPSS for data analysis and visualisation.

6.1.5 *Case Studies*

We developed detailed case studies of selected properties and projects implementing PropTech solutions for sustainability. These case studies allowed us to:

- Examine PropTech implementations in real-world contexts
- Analyse the effectiveness of various sustainable technologies
- Assess the long-term impacts on energy efficiency, resource management, and overall sustainability
- Identify best practices and lessons learned from successful implementations
- Explore challenges and barriers to adoption

Each case study involved a comprehensive analysis of project documentation, performance data, and interviews with key personnel who implemented and managed the PropTech solutions.

6.2 Real-World Impact of PropTech

6.2.1 *Conceptual Framework*

In Section 2.4.2, we proposed a conceptual framework that integrates several key theories and models to understand the complex interplay between PropTech and sustainability in the real estate sector. In the following sections, we will further evaluate a set of quantifiable metrics to empirically assess PropTech's impact on real estate sustainability.

6.2.2 *Quantifiable Metrics*

We have identified a set of quantifiable metrics (shown in Figure 6.1) to assess PropTech's impact on real estate sustainability empirically. These metrics are categorised according to the three pillars of sustainability and PropTech.

These metrics provide a comprehensive framework for quantifying PropTech's impact on sustainability across various real estate operations and performance dimensions.

6.2.3 *Comparative Analysis: PropTech-Driven Versus Traditional Office Buildings*

This analysis compares two office buildings in London's Holborn area: 150 Holborn,[1] a PropTech-driven building, and 280 High Holborn,[2] a more traditional office building. The

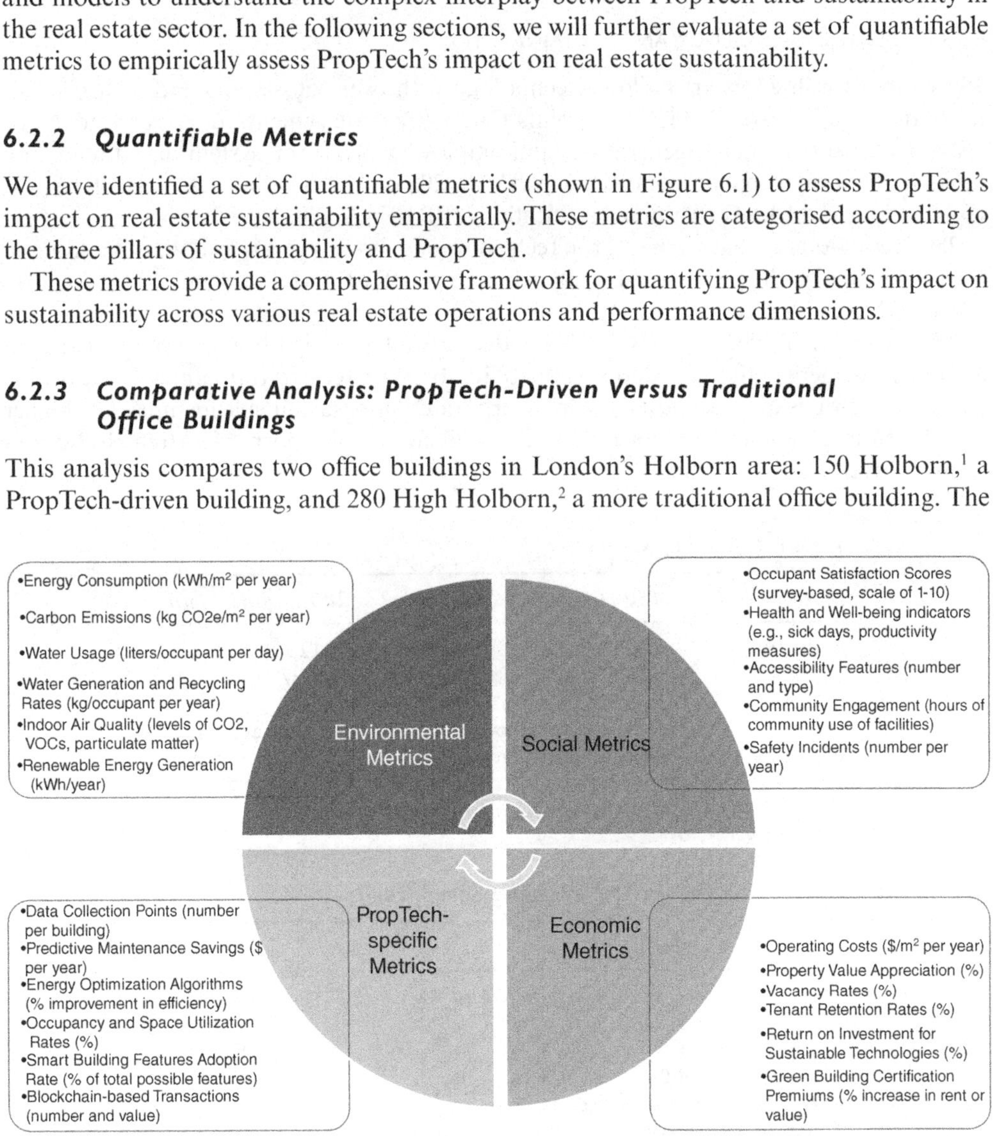

Figure 6.1 Measurable Metrics Assessing the Impact of PropTech on Sustainability.
By author (2024).

comparison aims to empirically investigate PropTech's impact on sustainability, efficiency, and overall building performance. Due to data availability, only part of the indicators in the metrics in Figure 6.1 have been examined and compared.

6.2.3.1 Building Overview

Table 6.1 presents the overview of the two buildings. Data is collected from CoStar (2024).

6.2.3.2 Comparative Analysis

6.2.3.2.1 ENERGY EFFICIENCY AND SUSTAINABILITY

150 Holborn utilised smart building technology with over 650 sensors and controls, collects, and analyses over 60,000 data points for energy management, features photovoltaic cells for renewable energy generation, and employs a blue roof system and attenuation tanks for rainwater management (Perkins&Will, 2024). The building achieved BREEAM Outstanding and LEED Platinum certifications (BREEAM, 2021).

280 High Holborn hasn't used PropTech solutions; however, it is a modern building in the same area with a similar size. It achieved a BREEAM Excellent rating in 2013 (BREEAM) and has the Energy Performance Certificate (EPC) rating of B (CoStar).

As shown in Table 6.2, 150 Holborn demonstrates a significantly more advanced approach to energy efficiency and sustainability. Its PropTech features allow for real-time energy use optimisation, while its design incorporates more sustainable features. The higher BREEAM target for 150 Holborn suggests it is likely to outperform 280 High Holborn in terms of overall sustainability.

Table 6.1 Building Overview

Feature	*150 Holborn*	*280 High Holborn*
Built year	2022	2012
Size	185,000 sq ft	70,097 sq ft
Floors	9 (including basement)	9
Owner	Sidara (Dar Group)	Citi Private Bank
Occupier	Owner-occupied	Multiple tenants

Table 6.2 Comparison on Energy Efficiency and Sustainability

Feature	*150 Holborn*	*280 High Holborn*
BREEAM rating	Outstanding with score of 89.9% (2021)	Excellent with score of 70.83% (2013)
Smart sensors	650+	N/A
Data points analysed	60,000+	N/A
Renewable energy	Photovoltaic cells	N/A
Water management	Blue roof system and attenuation tanks	N/A
EPC rating	Not specified	B

6.2.3.2.2 SPACE UTILISATION AND FLEXIBILITY

The space utilisation and flexibility approach differ significantly between 150 Holborn and 280 High Holborn. 150 Holborn encourages interaction and collaboration across firms, featuring flexible working zones and shared hubs. It also uses space utilisation analytics for optimal use of areas (Perkins&Will, 2024). In contrast, 280 High Holborn adheres to a more traditional office layout without specific flexible working or space optimisation (CoStar) features. The PropTech solutions implemented in 150 Holborn likely contribute to more efficient space utilisation and adaptability to changing work patterns. This advanced approach could lead to better occupancy rates and lower costs per employee, highlighting the potential benefits of PropTech in modern office design and management.

6.2.3.2.3 OCCUPANT WELL-BEING AND SATISFACTION

The approach to occupant well-being and satisfaction varies considerably between 150 Holborn and 280 High Holborn. 150 Holborn leverages PropTech to tailor environmental conditions to occupant needs using real-time data, features a roof terrace and pavilion for employee well-being, and incorporates biophilic design elements (Perkins&Will, 2024). In contrast, 280 High Holborn offers good natural light with floor-to-ceiling glazing and provides basic amenities like bicycle storage and shower facilities (CoStar). The PropTech systems in 150 Holborn allow for a more personalised and responsive environment, likely to result in higher occupant satisfaction and well-being. These advanced features, combined with the additional well-being amenities in 150 Holborn, may contribute to lower absenteeism and higher productivity, highlighting the potential benefits of PropTech in enhancing the overall workplace experience.

6.2.3.2.4 ECONOMIC PERFORMANCE

The economic performance of 150 Holborn and 280 High Holborn (outlined in Table 6.3) reveals interesting contrasts and insights into PropTech's potential impact on commercial real estate.

150 Holborn demonstrates superior occupancy with only 1.6% vacancy, compared to 280 High Holborn's 13.6% vacancy rate. This significant difference suggests that the PropTech-driven building may be more attractive to tenants or that the owner-occupier model employed at 150 Holborn is more space-efficient. However, it's important to note

Table 6.3 Comparison on Economic Performance

Metric	*150 Holborn*	*280 High Holborn*
Total size	185,000 sq ft	70,097 sq ft
Vacancy rate	1.6%	13.6%
Largest tenant	Sidara (owner-occupier)	Gilead Sciences Ltd (27,683 sq ft)
Number of tenants	3 (including owner)	5+
Longest lease	N/A (owner-occupied)	Until October 2033
Recent lease rate	N/A	£59.99 per sq ft (March 2022)

Source: Data collected from CoStar.

that 150 Holborn's recent completion and primary use as an owner-occupied building may contribute to this high occupancy rate.

While a direct comparison is challenging regarding rental values due to 150 Holborn's owner-occupier status, its estimated rental value (£70 - £85 per sq ft) is competitive with 280 High Holborn's current asking rent (£75.89 per sq ft). This suggests that the PropTech features and newer construction of 150 Holborn may command a premium in the market. However, 280 High Holborn has demonstrated strong rental growth, with increased rents from £37.82 to £47.70 per sq ft in May 2021 to the current asking rent of £75.89 per sq ft in 2024. This significant growth indicates robust market performance, although it's unclear how much of this increase is due to building-specific factors versus overall market trends.

The lease structure and tenant mix of the two buildings also differ notably. 280 High Holborn shows evidence of long-term leases, with some extending to 2032 and 2033, indicating strong tenant confidence in the building. It also demonstrates a diverse tenant mix, spreading risk across multiple industries. In contrast, 150 Holborn, being primarily owner-occupied, may have less diversity in its tenant base but potentially more stability. This owner-occupier model likely provides more stable income for 150 Holborn, as it's less subject to market fluctuations in rent and occupancy. However, this stability may come at the cost of potential upside from rising market rents.

While specific operating cost data is not provided for either building, it's reasonable to infer that 150 Holborn's PropTech features will likely result in lower operating costs due to improved energy efficiency and predictive maintenance capabilities. These advantages could lead to higher net operating income for 150 Holborn, even if the two buildings' gross rental incomes were similar. This potential for reduced operating costs represents a significant long-term economic benefit of PropTech integration in commercial real estate.

In conclusion, while both buildings show strong economic performance in their respective models, 150 Holborn's high occupancy rate, competitive rental values, and potential for lower operating costs suggest that its PropTech features and sustainable design provide certain economic advantages. However, 280 High Holborn's diverse tenant mix, long-term leases, and demonstrated rental growth indicate that it is a valuable and resilient asset. A full economic comparison would require more detailed data on operating costs, capital expenditures, and long-term value appreciation for both properties.

6.2.3.2.5 SMART BUILDING FEATURES

The differences in smart building features between 150 Holborn and 280 High Holborn are simple. 150 Holborn boasts an extensive Internet of Things (IoT) network for comprehensive data collection and analysis, employs digital twin technology for ongoing performance optimization, and utilises smart lighting and environmental controls (Perkins&Will, 2024). These advanced PropTech features provide 150 Holborn with significant advantages in operational efficiency, predictive maintenance, and adaptability to changing needs. In contrast, 280 High Holborn appears to rely on more basic building management systems, with no mention of advanced smart building features in its property description (CoStar). This technological gap underscores the potential impact of PropTech on building performance and management capabilities.

6.2.3.3 *Conclusions and limitations*

Based on the available information, 150 Holborn appears to outperform 280 High Holborn regarding sustainability, energy efficiency, space utilisation, and occupant well-being. The PropTech features integrated into 150 Holborn enable more responsive and efficient building management, likely resulting in environmental and economic benefits.

However, the analysis is limited by the data availability and the differences in size and age of the buildings, which may affect the direct comparisons. To fully quantify PropTech's impact, a more detailed analysis with specific performance data (e.g., actual energy consumption, occupant satisfaction surveys, and long-term economic performance) would be necessary. If the data were available, the higher initial investment in PropTech solutions should be considered against the long-term benefits for a comprehensive evaluation.

6.3 Environmental Performance Metrics

In Chapter 2, we highlighted the growing importance of sustainability in real estate. This section examines the empirical data on energy efficiency gains and resource consumption reduction achieved through implementing PropTech solutions. By analysing these metrics, we can better understand the tangible environmental benefits that PropTech brings to the real estate sector.

6.3.1 *Energy Efficiency Gains*

Energy efficiency has been at the forefront of sustainable real estate initiatives, and PropTech has played a crucial role in driving substantial improvements in this area. Integrating smart building technologies, IoT sensors, and artificial intelligence (AI)-driven analytics has enabled real estate owners and managers to optimise energy consumption in previously impossible ways (Figure 6.2).

One of PropTech's significant contributions to energy efficiency is through advanced building management systems (BMS). These systems leverage a network of sensors and actuators to monitor and control various building systems, including HVAC, lighting, and plug loads. Data from Johnson Controls (2017) shows that buildings using advanced BMS

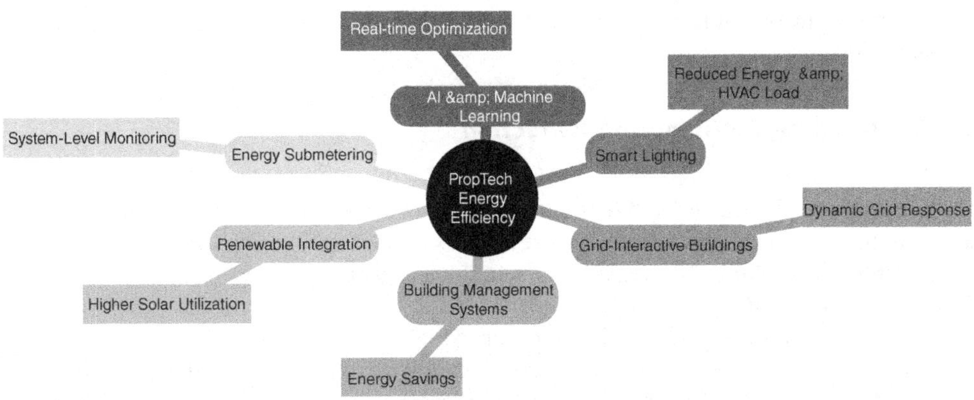

Figure 6.2 Energy Efficiency Gains Through PropTech.
By author (2024).

achieved 15–25% energy savings compared to conventional buildings. This significant reduction in energy consumption not only lowers operational costs but also substantially reduces the carbon footprint of these buildings.

Integrating machine learning algorithms further enhances BMS's effectiveness in improving energy efficiency. These AI-driven systems can analyse patterns in energy consumption, occupancy, and external factors like weather conditions to optimise building operations in real time.

Smart lighting systems represent another area where PropTech has driven significant energy efficiency gains. These systems use a combination of occupancy sensors, daylight harvesting, and LED technology to minimise lighting energy consumption. The impact of smart lighting goes beyond just the energy saved by the lighting system itself. By reducing the heat output from lighting fixtures, smart lighting systems also decrease the cooling load on HVAC systems, leading to additional energy savings. This synergistic effect highlights the importance of considering building systems holistically when implementing PropTech solutions.

Energy submetering, enabled by IoT sensors and advanced analytics platforms, has also significantly improved energy efficiency. By providing granular data on energy consumption at the level of individual systems or even specific pieces of equipment, submetering allows facility managers to identify inefficiencies and optimise operations.

PropTech has also enabled more effective implementation of renewable energy systems in buildings. Smart inverters and energy management systems can optimise on-site renewable energy, balancing it with grid electricity to minimise energy costs and carbon emissions. A study of commercial buildings with solar PV installations found that those using smart energy management systems achieved 20–30% higher solar energy utilisation than buildings without such systems (Margolis, 2021).

The advent of grid-interactive efficient buildings (GEBs) represents another frontier where PropTech is driving energy efficiency gains. These buildings use smart technologies to interact with the electric grid, adjusting their energy consumption in response to grid conditions. This capability reduces energy costs for the building and supports grid stability and the integration of renewable energy at the grid level.

The impact of PropTech on energy efficiency is not limited to large commercial buildings. In the residential sector, smart home technologies are driving significant improvements in energy performance. Smart thermostats, which learn from occupant behaviour and optimise heating and cooling schedules, have been shown to reduce HVAC energy consumption.

6.3.2 Resource Consumption Reduction

While energy efficiency has been a primary focus of sustainable real estate initiatives, PropTech is also significantly contributing to reducing the consumption of other resources and minimising waste generation. This section examines how various PropTech solutions (Figure 6.3) are helping the real estate sector use resources more efficiently and move towards a more circular economy model.

Water conservation is one area in which PropTech has made substantial impacts. Smart water management systems, which use a combination of IoT sensors, real-time monitoring, and predictive analytics, have been shown to reduce water consumption in buildings significantly. The EPA's (2012) study of commercial buildings that implemented smart water

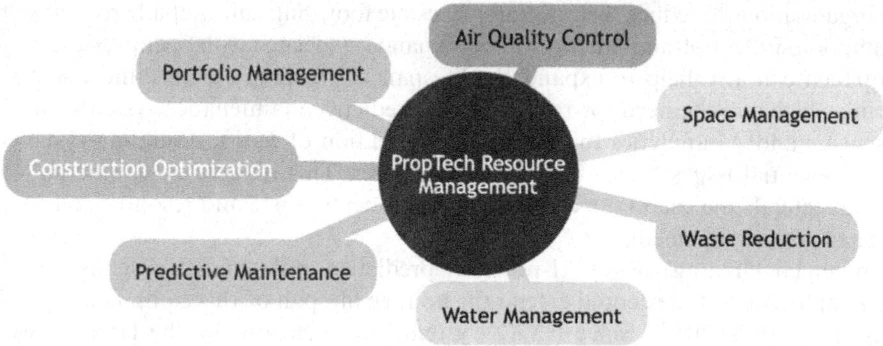

Figure 6.3 Resource Consumption Reduction Through PropTech.
By author (2024).

management systems found an average reduction in water consumption of 15–20%. These savings were achieved through a combination of leak detection, optimisation of water-using systems, and behaviour change prompted by real-time feedback.

In terms of waste reduction, PropTech is driving improvements through better monitoring, optimisation of waste collection processes, and facilitation of recycling and composting efforts. Smart waste management systems use sensors in waste bins to monitor fill levels and optimise collection routes. This reduces the fuel consumption and emissions associated with waste collection and prevents overflow situations that can lead to litter and health hazards.

In the construction sector, PropTech is helping to reduce waste through more accurate planning and better on-site resource management. When combined with AI-powered analytics, building information modeling (BIM) technology can optimise material use in the design phase, reducing over-ordering and minimising off-cuts. Construction projects using advanced BIM and AI-powered resource optimisation found a reduction in material waste of 15–20% compared to conventional projects (Ribeirinho et al., 2020).

On-site, IoT-enabled tracking systems can monitor material use in real time, alerting project managers to any deviations from the plan that might lead to waste. These systems can also facilitate better inventory management, reducing the likelihood of materials being damaged or lost on site.

PropTech also facilitates the transition to a more circular economy in the real estate sector. Digital material passports, which provide detailed information about the materials used in a building, make it easier to reuse and recycle building components at the end of their life cycle. These passports, often implemented using blockchain technology for security and traceability, can provide information on the composition, origin, and potential for recycling or reuse of various building materials.

PropTech enables more efficient use of space in buildings' operational phases, which can indirectly lead to significant resource savings. Advanced occupancy monitoring systems and flexible space management platforms allow organisations to optimise their space use. This can reduce the need for new construction and all the associated resource consumption and waste generation.

Office buildings implementing smart space management systems found that they could increase space utilisation by 20–25% on average (JLL, 2020). This improved efficiency allowed some organisations to reduce their overall real estate footprint, substantially reducing energy consumption, water use, and other resource demands associated with operating excess space.

PropTech can also help to expand the lifespan of building systems and components, reducing waste and the need for replacement. Predictive maintenance systems, which use IoT sensors and AI analytics to monitor the condition of various building systems, can identify potential issues before they lead to failures. This reduces downtime and repair costs and extends the useful life of equipment, reducing waste and resource consumption associated with replacements.

Commercial buildings using AI-powered predictive maintenance systems for HVAC equipment found that they could extend the average lifespan of this equipment by 15–20% (Johnson Controls, 2017). Since HVAC systems are often one of the largest and most resource-intensive components of a building, this extension of useful life can significantly reduce resource consumption and waste generation over time.

Regarding indoor environmental quality, PropTech enables more efficient resource use while improving occupant comfort and well-being. Advanced air quality monitoring systems and demand-controlled ventilation can optimise fresh air delivery based on actual occupancy and pollutant levels. This approach ensures good air quality and minimises the energy consumption associated with overventilation.

Office buildings implementing smart air quality management systems found that they could reduce ventilation-related energy consumption by 20–25% while maintaining or improving indoor air quality (Chaouch et al., 2021). These systems used a combination of CO_2 sensors, volatile organic compound (VOC) sensors, and occupancy data to dynamically adjust ventilation rates, ensuring that resources were used efficiently without compromising occupant health and comfort.

PropTech has the potential to significantly reduce resource consumption and waste generation in the real estate sector. From smart water management systems and AI-powered waste sorting to digital material passports and predictive maintenance, these technologies enable unprecedented levels of efficiency and circularity in buildings. As they continue to evolve and become more widely adopted, they have the potential to dramatically reduce the built environment's environmental footprint, contributing to broader sustainability goals and the transition to a more circular economy.

6.4 Social Impact Assessments

This section examines the social impact of PropTech in sustainable real estate projects, focusing on community engagement and social equity. We present empirical evidence from a case in the UK to assess how PropTech influences these key social dimensions. The PropTech Innovation Fund, launched by the UK's Department for Levelling Up in 2021, Housing and Communities (DLUHC), has provided valuable insights into the social impact of PropTech in sustainable real estate projects.

6.4.1 *Funding Distribution*

Figure 6.4 shows that the funding amounts for PropTech projects range from £100,000 to £285,000, indicating a significant variation in project scales and resource requirements.

Plymouth & South West Devon received the highest funding at £285,000, followed by Harlow & Gilston Garden Town at £228,800. The average funding is approximately

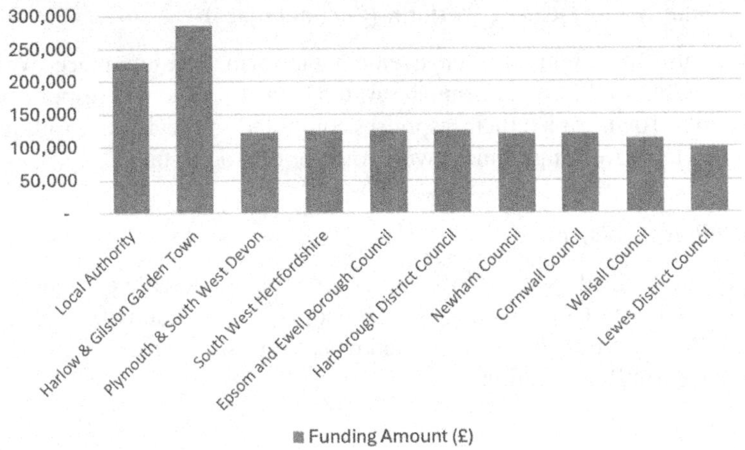

Figure 6.4 Funding Distribution.

By author (2024), data source: UK Ministry of Housing, Communities and Local Government (MHCLG, 2024).

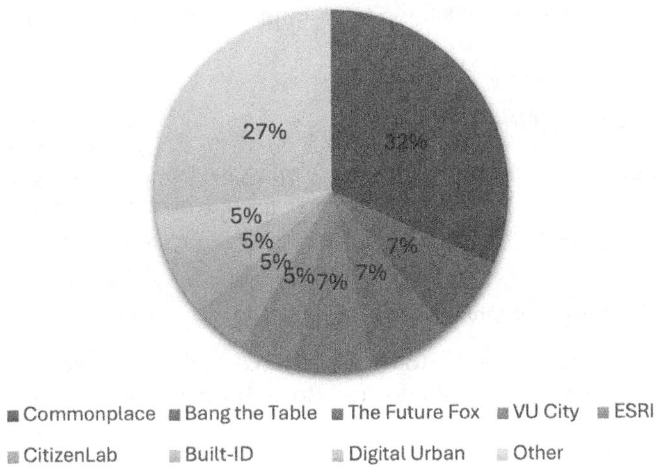

Figure 6.5 Digital Tools Used.

By author (2024); data source: UK Ministry of Housing, Communities and Local Government (MHCLG, 2024).

£148,805, suggesting substantial investment is required to implement digital engagement tools effectively.

6.4.2 *Digital Tools Utilised*

Figure 6.5 illustrates the various digital tools used across the projects. The most common platforms are commonplace, VR/AR technologies, and specialised engagement platforms like CitizenLab and Bang the Table. This diversity in tools reflects the comprehensive approach taken to address different aspects of community engagement.

Based on the cases, the effectiveness of the tools is assessed.

6.4.2.1 *Community Engagement Platform (Commonplace)*

Many councils use the community engagement platform Commonplace[3] with generally positive results. Cotswold District Council saw 6,532 visitors and 755 respondents. Harlow & Gilston Garden Town tripled their responses compared to previous consultations. Lewes District Council improved engagement with younger demographics.

6.4.2.2 *VR/AR Technologies*

Birmingham City Council used VR/AR and other tools to increase engagement from 406 to 2,216. Surrey County Council engaged 400 people with VR activities for the Farnham Infrastructure Programme. Cornwall Council received positive feedback on its VR experience for Town Centre Regeneration.

6.4.2.3 *Interactive Web-Based Tools*

Bolsover District Council's PlaceBuilder tool reported increased responses and more diverse participants, including younger residents. Leicester City Council's Engaged Space platform enabled wider participation, with 90% of respondents accessing via mobile phone. Greater Cambridge analysed 80,000 tweets to improve understanding of public sentiment on planning issues.

6.4.2.4 *Specialised Platforms*

South West Hertfordshire used Bang the Table and saw a significant increase in responses (3,122 vs. 1,057 previously). Newham Council used CitizenLab and achieved 54% of respondents from ethnic minority backgrounds.

6.4.2.5 *Multi-language Support*

Regarding social equity outcomes, tools that offered multi-language support or targeted specific demographics seemed most effective. Royal Borough of Kensington and Chelsea's WeAreW11 app improved access for non-English speaking communities. Newham Council's use of CitizenLab resulted in 54% of respondents from ethnic minority backgrounds. Havant Borough Council's use of CitizenLab and SmartSurvey improved engagement with younger demographics, with 52% of respondents aged 40 and under.

Community engagement platforms like Commonplace, VR/AR technologies, and specialised engagement platforms like CitizenLab and Bang the Table are associated with the highest engagement rates and best social equity outcomes. The effectiveness of these tools appears to be enhanced when combined with targeted strategies to reach specific demographics or address language barriers.

6.4.3 Project Focus Areas

Figure 6.6 categorises the projects into their primary focus areas. The main categories include Local Plan/Strategic Plan Development, Regeneration/Masterplan Projects, Specific Area Consultations, and General Planning Engagement Tools. This distribution highlights PropTech's versatility in addressing various planning and development needs.

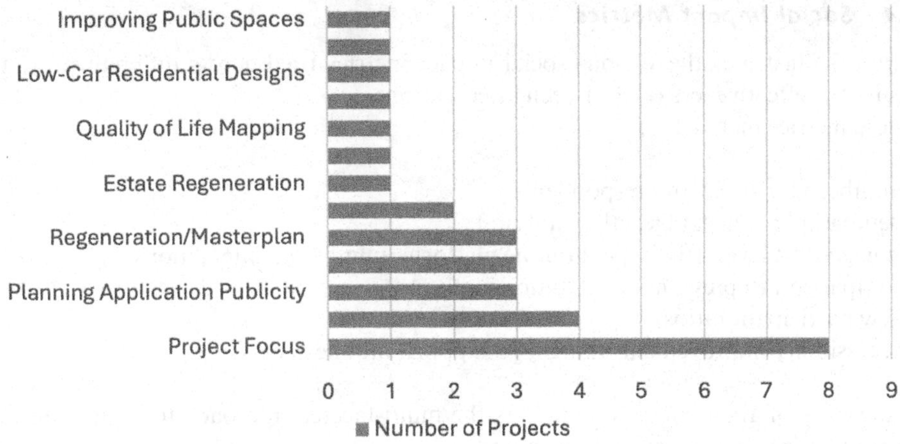

Figure 6.6 Project Focus.

By author (2024); data source: UK Ministry of Housing, Communities and Local Government (MHCLG, 2024).

6.4.3.1 *Local Plan/Strategic Plan Development*

These projects generally saw significant increases in engagement compared to previous consultations. South West Hertfordshire reported 3,122 responses versus 1,057 in a previous consultation. Cotswold District Council saw 6,532 visitors and 755 respondents, a 119% increase from 2015. Plymouth & South West Devon captured 99.9% of responses online, allowing for better data analysis.

6.4.3.2 *Regeneration/Masterplan Projects*

Birmingham City Council increased engagement from 406 in 2017 to 2,216 in 2022. Leicester City Council enabled broader participation, especially in areas of digital poverty. In Epsom and Ewell, 55% of respondents stated they had never responded to a planning consultation.

6.4.3.3 *Specific Area Consultations*

Lambeth Council achieved 60.13% and 72.98% positive responses across two consultations on improving public spaces. Wandsworth and Runnymede both received over 1,000 contributions for their Night Time Strategy consultations. Havant Borough Council improved engagement with younger demographics regarding play area design.

6.4.3.4 *General Planning Engagement Tools*

Watford Borough Council found that 37% of respondents wouldn't engage without their new platform, and 83% said QR codes enhanced engagement. Stevenage Borough Council increased responses from 487 to over 1,900 while reducing face-to-face engagement time.

Projects focused on local plan development and general planning engagement tools have significantly increased participation and engagement rates.

6.4.4 *Social Impact Metrics*

Figure 6.7 illustrates the various social impact metrics used across different projects to measure the effectiveness of PropTech interventions.

These metrics include:

- Number of visitors and respondents
- Demographic reach (age, ethnicity, etc.)
- Engagement levels (time spent on the platform, number of interactions)
- Comparison to previous consultation methods
- New participant ratios
- Accessibility improvements (language support, mobile access)

The diversity of these metrics highlights the multi-faceted approach to evaluating social impact in PropTech projects. For instance:

- Cotswold District Council tracked visitor numbers (6,532) and respondents (755), allowing them to calculate engagement rates.
- Waltham Forest measured the increase in engagement from specific age groups (17% increase in 18–24-year-olds).
- Epsom and Ewell Borough Council focused on new participant ratios, reporting that 55% of respondents had not participated in planning consultations before.
- The Royal Borough of Kensington and Chelsea evaluated accessibility improvements through language support for non-English speaking communities.

These varied metrics allow for a comprehensive understanding of how PropTech impacts different aspects of community engagement and social equity in planning processes.

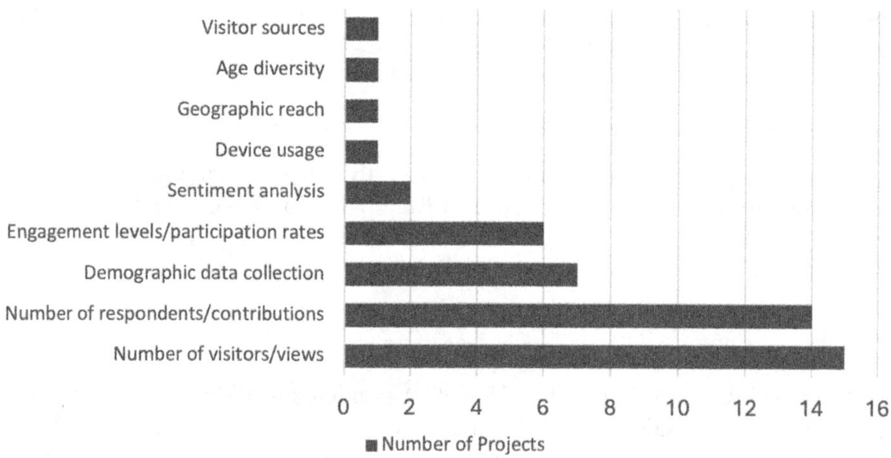

Figure 6.7 Social Impact Metrics.

By author (2024); data source: UK Ministry of Housing, Communities and Local Government (MHCLG, 2024).

6.4.5 Community Engagement

6.4.5.1 Engagement Outcomes

Figure 6.8 provides a visual representation of the engagement outcomes across different projects.

Plymouth & South West Devon had the highest number of visitors at 10,471, significantly higher than other projects. This suggests their digital platform was particularly successful in attracting attention. Among the projects with visitor and respondent data, Cotswold District Council had the highest number of respondents (755) despite having a lower visitor count. This indicates a high engagement rate among those who visited their platform. The ratio of visitors to respondents varies across projects. For example, Harlow & Gilston Garden Town had 7,000 visitors but only 554 respondents, while Lewes District Council had 5,096 visitors and 609 respondents. This suggests differences in how effectively each project converted visitors into active participants. Some projects (Wandsworth Council and Runnymede Borough Council) show identical numbers, which could indicate similar strategies or tools used.

6.4.5.2 Youth Engagement Impact

Figure 6.9 highlights the success in engaging younger demographics.

Waltham Forest saw a 17% increase in engagement from 18–24-year-olds, demonstrating the effectiveness of their digital strategy in reaching younger demographics. Wirral Council and Havant Borough Council reported 33% and 52% of participants under 40, respectively, successfully engaging a younger audience. East Hampshire District Council found that

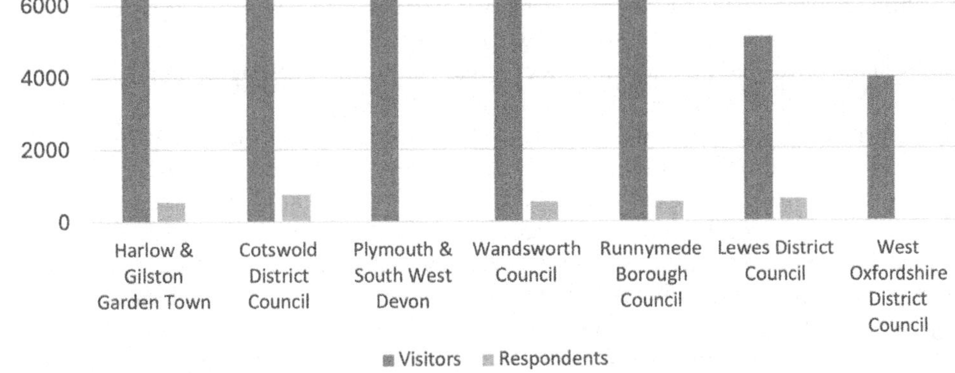

Figure 6.8 Engagement Comparison.

By author (2024); data source: UK Ministry of Housing, Communities and Local Government (MHCLG, 2024).

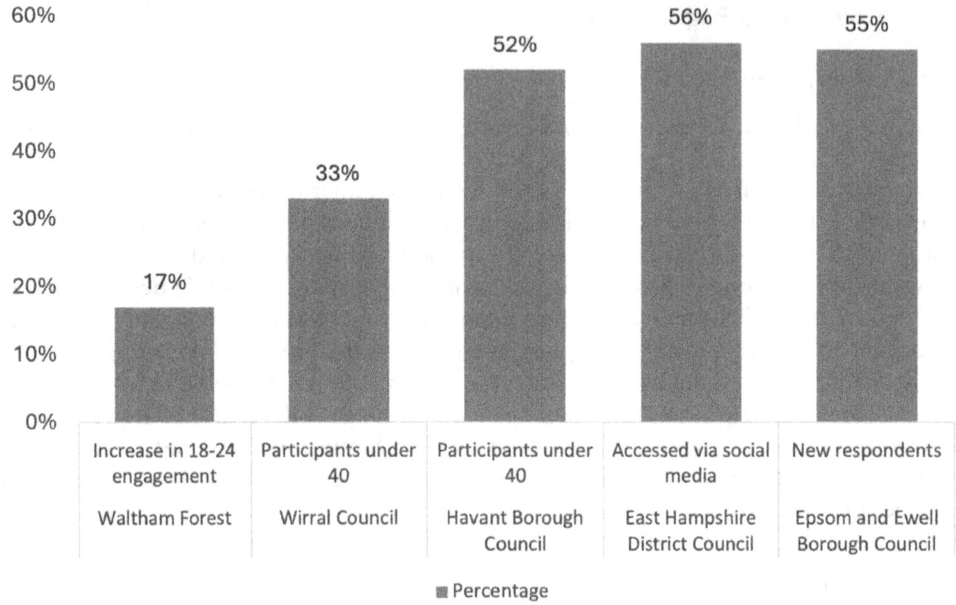

Figure 6.9 Youth Engagement Impact.

By author (2024); data source: UK Ministry of Housing, Communities and Local Government (MHCLG, 2024).

56% of visitors accessed the consultation via social media or online referrals, highlighting the importance of these channels for reaching broader audiences. Epsom and Ewell Borough Council reported that 55% of respondents had not participated in planning consultations before, indicating success in broadening civic engagement. The different metrics used by each council make direct comparisons challenging, but all show positive trends in engaging younger or previously unengaged demographics.

6.4.6 Conclusions and Implications

The PropTech Innovation Fund has demonstrated the significant potential of property technology to enhance community engagement and promote social equity in sustainable real estate projects. These outcomes suggest that PropTech is reshaping community interaction with planning processes more inclusively and efficiently.

However, several challenges remain that need to be addressed for the widespread and equitable adoption of PropTech in planning processes. The digital divide presents a significant concern, as there is a risk of excluding those with limited digital access or skills. This highlights the need for complementary strategies to ensure digitalisation does not inadvertently marginalise specific population segments. LPAs and policymakers must consider bridging this gap, possibly through digital literacy programs or by maintaining alternative engagement methods alongside digital ones. The regional disparities observed in PropTech adoption, with less representation from certain regions, indicate potential gaps that need to be addressed. These disparities could stem from various factors, including resource availability, local priorities, or varying levels of digital readiness. Addressing these regional imbalances is crucial to ensure that the benefits of PropTech are evenly distributed across the country.

PropTech has shown its ability to increase participation rates, engage previously underrepresented groups, and provide planners and developers with more prosperous and diverse community input. As technology evolves and best practices emerge from pioneering projects, many of the challenges may be mitigated. The key to maximising the positive impact of PropTech in sustainable real estate projects lies in a balanced approach. This approach should leverage the advantages of digital engagement while being mindful of potential exclusionary effects. It should also focus on knowledge sharing between the stakeholders, continued innovation in engagement methods, and policy support to ensure equitable access to these technologies.

6.5 Sustainability in PropTech Start-Ups

The intersection of PropTech and sustainability has given rise to a new wave of innovative start-ups reshaping the real estate industry. These emerging companies leverage cutting-edge technologies to address environmental challenges, improve energy efficiency, and promote sustainable practices across the built environment.

In recent years, the PropTech sector has seen a surge of start-ups focused on sustainability, reflecting the growing importance of environmental concerns in the real estate industry. These companies address various sustainability challenges, from energy efficiency and carbon reduction to waste management and sustainable materials. Key focus areas for sustainability-oriented PropTech start-ups include energy management and optimisation, with companies like Envio Systems (acquired by JLL) offering AI-powered building management systems using IoT sensors and GridPoint[4] providing energy management platforms for commercial buildings. CarbonCure[5] Technologies injects recycled CO_2 into concrete in sustainable construction materials to reduce its carbon footprint, while Made of Air[6] creates carbon-negative thermoplastic materials from waste biomass.

Other notable start-ups include Rheaply,[7] a resource-sharing platform that reduces waste in commercial real estate, and WINT Water Intelligence,[8] which uses AI-driven technology to detect and prevent water leaks. Urban Footprint[9] is transforming sustainable urban planning with its data-driven platform that assesses the sustainability of development scenarios.

These sustainability-focused PropTech start-ups are not merely introducing new technologies but fundamentally reshaping how the real estate industry approaches sustainability. Their innovations disrupt traditional practices across the entire value chain. Companies like Urban Footprint integrate sustainability considerations from the earliest design stages in design and construction. Innovative low-carbon materials, such as those developed by CarbonCure and Made of Air, transform buildings into carbon sinks.

PropTech start-ups are also revolutionising building operations and facility management. AI-powered, predictive building management systems from companies like Envio and GridPoint are optimising energy use in real time. WINT's proactive water leak detection and management solutions prevent water waste and damage, while Rheaply facilitates circular economy practices through its resource-sharing platform. Energy management is another area ripe for disruption, with start-ups like Span enabling grid-interactive efficient buildings and decentralised energy systems seamlessly integrating renewable energy sources.

In financing, new models like C-PACE, offered by Nuveen Green Capital,[10] provide innovative ways to fund sustainable building improvements. Sustainability-linked loans, tied to real-time building performance data, are also emerging to incentivise and reward sustainability efforts. PropTech is also driving change in urban planning and development, with Urban Footprint's data-driven, evidence-based approach to sustainable urban design helping cities and developers make more informed decisions.

As these disruptive innovations converge, they enable new paradigms such as grid-interactive efficient buildings that actively balance energy demand, lifecycle building management that optimises sustainability across a building's lifespan, and a fundamental shift towards decentralised, service-based business models. The growth of these start-ups is fuelled by increasing investor interest. In 2019, PropTech investment reached USD 14 billion, with a significant portion going to sustainability-focused companies (PwC & ULI, 2022).

However, PropTech disruption also faces challenges around data privacy and security, interoperability of solutions, and the overall complexity and inertia of the real estate industry. The industry's conservative nature and fragmented stakeholder landscape pose challenges for widespread adoption. Overcoming these barriers will be crucial for PropTech start-ups to scale their impact and drive widespread adoption of sustainable practices.

Sustainability-focused PropTech start-ups drive significant disruption and are poised to reduce the built environment's carbon footprint dramatically. As the demand for sustainability grows, opportunities will continue to emerge for innovative PropTech companies to scale their solutions and impact. Overcoming barriers to adoption will be key to realising the full potential of this transformation towards a sustainable real estate industry.

6.6 Statistical Analysis on PropTech Impact

PropTech has emerged as a significant disruptor in the real estate industry, similar to how FinTech has transformed the financial sector. This section examines the relationship between PropTech adoption and environmental, social, and governance (ESG) performance in UK real estate investment trusts (REITs) from 2019 to 2023.

6.6.1 *Methodology*

6.6.1.1 Data Collection and Sample Selection

The London Stock Exchange lists 48 REITs with a market capitalisation of over USD 57 billion. These REITs invest in industrial, office, residential, retail, speciality, hotel, and lodging real estate. From 2019 to 2023, we collected data on PropTech adoption and ESG performance for 34 UK REITs. The PropTech adoption data was gathered through textual analysis of annual reports, similar to the method used by Cheng and Qu (2020) and He et al. (2023) for measuring FinTech adoption in banks. The ESG score assessed by Refinitiv (LSEG, 2024) represents the ESG performance for each selected REIT.

6.6.1.2 PropTech Adoption Index

To proxy PropTech adoption in REITs, we follow Cheng and Qu (2020) and He et al. (2023) use the following steps to build REIT-level PropTech adoption measures by performing a textual analysis of REITs' annual reports.

First, since PropTech refers to applying information technology and platform economics to the real estate industry (Shaw, 2020), we set up PropTech-related keywords using three dimensions: technical foundations, platform applications, and general terminologies. The first dimension involves modern technologies such as AI, blockchain, cloud computing, big data, machine learning virtual reality, and augmented reality; the second focuses on PropTech platform applications in tenant management, facility management, energy management, and information transmission; the last dimension refers to the general terminologies, such as "PropTech" and "real estate technology," in the annual report.

Second, with the help of Python, we identify the 36 keywords in Table 6.4 from an REIT's annual report.

Finally, we use the natural logarithm of one plus the total number of these keywords to quantify the PropTech adoption (*PT*) level for a REIT in one year. As Table 6.5 shows, *PT* ranges from a minimum of 0.0000 to a maximum of 4.2627, with a mean of 1.2735 and a standard deviation of 1.2188.

6.6.1.3 *ESG Performance Measure*

ESG performance was measured using a comprehensive ESG score from Refinitiv (LSEG, 2024). The descriptive statistics of ESG are presented in Table 6.5.

Table 6.4 Keywords of PropTech Adoption

Dimensions	*Keywords*
Technical foundations	
AI technology	AI, machine learning, Real estate automation
Blockchain technology	Blockchain, smart contracts, distributed ledger
Cloud technology	Cloud computing, cloud service
Data technology	Big data, data analytics, data mining, data visualization, cybersecurity
Internet technology	4G, 5G, mobile broadband, IOT
Manufacturing technology	3D printing
VR and AR	Virtual reality, augmented reality, digital twins
Platform applications	
Tenant management	Software-as-a-service (SaaS), tenant experience
Energy management	Energy management systems
Facility management	Smart building, online property management, facility management software
Information transmission	Smart meters, mobile applications, building information modeling
General terms	
Proptech, real estate technology, innovation, digital transformation, technology adoption, fintech integration	

Table 6.5 Descriptive Statistics

Variable	*Obs.*	*Mean*	*Std. Dev.*	*Min*	*Max*
PT	170	1.2735	1.2188	0.0000	4.2627
ESG	170	52.1073	21.4895	5.4180	87.5119

6.6.2 Descriptive Statistical Analysis

6.6.2.1 General Trends

Figure 6.10 illustrates the movements of PropTech adoption and ESG performance for selected UK REITs from 2019 to 2023. The graph shows a general upward trend in PropTech adoption and ESG scores over the five years.

6.6.2.2 Correlation Analysis

Figures 6.11a and 6.11b demonstrate a positive correlation between PropTech adoption and ESG scores across 25 UK REITs.[11] This suggests that companies investing more in property technology are inclined to have better environmental, social, and governance performance.

The strength of the correlation varies significantly among different REITs. Some companies show very strong positive correlations (e.g., Assura PLC[12] and Derwent London PLC[13]), while others show weak or even negative correlations (e.g., Safestore Holdings PLC[14] and Regional REIT Ltd[15]). Many REITs show similar correlation percentages in both graphs, indicating the consistent relationship between PropTech adoption and ESG scores across different measurement methods or time periods. A few REITs display negative correlations, most notably UK Commercial Property REIT Ltd[16] and Abrdn Property Income Trust Ltd.[17] This could indicate that increased PropTech adoption is associated with lower ESG scores for these companies or vice versa. It would be worth investigating the specific circumstances of these outliers. Several REITs show strong positive correlations, which implies that these REITs could be seen as leaders in successfully integrating PropTech and ESG initiatives.

The positive correlation could be due to several factors:

- PropTech solutions often focus on energy efficiency and resource management, directly improving environmental scores.
- Companies investing in technology may be more forward-thinking and thus more likely to prioritise ESG factors.

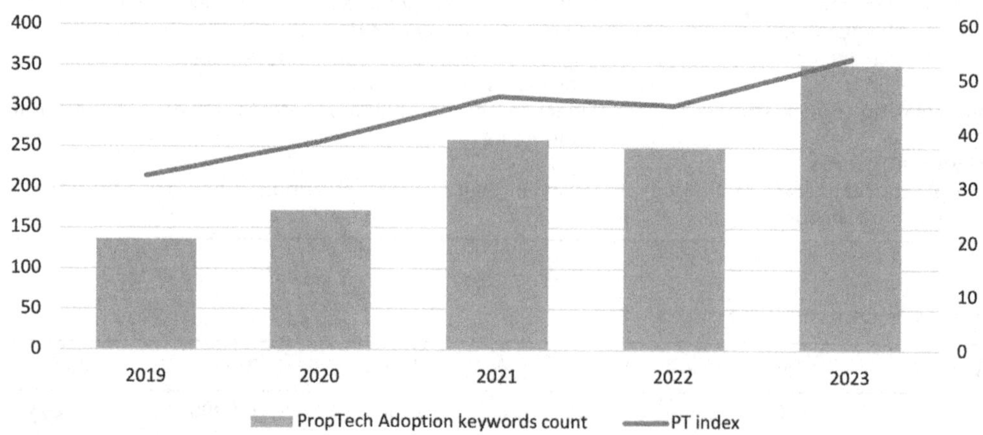

Figure 6.10 PropTech Adoption Keywords Counts Versus PT Index (UK REITs, 2019–2023).

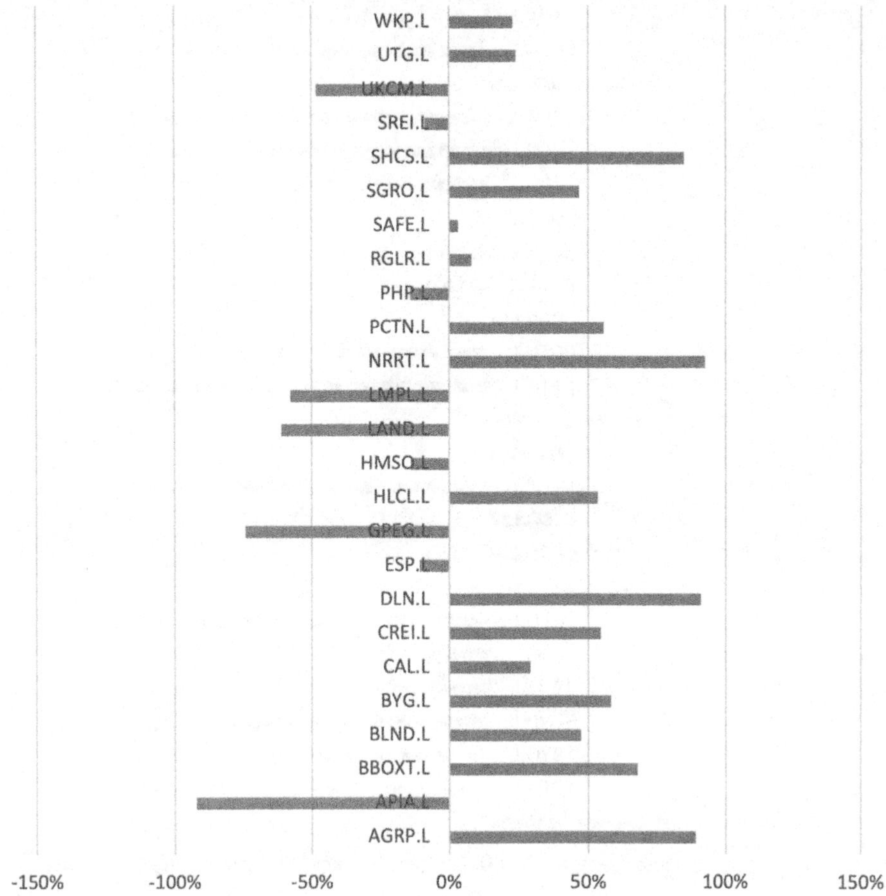

Figure 6.11a Correlation Between PropTech Keywords Counts and ESG Score UK REITs (2019–2023).

- PropTech can improve transparency and data collection, potentially boosting governance scores.

These findings suggest that investment in PropTech could be a pathway for REITs to improve their ESG performance. However, the varying correlation strengths indicate that PropTech adoption alone does not guarantee better ESG scores.

6.6.2.3 Heterogeneity Analysis

Following Guo et al. (2024), we conducted a heterogeneity analysis to examine whether the impact of PropTech adoption on ESG performance varies across different types of REITs. We considered these subgroups: diversified REITs, healthcare REITs, retail REITs, office REITs, residential REITs, industrial REITs, and specialised REITs.

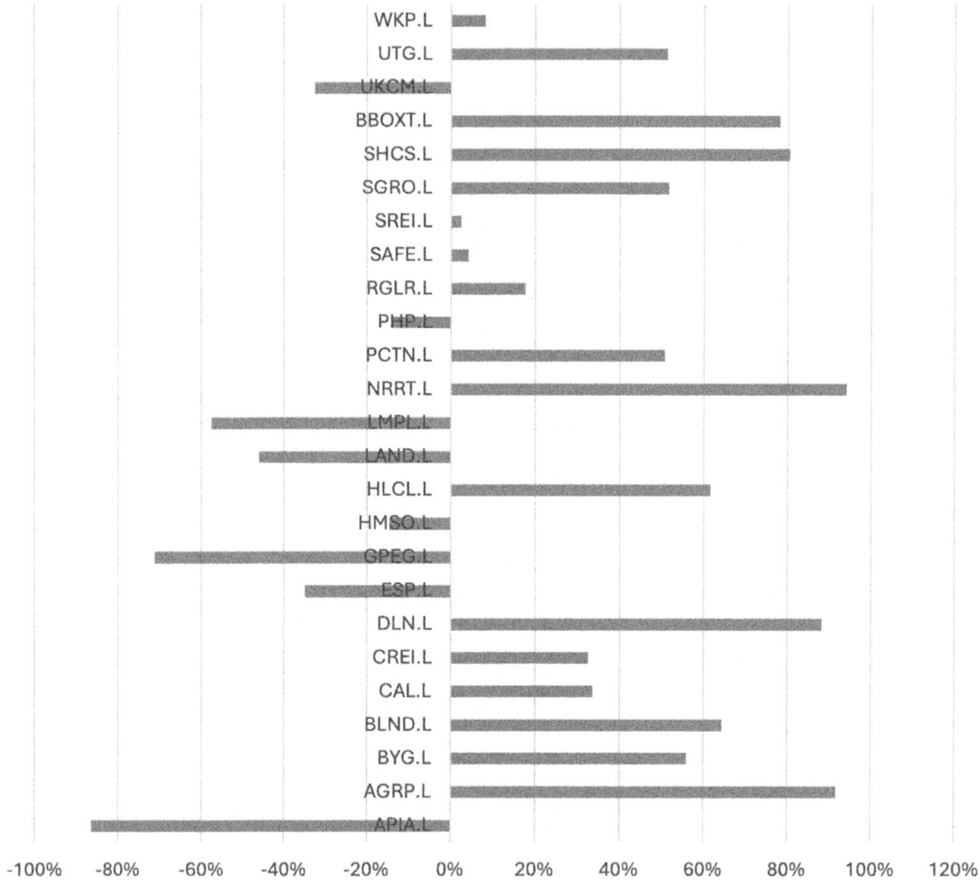

Figure 6.11b Correlation Between PropTech Adoption Index and ESG Score UK REITs (2019–2023).

6.6.2.3.1 DIVERSIFIED REITs

Figure 6.12 shows that the diversified REITs show heterogeneous patterns in their PropTech adoption and ESG performance. While some companies, like Picton Property Income Ltd and British Land Company PLC, show a generally positive relationship between PropTech adoption and ESG scores, others, like Land Securities Group PLC, display an inverse relationship. Consistent patterns across these REITs suggest that factors beyond PropTech adoption, such as company-specific strategies, market conditions, or other external factors, may significantly influence ESG performance. Further investigation into these company-specific factors would be necessary to fully understand the relationship between PropTech adoption and ESG performance in diversified REITs.

6.6.2.3.2 HEALTHCARE REITs

Healthcare REITs in Figure 6.13 show a general upward trend in ESG scores over the five years. This indicates an increased focus on environmental, social, and governance factors

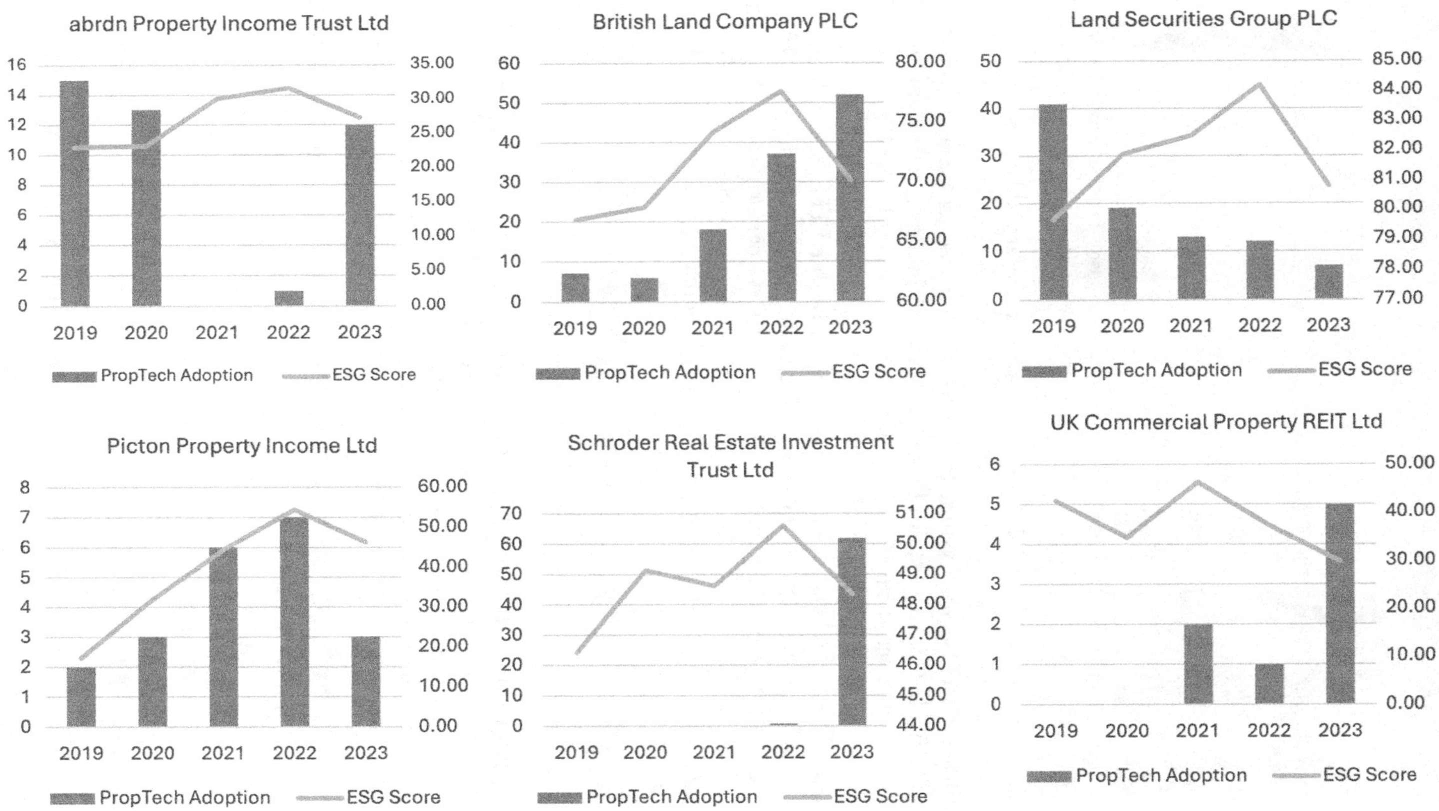

Figure 6.12 PropTech Adoption and ESG Performance (Diversified REITs).

By author (2024).

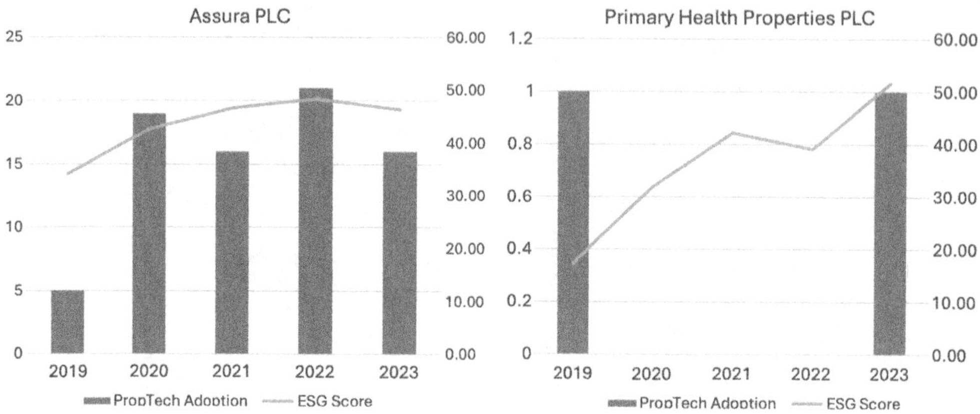

Figure 6.13 PropTech Adoption and ESG Performance (Healthcare REITs).

in the healthcare REIT sector. Assura PLC shows more frequent changes in PropTech adoption levels, while Primary Health Properties PLC maintains a consistent level (based on available data). The healthcare REITs maintained or improved their ESG scores during the pandemic years (2020–2021), possibly reflecting the essential nature of their properties and the increased focus on health-related technologies. Both REITs show a slight decrease in ESG scores in 2023, despite maintaining or increasing PropTech adoption. This could indicate that factors beyond PropTech are influencing ESG performance. The healthcare REIT sector has maintained a focus on both technological adoption and ESG improvements over the studied period, possibly driven by the increased importance of healthcare facilities and the need for technological solutions in the face of recent global health challenges.

6.6.2.3.3 RETAIL REITs

As shown in Figure 6.14, the retail REITs show diverse approaches to PropTech adoption, ranging from steady increases (Shaftesbury) to volatile patterns (Capital & Regional). Despite differences in PropTech adoption, all four companies show an overall upward trend in ESG scores, indicating a sector-wide focus on improving environmental, social, and governance factors. Most companies have shown continued improvement in both PropTech adoption and ESG scores in recent years, suggesting an ongoing commitment to technological integration and sustainability in the retail REIT sector. While some companies show a clearer positive correlation between PropTech adoption and ESG scores, others display more complex patterns. This suggests that factors beyond PropTech adoption, such as company-specific strategies, market conditions, or other external factors, play significant roles in determining ESG performance in the retail REIT sector.

The overall trend of improving ESG scores across all four companies, despite varying PropTech adoption patterns, indicates a sector-wide commitment to sustainability and governance improvements. This could be driven by increasing regulatory pressures, investor demands, or recognising the long-term benefits of strong ESG performance in the retail real estate sector.

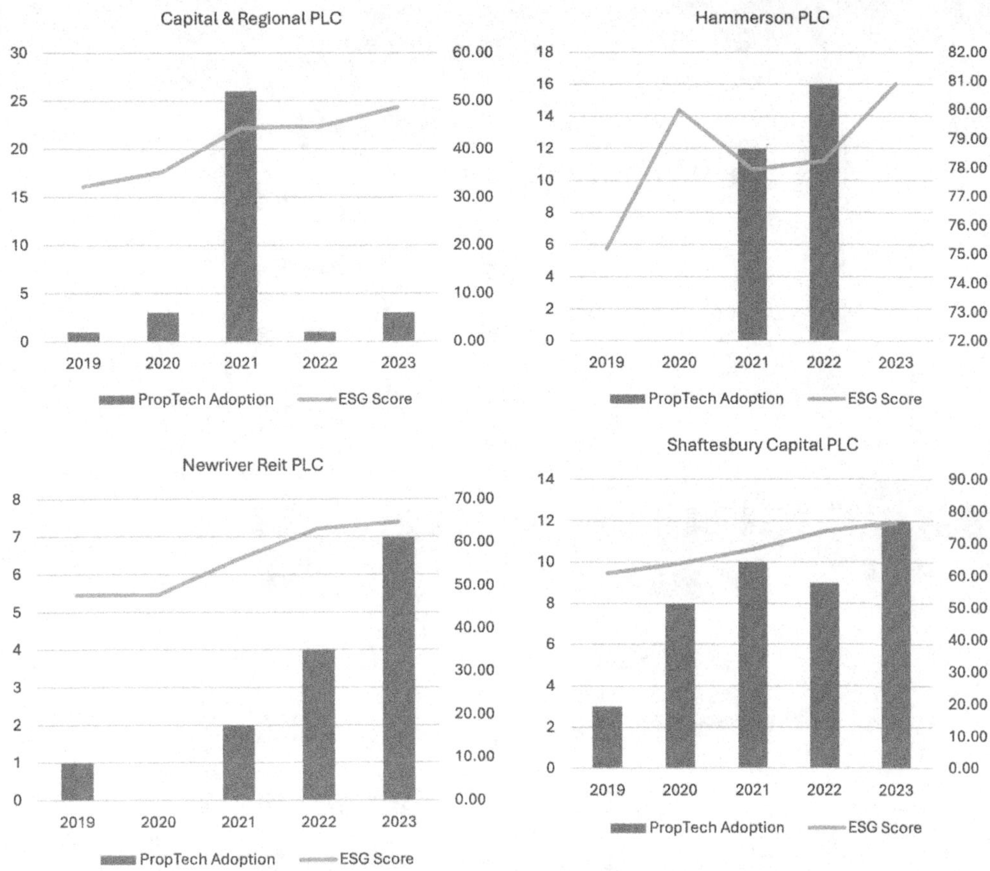

Figure 6.14 PropTech Adoption and ESG Performance (Retail REITs).

6.6.2.3.4 OFFICE REITs

The office REITs show varied approaches to PropTech adoption, ranging from steady increases (Derwent London) to more volatile patterns (Regional REIT). The relationship between PropTech adoption and ESG scores varies significantly among office REITs. Some show positive correlations (Derwent London, Helical), while others display negative or unclear relationships (Great Portland Estates, Regional REIT). The varied patterns observed suggest that factors beyond PropTech adoption, such as company-specific strategies, changing work patterns due to the pandemic, and other external factors, play significant roles in determining ESG performance in the office REIT sector.

The overall trend of increasing PropTech adoption across most companies, despite varying ESG score patterns, indicates a sector-wide recognition of the importance of technological integration in office spaces (Figure 6.15). This could be driven by the need to adapt to changing work environments, enhance building efficiency, and improve tenant experiences in a competitive market.

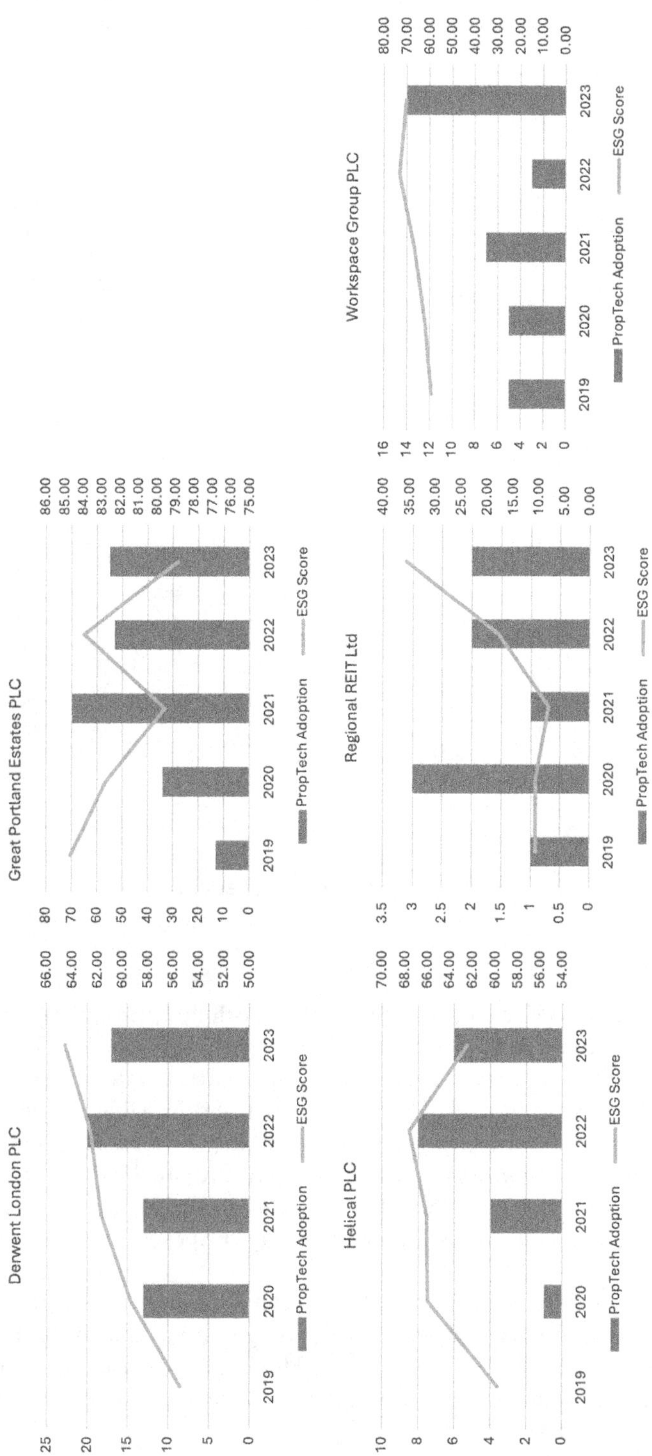

Figure 6.15 PropTech Adoption and ESG Performance (Office REITs).

6.6.2.3.5 RESIDENTIAL REITs

Two residential REITs are shown in Figure 6.16. Empiric Student Property shows a more consistent positive relationship between PropTech adoption and ESG performance, suggesting that their technological investments may be aligned with or contribute to their ESG goals. Unite Group, on the other hand, displays a more complex relationship, with their recent substantial increase in PropTech adoption not yet reflected in their ESG scores. This could indicate a lag between technology implementation and ESG impact or suggest that other factors are influencing their ESG performance more strongly. Both companies' focus on student accommodation adds an interesting dimension to the analysis. The student housing market may have unique technological needs and ESG considerations, which could explain some of the observed patterns.

6.6.2.3.6 INDUSTRIAL REITs

The three industrial REITs show different approaches to PropTech adoption (Figure 6.17). LondonMetric maintains a consistent level (based on available data), Tritax Big Box shows a peak followed by a gradual decrease, and SEGRO demonstrates an overall increase with a recent decline. Both Tritax Big Box and SEGRO show steady improvements in ESG scores over the five years, while LondonMetric's ESG score fluctuates significantly. Tritax Big Box's data suggests a possible lag between PropTech adoption and ESG score improvements, indicating that the benefits of technological investments may take time to reflect in ESG performance.

The overall trend of improving ESG scores for Tritax Big Box and SEGRO, despite recent decreases in PropTech adoption, suggests that the benefits of earlier technological investments continue to contribute to ESG performance. This could indicate a maturation phase where the focus shifts from adopting to optimising existing technologies. The industrial REIT sector's unique characteristics, such as the importance of logistics and supply chain efficiency, may influence the relationship between PropTech adoption and ESG performance. For instance, technologies that improve energy efficiency in large warehouse spaces or optimise transportation routes could significantly impact environmental scores.

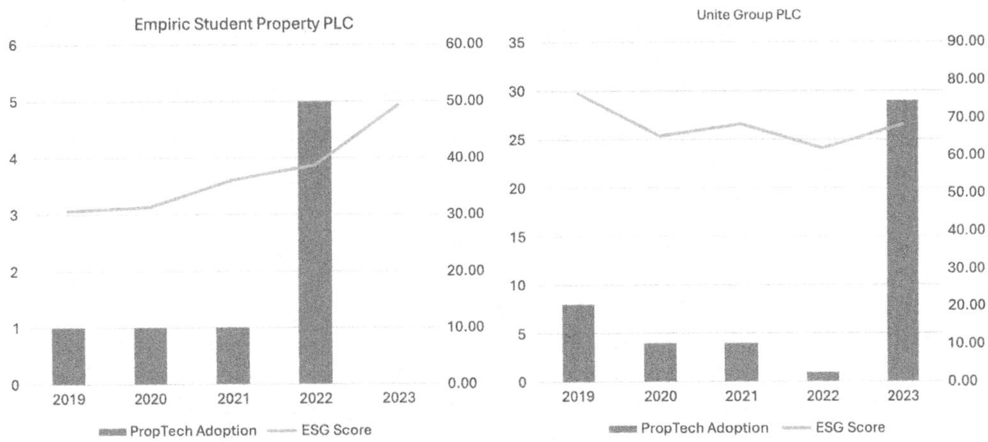

Figure 6.16 PropTech Adoption and ESG Performance (Residential REITs).

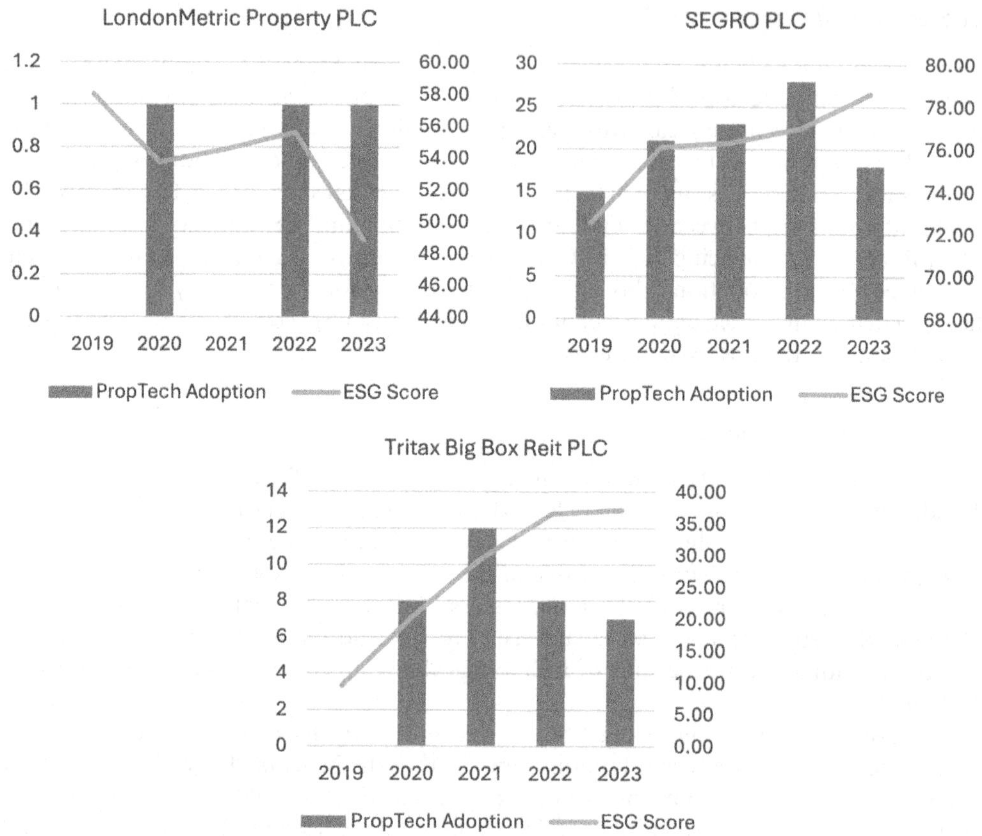

Figure 6.17 PropTech Adoption and ESG Performance (Industrial REITs).

Further research into sector-specific PropTech solutions, such as warehouse automation, IoT-enabled inventory management, or sustainable building technologies, could provide additional insights into how different technologies contribute to ESG improvements in the industrial real estate sector. Additionally, investigating the impact of e-commerce growth and changing supply chain dynamics on PropTech adoption and ESG performance in industrial REITs could offer a more comprehensive understanding of the observed trends.

6.6.2.3.7 SPECIALISED REITs

As exhibited in Figure 6.18, both specialised REITs show overall increases in PropTech adoption. Safestore's pattern is more volatile compared to Big Yellow's steadier increase. Both companies' high levels of PropTech adoption in 2023 suggest a continued focus on technological integration in the self-storage sector.

The self-storage sector's unique characteristics, such as the need for efficient space management, security systems, and customer self-service options, may influence the relationship between PropTech adoption and ESG performance. For instance, technologies that improve energy efficiency in storage units, enhance security measures, or streamline customer interactions could significantly impact operational efficiency and ESG scores.

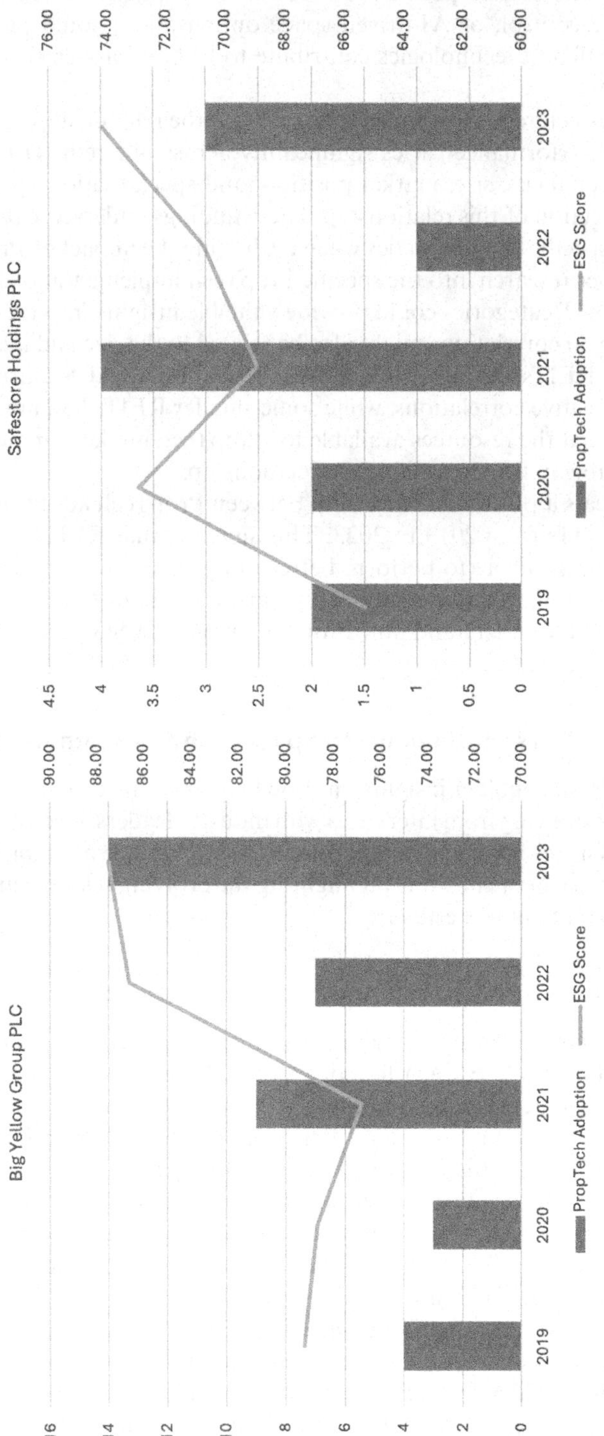

Figure 6.18 PropTech Adoption and ESG Performance (Specialised REITs).

Further research into sector-specific PropTech solutions, such as smart access systems, IoT-enabled climate control, or AI-driven space optimisation, could provide additional insights into how different technologies contribute to ESG improvements in the self-storage real estate sector.

In conclusion, our heterogeneity analysis reveals that the relationship between PropTech adoption and ESG performance varies significantly across different REIT types. Factors such as property type focus, size, market position, and specialisation appear to influence the strength and direction of this relationship. These findings underscore the importance of considering REIT-specific characteristics when evaluating the impact of PropTech on ESG performance. Further research into the specific PropTech implementations and ESG strategies of different REIT categories could provide valuable insights into these variations.

The heterogeneity in correlations might also be related to the size and market position of the REITs. Larger REITs like British Land Co PLC and Land Securities Group PLC showed moderate positive correlations, while some smaller REITs had more varied results. This could suggest that the resources available for PropTech implementation and ESG initiatives may play a role in the strength of the relationship.

Our analysis reveals a positive relationship between PropTech adoption and ESG performance in UK REITs from 2019 to 2023. This suggests that REITs investing in property technology are also likely to perform better in terms of environmental, social, and governance factors. However, this relationship varies across different REIT types, indicating the need to better understand how PropTech impacts ESG performance in the real estate sector.

6.7 Stakeholder Perspectives on PropTech and Sustainability

This section analyses stakeholder insights on PropTech's role in advancing sustainability in the real estate sector, drawing from interviews with industry leaders, investors, and PropTech founders. The analysis reveals key themes around adoption patterns, implementation challenges, and impact measurement while highlighting the evolving relationship between technology and sustainability in real estate.

6.7.1 Current State of PropTech Adoption for Sustainability

Stakeholders consistently identify energy management as a primary focus for PropTech adoption. Participant 1, a PropTech founder, noted, "We can understand the real-time occupancy of parking lots, buildings and overall traffic situation in the area … to maximise space utilisation for real estate owners." This exemplifies how PropTech solutions are addressing both operational efficiency and sustainability goals.

Participant 2, a senior executive at a storage solutions company, emphasises that "IoT-based smart storage management system, we're able to reduce energy consumption by optimising climate control in our facilities." This demonstrates the practical application of technology in achieving tangible sustainability outcomes.

Interview data reveals a strong focus on building performance optimisation through technology. Participant 3, an experienced sustainability investor, highlights that "retrofitting existing assets is crucial because there are so many buildings operating now that do not support 1.5° decarbonisation." This perspective underscores the importance of addressing existing building stock rather than focusing solely on new construction.

6.7.2 *Implementation Challenges and Barriers*

- *Integration with Legacy Systems*: A significant challenge identified across interviews is the integration of new technologies with existing infrastructure. Participant 4, a PropTech CEO, noted, "One of the key challenges is integration. Basically, to make it very effective, we need to compute the space usage and convert that into an ideal temperature, but the challenge is how we can integrate it with the air con system."
- *Stakeholder Resistance*: Multiple interviewees highlight resistance from traditional industry stakeholders as a major barrier. Participant 5, an industry expert, observes that "traditional real estate enterprises are cautious about new technology adoption; internal personnel aging and conservatism" pose significant challenges to implementation.
- *Data Management and Standardization*: The complexity of data management emerges as a crucial challenge. Participant 6, a data solutions provider, notes that "the data is very fragmented, data is spread out across different government and private organisations ... they need to be matched, they need to be integrated." This highlights the need for better data standardisation and integration capabilities.

6.7.3 *Measuring Impact and Success*

- *Quantitative Metrics*: Stakeholders employ various metrics to measure the impact of PropTech on sustainability, including energy consumption reduction, carbon emission reductions, space utilisation efficiency, and operational cost savings. Participant 7, a PropTech founder, emphasises the importance of concrete metrics: "We can calculate how much energy is generated through heat or waste heat ... we can convert into a monetary value, so it's easier for people to measure."
- *Qualitative Assessment*: Beyond quantitative metrics, stakeholders also consider qualitative impacts enhanced user experience, improved operational efficiency, better decision-making capabilities, and environmental impact reduction.

6.7.4 *Future Opportunities*

- *Technology integration*: Interviewees anticipate greater integration of different technologies and solutions. As Participant 8, a venture capital investor, notes, "Data centers are very ripe for disruption ... the amount of heat that's rejected from having to cool the chips down is huge ... anything around that would make such a big difference."
- *Market Evolution*: The interviews reveal expectations for market evolution towards:

  a More comprehensive end-to-end solutions
  b Greater focus on existing building retrofitting
  c Increased emphasis on data-driven decision making
  d Enhanced integration between different sustainability solutions

- *Regulatory Influence*: Stakeholders anticipate increasing regulatory pressure driving adoption. Participant 9, a PropTech founder, observes that "certain regions have a lot of innovation coming out of them because they saw scope to expand within their market in general."

6.7.5 Key Success Factors for Implementation

- *Clear Value Proposition*: Successful implementation requires demonstrating clear value to stakeholders. As Participant 10 emphasises, "everybody gives lip service to sustainability, but it needs to be sort of in context of not just sustainability on its own but in terms of sort of a bigger business imperative."
- *Stakeholder Engagement*: Effective stakeholder engagement emerges as crucial for success. Participant 11 highlights the importance of "having different stakeholders inside any like real estate company involved and responsible for sustainability."
- *Technology Readiness*: The maturity and reliability of technology solutions play a vital role. Interviewees stress the importance of:

  a Proven technology solutions
  b Scalable implementation capabilities
  c Integration with existing systems
  d Robust data management

The analysis of stakeholder perspectives reveals a complex landscape where technology, sustainability, and business imperatives intersect. While PropTech offers significant potential for advancing sustainability in real estate, successful implementation requires addressing multiple challenges.

The future success of PropTech in driving sustainability will depend on:

a Better integration of solutions
b More comprehensive approach to implementation
c Stronger focus on existing building stock
d Enhanced stakeholder engagement and education
e Clear regulatory frameworks and incentives

These insights provide valuable guidance for industry stakeholders seeking to leverage PropTech for sustainability improvements in the real estate sector.

6.8 Concluding Remarks

This chapter has presented the empirical analysis of PropTech's impact on sustainability in the real estate sector, examining environmental performance, social engagement, and economic outcomes. Our mixed-method research approach, combining quantitative analysis, case studies, and stakeholder interviews, reveals several key findings.

First of all, the case study of 150 Holborn versus 280 High Holborn shows that PropTech integration can lead to superior building performance across multiple sustainability metrics, though the relationship between technology investment and outcomes requires careful consideration of implementation costs and long-term benefits.

Second, our statistical analysis of UK REITs from 2019 to 2023 reveals a positive correlation between PropTech adoption and ESG performance, though with significant variations across different REIT types. This heterogeneity suggests that the effectiveness of PropTech in improving sustainability outcomes depends on factors such as property type, company size, and market position.

Third, assessing social impact through the PropTech Innovation Fund initiatives demonstrates technology's potential to enhance community engagement and social equity in real estate development. Digital tools have shown particular effectiveness in increasing participation from traditionally under-represented groups and improving accessibility to planning processes.

Finally, stakeholder perspectives highlight both the opportunities and challenges in PropTech implementation. While technology offers significant potential for advancing sustainability, successful implementation requires addressing integration challenges, stakeholder resistance, and data management issues.

These findings have several important implications for practice and policy. The real estate industry ought to adopt a more nuanced approach to PropTech implementation, taking into account property-specific characteristics and sustainability goals rather than pursuing a one-size-fits-all strategy. Policymakers should reflect on how regulatory frameworks can better support PropTech adoption while ensuring equitable access to digital tools and preventing the exacerbation of existing digital divides. Furthermore, investment decisions in PropTech should be guided by clear sustainability metrics and the potential for long-term value creation rather than solely by technological innovation.

As the real estate industry continues its digital transformation, integrating PropTech and sustainability will likely become increasingly important. The evidence presented in this chapter suggests that while PropTech offers significant potential for advancing sustainability goals, success requires careful consideration of implementation strategies, stakeholder needs, and market-specific factors.

Notes

1 https://150holborn.com/.
2 https://www.hines.com/properties/280-high-holborn-london.
3 https://www.commonplace.is/.
4 https://www.gridpoint.com/.
5 https://www.carboncure.com/.
6 https://www.madeofair.com/.
7 https://rheaply.com/.
8 https://wint.ai/.
9 https://urbanfootprint.com/.
10 https://www.nuveen.com/greencapital.
11 Due to the ESG data availability, only 25 UK REITs can be included in the correlation analysis.
12 Ticker: AGRP.L.
13 Ticker: DLN.L.
14 Ticker: SAFE.L.
15 Ticker: RGLR.L.
16 Ticker: UKCM.L.
17 Ticker: APIA.L.

References

BREEAM. 2021. Best of BREEAM 2021. https://files.bregroup.com/breeam/Winners_Brochure_2021.pdf

Chaouch, H., Çeken, C., & Ari, S. 2021. Energy management of HVAC systems in smart buildings by using fuzzy logic and M2M communication. *Journal of Building Engineering*, 44, 102606.

Cheng, M., & Qu, Y. 2020. Does bank FinTech reduce credit risk? Evidence from China. *Pacific-Basin Finance Journal*, 63, 101398.

CoStar Property Data. 2024. CoStar Group, C. https://costar.co.uk/home/demo

EPA. 2012. *WaterSense at work: Best management practices for commercial and institutional facilities*. Office of Water, U.S. Environmental Protection Agency.

Guo, P., Cheng, M., & Shen, Y. 2024. FinTech adoption and bank risk-taking: Evidence from China. *Applied Economics Letters*, 31, 594–602.

He, M., Song, G., & Chen, Q. 2023. Fintech adoption, internal control quality and bank risk taking: Evidence from Chinese listed banks. *Finance Research Letters*, 57, 104235.

JLL. 2020. *The impact of sustainability on value*. Author.

Johnson Controls. 2017. *Energy efficiency indicator survey*. Author.

LSEG. 2024. *Workspace data & analytics*. London Stock Exchange Group.

Margolis, R. 2021. Solar futures study databook. In *National Renewable Energy Laboratory—data (NREL-DATA)*. NREL.

MHCLG. 2024. Digital planning case studies. https://www.localdigital.gov.uk/digital-planning/case-studies/?filters=.proptech

Perkins&Will. 2024. 150 Holborn. https://perkinswill.com/project/150-holborn/

PwC & ULI. 2022. Emerging trends in real estate. https://www.pwc.com/gx/en/industries/financial-services/real-estate/emerging-trends-real-estate/global-outlook-2022.html

Ribeirinho, M. J., Mischke, J., Strube, G., Sjödin, E., Blanco, J. L., Palter, R., Biörck, J., Rockhill, D., & Andersson, T. 2020. *The next normal in construction: How disruption is reshaping the world's largest ecosystem*. McKinsey & Company.

Shaw, J. 2020. Platform real estate: Theory and practice of new urban real estate markets. *Urban Geography*, 41, 1037–1064.

Chapter 7

Drivers and Challenges for the PropTech Ecosystem

7.1 Introduction to the PropTech Ecosystem

The PropTech ecosystem is a complex network of stakeholders, technologies, and business models transforming the traditional real estate industry. This section systematically analyses the ecosystem's components and their interrelationships, drawing from academic literature and industry practice. PropTech's rapid evolution has created a need for structured frameworks to understand its impact on real estate markets, stakeholder relationships, and value creation mechanisms.

7.1.1 *Theoretical Framework for PropTech Ecosystem Analysis*

This section presents three theoretical perspectives that help explain PropTech's development, dynamics, and impact on the real estate industry.

7.1.1.1 Innovation Ecosystems Theory

Moore (1993) developed the innovation ecosystems theory, which provides valuable insights into how PropTech stakeholders cooperate and compete to create value. The theory suggests that innovation success depends on individual company capabilities and the alignment and coordination of multiple participants.

In the PropTech context, the theory explains how PropTech companies must coordinate with complementors to deliver value. For example, a property management platform needs to work with payment processors, maintenance service providers, and communication tools to create a comprehensive solution. This interdependence creates both opportunities and challenges for innovation. The ecosystem perspective illuminates the role of platform leaders in shaping industry development. Major PropTech platforms often act as keystone organisations, providing the infrastructure and standards that enable other ecosystem participants to create value. These platforms must balance their own interests with the need to foster a healthy ecosystem that attracts and retains participants. The theory also highlights the importance of co-evolution among ecosystem participants. The stakeholders must simultaneously adapt their capabilities and business models. For instance, when property listing platforms introduced virtual tours, real estate agents needed to develop new skills and adjust their services accordingly.

DOI: 10.1201/9781003504436-7

7.1.1.2 Digital Platform Economics

The principles of digital platform economics, as articulated by Parker (2016) and Eisenmann (2006), are fundamental to understanding how PropTech platforms create and capture value. The concept of two-sided markets shows how PropTech platforms must simultaneously attract and serve different user groups. A property listing platform, for example, must attract both property sellers and potential buyers to create value. The theory provides insights into how platforms can overcome the "chicken-and-egg" problem of simultaneously building both sides of the market. Network effects play a crucial role in platform success. As more users join a PropTech platform, its value to other users increases, creating a virtuous growth cycle. Understanding these network effects is crucial for platform strategy and investment decisions. For instance, property valuation platforms become more accurate as they accumulate more transaction data, attracting more users and generating even more data. The theory also explains pricing strategies in platform markets. PropTech platforms often subsidise one side of the market while generating revenue from the other. For example, many property search platforms offer free access to property seekers while charging real estate agents and developers for listings and enhanced features.

7.1.1.3 Institutional Change Theory

Institutional change theory, developed by North (1990), provides a framework for understanding how PropTech is transforming long-established real estate practices and institutions. Both formal institutions (including laws, regulations, and property rights) and informal institutions (such as industry norms, professional practices, and cultural expectations) critically impact PropTech adoption and innovation. For example, digital signature laws and electronic transaction regulations have been crucial enablers of online real estate transactions. Accepting virtual property tours required changes in formal regulations and informal industry practices.

Understanding these institutional constraints helps explain why PropTech adoption varies across different jurisdictions. The theory explains how PropTech innovations can lead to institutional changes, enabling further innovation. This recursive relationship between technology and institutions helps explain the pace and pattern of PropTech adoption.

7.1.2 Stakeholder Analysis and Value Networks

The PropTech ecosystem comprises a complex network of stakeholders whose interactions drive innovation and value creation in the real estate industry. Understanding these stakeholders and their relationships is crucial for analysing the ecosystem's development and potential trajectories. This section systematically analyses key stakeholders, their roles, and the value networks they create.

7.1.2.1 Core Technology Providers: The Innovation Engine

Core technology providers serve as the primary drivers of innovation within the PropTech ecosystem. These organisations develop and deploy the technological solutions that enable digital transformation in real estate. PropTech start-ups and established technology companies form the backbone of innovation in the ecosystem. These organisations typically focus on specific real estate value chain pain points, developing specialised solutions that address particular market needs. The success of these technology providers often depends on their ability to integrate with existing systems and workflows.

7.1.2.2 Real Estate Industry Players: The Traditional Innovators

Traditional real estate industry players remain central to the PropTech ecosystem. They often adopt new technologies and drive innovation. Their deep industry knowledge and established market positions make them crucial partners in developing and deploying PropTech solutions.

Property developers and investors have emerged as significant drivers of PropTech adoption. These organisations increasingly leverage technology to optimise their operations and enhance returns. Real estate agents and brokers represent a critical stakeholder group undergoing significant transformation through PropTech adoption. They started using digital tools to enhance their service offerings and maintain competitiveness. Customer relationship management (CRM) systems tailored to real estate help agents manage client relationships more effectively. Marketing automation platforms enable them to reach broader audiences with sophisticated digital campaigns. However, this group also often faces disruption from PropTech innovations, necessitating continuous adaptation of their business models. Property managers and facility operators leverage PropTech solutions to enhance operational efficiency and tenant satisfaction. Their adoption of smart building technologies, automated maintenance systems, and tenant engagement platforms demonstrates how technology can transform traditional property management practices. These stakeholders are crucial in validating new technologies and driving their widespread adoption.

7.1.2.3 Financial Institutions and Investors: The Capital Providers

Financial institutions and investors play a dual role in the PropTech ecosystem as technology adopters and providers of capital that fuels innovation. Their involvement shapes both the development of PropTech solutions and their market adoption.

Traditional banks and lending institutions have become significant adopters of PropTech solutions, particularly in mortgage lending and real estate financing. These organisations increasingly employ digital platforms to streamline loan origination, automate underwriting processes, and enhance risk assessment. Their technology adoption improves operational efficiency and expands access to real estate financing.

Venture capital and private equity firms provide crucial funding for PropTech innovation. These investors supply capital and contribute valuable expertise and network connections that help PropTech companies scale. Their investment decisions often signal market trends and influence the direction of PropTech development. For example, increased investment in sustainability-focused PropTech solutions reflects a growing market interest in environmental considerations.

Real estate investment trusts (REITs) are another crucial stakeholder group that both adopts and influences PropTech development. These organisations increasingly leverage technology to optimise portfolio management, enhance operational efficiency, and improve investor communications. Their scale and market influence make them important drivers of PropTech adoption and innovation.

7.1.2.4 End Users: The Value Recipients

End users represent the ultimate beneficiaries of PropTech innovation, and their needs and preferences significantly influence ecosystem development. Property buyers and sellers increasingly expect digital-first experiences in their real estate transactions. Their demand for virtual property tours, online transaction management, and digital documentation has driven significant innovation in these areas. The COVID-19 pandemic accelerated this trend, making digital solutions essential rather than optional for many real estate transactions. Tenants and occupiers benefit from PropTech solutions that enhance their living or working experiences. Smart home technologies, digital payment systems, and communication platforms have transformed the way people interact with properties and property managers. Their feedback and adoption patterns provide valuable insights for PropTech development and refinement.

7.1.3 Value Creation and Capture Mechanisms in PropTech

The PropTech ecosystem generates value through various mechanisms that leverage technology to transform traditional real estate practices. Understanding these mechanisms is crucial for stakeholders to effectively participate in and benefit from the ecosystem. This section demonstrates the primary value creation mechanisms, their implementation, and how stakeholders capture this value.

7.1.3.1 Primary Value Creation Mechanisms

Process automation and efficiency improvements represent a fundamental source of value creation in PropTech. PropTech solutions reduce operational costs and minimise human error by digitising and automating manual processes. The automation extends beyond simple task management, including complex processes such as lease administration and facility maintenance scheduling.

Data analytics and insights generation create value by enabling more informed decision-making across the real estate industry. PropTech solutions aggregate and analyse vast amounts of property data, market information, and operational metrics to generate actionable insights. Real estate investors, for example, can leverage predictive analytics to identify emerging market opportunities and optimise portfolio performance. Property managers use occupancy analytics and tenant behaviour data to enhance building operations and improve service delivery.

Enhanced user experience represents another significant value-creation mechanism. PropTech solutions improve accessibility, transparency, and convenience in real estate transactions and operations. Virtual property tours, for instance, save time and resources for both agents and potential buyers while providing a more flexible viewing experience. Digital transaction platforms streamline the purchase process by integrating document management, payment processing, and communication tools into a single interface.

7.1.3.2 Implementation of Value Creation Mechanisms

Implementing value-creation mechanisms in PropTech requires careful consideration of several key factors. On the technology infrastructure side, PropTech solutions demand robust technical foundations to deliver value effectively, including cloud computing platforms for scalable service delivery, data storage and processing capabilities, integration frameworks for connecting different systems, and security protocols for protecting sensitive information. While developing this infrastructure often requires significant investment, it creates the foundation for sustainable value creation. Equally important is stakeholder adoption and integration, as successful value creation depends on effectively incorporating PropTech solutions into existing workflows. This integration process encompasses user training and education programmes, change management initiatives to overcome resistance, technical support and maintenance services, and continuous feedback loops for solution improvement. Organisations must carefully manage these aspects to ensure maximum value realisation from their PropTech investments.

7.1.3.3 Value Capture Strategies

Different stakeholders in the PropTech ecosystem employ various strategies (shown in Figure 7.1) to capture the value created by technological innovations.

PropTech's value capture mechanisms benefit multiple stakeholder groups. Technology providers capture value through diverse revenue streams, including subscription fees for software services; transaction-based charges for platform usage, premium features, and add-on services; implementation and consulting services; and data monetisation opportunities. Real estate organisations, in turn, capture value through several operational and strategic advantages. These include significant operational cost reductions, enhanced asset performance, improved customer satisfaction, market differentiation, and competitive advantage in service delivery. Their adoption of PropTech solutions enables them to streamline operations while delivering superior customer value. Financial institutions represent another key stakeholder group that benefits substantially from PropTech adoption. They capture value through reduced risk enabled by better data analysis, improved loan processing efficiency, enhanced customer acquisition capabilities, new revenue streams from digital services, and better portfolio management capabilities. This integration of PropTech solutions allows them to optimise their operations while offering more sophisticated services to their clients.

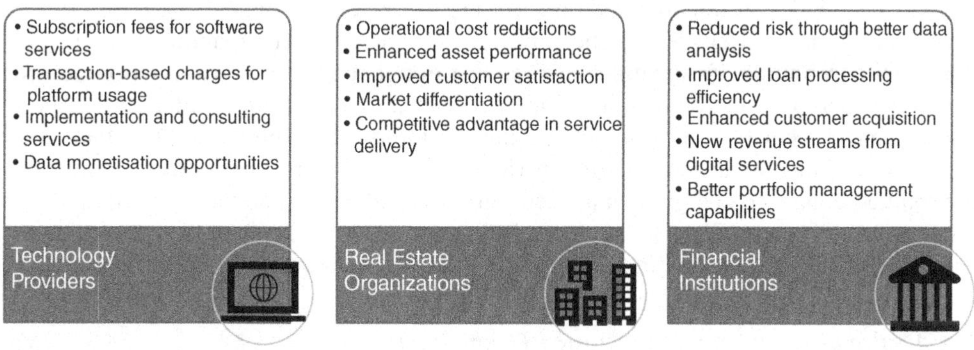

Figure 7.1 PropTech Value Capture Strategies.

By author (2024).

7.1.3.4 Value Distribution and Network Effects

Network effects and platform dynamics influence value distribution within the PropTech ecosystem. Direct network effects occur when the value of a PropTech solution increases with the number of users on the same side of the market. For example, property listing platforms become more valuable to buyers as more properties are listed, creating a self-reinforcing growth cycle. Indirect network effects arise when the value increases due to complementary users on different *market sides*. Property management platforms, for instance, become more valuable to tenants as more service providers join the network and vice versa.

7.1.3.5 Creating Sustainable Value

Long-term value creation in PropTech requires careful attention to innovation and risk management. On the innovation side, successful PropTech initiatives must maintain a delicate balance between continuous improvement of existing solutions, development of new capabilities, response to emerging market needs, and integration of new technologies. This balanced approach ensures sustained value creation over time. Implementing robust risk management strategies that address cybersecurity threats, data privacy concerns, regulatory compliance, technical reliability, and market adoption risks is equally critical. Organisations must use comprehensive risk management frameworks to protect and sustain their value creation efforts, ensuring long-term success in the PropTech ecosystem.

7.1.3.6 Measuring Value Creation

The assessment of value creation in PropTech involves quantitative and qualitative metrics to provide a comprehensive evaluation of success. Quantitative metrics include measurable factors such as return on investment (ROI), cost reduction percentages, process time savings, user adoption rates, and transaction volume increases. Qualitative metrics complement these and encompass less tangible but equally important factors like user satisfaction levels, service quality improvements, market positioning, innovation capabilities, and ecosystem relationships. Together, these metrics provide a holistic view of how effectively PropTech solutions create and deliver value.

The evolution of value creation in PropTech continues to be shaped by key emerging technologies. Artificial intelligence (AI) and machine learning enable more sophisticated predictive analytics, automated decision-making processes, personalised user experiences, enhanced operational efficiency, and advanced risk assessment capabilities. Meanwhile, blockchain technology is creating new value through transparent transaction records, smart contracts automation, tokenisation of real estate assets, reduced intermediary costs, and enhanced security and trust. Integrating the Internet of Things (IoT) further drives value by enabling real-time building performance monitoring, predictive maintenance capabilities, enhanced energy management, improved tenant experiences, and data-driven facility optimisation. These technological advancements fundamentally transform how value is created and captured in the property technology sector.

7.1.4 Regulatory and Compliance Frameworks

The regulatory landscape for PropTech operates at the intersection of real estate law, technology regulation, and financial services oversight. Understanding and navigating this complex regulatory environment is crucial for PropTech stakeholders to ensure compliance while fostering innovation.

7.1.4.1 *Global Regulatory Landscape*

The regulatory framework for PropTech varies significantly across jurisdictions, creating a complex environment for international organisations. A comparative analysis of major data protection regulations reveals substantial differences in scope, requirements, and enforcement mechanisms across regions. Table 7.1 illustrates these key differences among three significant regulatory frameworks: the EU's General Data Protection Regulation (GDPR), California's Consumer Privacy Act (CCPA), and Singapore's Personal Data Protection Act (PDPA).

As shown in Table 7.1, the varying requirements across jurisdictions create significant compliance challenges for PropTech companies operating globally. The GDPR's strict requirements and substantial penalties make it particularly impactful for PropTech operations in or targeting the European market.

7.1.4.2 *Business Model–Specific Compliance Requirements*

Different PropTech business models face varying regulatory requirements based on their specific activities and services. Understanding these differences is crucial for developing appropriate compliance strategies. Table 7.2 outlines the primary regulations, key compliance requirements, and necessary technological infrastructure for major PropTech business models.

As illustrated in Table 7.2, compliance requirements vary significantly across different PropTech business models. Property listing platforms must focus primarily on data protection and fair housing compliance, while real estate investment platforms face more stringent financial regulations and security requirements.

Table 7.1 Comparative Analysis of Major Data Protection Regulations Affecting PropTech

Aspect	*GDPR (EU)*	*CCPA (California)*	*PDPA (Singapore)*
Scope	All personal data processing	Consumer data	Personal data
Consent requirements	Explicit opt-in	Opt-out rights	Explicit consent
Data subject rights	Comprehensive (access, portability, erasure)	Limited (access, deletion)	Moderate (access, correction)
Penalties	Up to €20 million or 4% global revenue	Up to USD 7,500 per violation	Up to SGD 1 million
Cross-border transfers	Strict requirements	Limited restrictions	Moderate restrictions
Impact on PropTech	High—requires significant compliance measures	Medium—affects customer data handling	Medium—focuses on data protection

Table 7.2 PropTech Compliance Requirements by Business Model

Business Model	*Primary Regulations*	*Key Compliance Requirements*	*Technology Requirements*
Property listing platforms	Data protection laws, real estate advertising regulations	• User data protection • Accurate listing information • Fair housing compliance	• Secure data storage • Content moderation tools
Real estate investment platforms	Securities laws, investment regulations, AML/KYC requirements	• Investor verification • Risk disclosure • Transaction monitoring	• Identity verification systems • Risk assessment tools
Property management software	Landlord–tenant laws, payment processing regulations	• Secure payment handling • Privacy protection • Document management	• Payment processing systems • Document encryption
Smart building solutions	IoT security standards, building codes, privacy regulations	• Device security • Data protection • Performance monitoring	• IoT security protocols • Data encryption tools

7.1.4.3 Regulatory Innovation and Sandbox Initiatives

Regulatory authorities worldwide have recognised the need to support PropTech innovation while maintaining appropriate oversight. This has led to the development of various regulatory sandbox initiatives. Table 7.3 overviews major sandbox programs and their impacts on PropTech development.

The sandbox initiatives outlined in Table 7.3 demonstrate how different jurisdictions are approaching the balance between innovation and regulation. These programs have been particularly effective in facilitating the testing and implementation of new PropTech solutions while ensuring regulatory compliance.

7.1.4.4 Future Regulatory Trends

The regulatory landscape for PropTech continues to evolve rapidly, with new requirements and standards emerging regularly. Table 7.4 analyses key regulatory trends expected to impact the PropTech sector in 2024–2025 and the actions required to address them.

As shown in Table 7.4, PropTech companies must prepare for significant regulatory changes across multiple areas. The enhanced privacy regulations and digital transaction standards are expected to have particularly high impacts, requiring substantial updates to existing systems and processes.

These emerging trends highlight the need for PropTech companies to maintain flexible compliance frameworks that can adapt to changing regulatory requirements. Organisations must not only comply with current regulations but also anticipate and prepare for future changes in the regulatory landscape.

Table 7.3 Global Regulatory Sandbox Initiatives for PropTech

Country/ Region	*Sandbox Program*	*Focus Areas*	*Key Benefits*	*Notable Outcomes*
UK	FCA Sandbox	• Digital transactions • Property investment platforms	• Reduced regulatory burden • Fast-track testing	Several successful PropTech launches
Singapore	MAS FinTech Sandbox	• Property tokenisation • Digital payments	• Regulatory guidance • Market testing	Enhanced digital payment solutions
Australia	ASIC Innovation Hub	• Real estate investment • Digital platforms	• Regulatory support • Compliance guidance	New investment platform models
UAE	ADGM RegLab	• Property technology • Smart contracts	• Testing environment • Mentorship	Blockchain-based solutions

Table 7.4 Emerging Regulatory Trends and Their Impact on PropTech (2024–2025)

Trend	*Description*	*Expected Impact*	*Required Actions*
Enhanced privacy regulations	Stricter data protection requirements and user consent	High—requires significant system updates	• Update privacy policies • Enhance data security • Implement consent management
Sustainable technology requirements	Green building standards and environmental reporting	Medium—affects development and operations	• Implement sustainability metrics • Update reporting systems
Digital transaction standards	Standardization of digital real estate transactions	High—affects core business processes	• Update transaction systems • Enhance security protocols
AI governance	Regulations on AI use in real estate	Medium—affects automated systems	• Implement AI transparency • Enhance model documentation

7.1.5 Market Dynamics and Competition

The PropTech market displays complex and competitive dynamics shaped by rapid technological advancements, changing consumer preferences, and evolving industry structures. Understanding these dynamics is essential for stakeholders navigating this transformative landscape. At the heart of the PropTech market dynamics are strong network effects, where the value of platforms increases with user numbers. These effects reveal themselves in two key ways: (1) direct network effects, where growth in the user base enhances platform value through increased interactions and data generation, and (2) indirect network effects, where benefits accrue across different user groups, creating valuable ecosystem synergies.

Market concentration in PropTech varies significantly across segments. High concentration is evident in property listing platforms, real estate search engines, and property management systems, where network effects often lead to "winner-takes-most" scenarios. Conversely, more fragmented segments exist in specialised property services, local market solutions, and custom development services, where regional specificity and unique client needs prevent significant growth consolidation.

PropTech companies implement a range of differentiation strategies to gain competitive advantages. Technology-oriented differentiation focuses on crafting advanced analytics capabilities, proprietary algorithms, and user-friendly interfaces. Market-oriented differentiation highlights specialisation by geography, property type, and specific customer segments. PropTech innovation strategies usually take one of two routes: incremental innovation, which focuses on enhancing features and optimising performance, or disruptive innovation, which introduces new business models and groundbreaking technologies that transform industry standards.

Barriers to entry in the PropTech markets are significant and complex. Technical barriers involve high technology development costs, necessary infrastructure, and data acquisition challenges. Market barriers include network effects, brand awareness, and the costs associated with user switching. These challenges are further heightened by regulatory compliance needs and considerable capital investment, particularly in areas requiring ownership of real estate assets management.

Companies in the PropTech ecosystem pursue various growth strategies. Organic growth focuses on market penetration, geographic expansion, and product development. Inorganic growth often involves strategic acquisitions, partnerships, and joint ventures, which allow companies to quickly acquire technology capabilities or market presence. The selection of a growth strategy often depends on market maturity, competition intensity, and available resources.

Competitive responses in PropTech markets are characterised by both defensive and offensive strategies. Defensive strategies include feature matching, price competition, and customer retention programs, while offensive strategies focus on market expansion, innovation leadership, and aggressive marketing. The industry also exhibits significant competition, with competitors collaborating in areas such as standard setting and technology sharing while competing in others.

The PropTech market shows clear signs of evolution towards maturity, manifesting in several trends. Consolidation is increasing as platforms converge and standards emerge. Simultaneously, market specialisation occurs as companies focus on specific niches or vertical integration opportunities. This maturation is accompanied by new competitive patterns driven by emerging technologies such as AI, blockchain, and IoT integration.

Critical success factors in PropTech encompass multiple dimensions. Technological capabilities, including innovation capacity and integration ability, are fundamental. Market position, reflected in brand strength and user base size, remains crucial. Operational excellence in customer service and cost management provides a competitive advantage. Strategic adaptability, particularly in responding to market changes and managing risks, is increasingly important for long-term success.

The future evolution of PropTech markets will likely be characterised by increased platform dominance in certain segments while maintaining space for specialised players in others. The emergence of comprehensive ecosystems, where multiple services are integrated under single platforms, is expected to continue. Global competition will intensify, though local market focus will remain important due to the inherently local nature of real estate markets.

Collaboration dynamics in PropTech are evolving, with companies increasingly engaging in strategic alliances for technology integration, market access, and risk mitigation. These partnerships often accelerate innovation and create value through shared resources and capabilities. However, managing these relationships requires careful attention to intellectual property rights, data-sharing protocols, and aligned incentives.

Technological advancement, changing consumer expectations, and regulatory evolution will continue to shape PropTech market dynamics. Success in this environment will require companies to maintain technological leadership while building strong market positions and operational capabilities. Adapting to changing market conditions while focusing on core competencies will be crucial for sustained competitive advantage.

7.2 Empirical Research Methodology

In this chapter, we empirically examine the drivers propelling the growth of the PropTech ecosystem and the challenges it encounters in the real estate landscape. We employed a mixed-method research methodology that combines qualitative and quantitative approaches to provide a nuanced understanding of the factors influencing PropTech adoption and evolution. This section outlines our data collection strategies, interview processes, sampling techniques, quantitative analysis, and case study methods.

7.2.1 *Data Collection*

Our empirical research relied on a wide range of data from multiple sources to capture the complexity of the PropTech ecosystem.

7.2.1.1 Primary Data Collection

We conducted semi-structured interviews with key PropTech stakeholders to gather first-hand information and insights. The interviews provided qualitative data on PropTech's drivers, challenges, opportunities, and real-world impacts on the real estate industry.

7.2.1.2 Secondary Data Collection

We utilised secondary data sources to quantitatively analyse the PropTech landscape. This included:

- Reports from leading real estate firms, technology companies, and consultancies specialising in PropTech
- Listed companies' (REITs) annual reports and technology adoption statements
- Government publications on digital transformation in real estate
- Statistical datasets on PropTech investments and adoption

This secondary data allowed us to gather comprehensive information on PropTech drivers and challenges from various perspectives.

7.2.2 *Interviews*

We conducted 15 in-depth semi-structured interviews to allow for standardised questions and open-ended discussions.

7.2.2.1 *Participant Selection*

Interviewees were chosen based on their expertise and experience in the PropTech ecosystem. Participants included:

- PropTech start-up founders
- Real estate developers and asset managers
- Institutional real estate investors
- Technology consultants specialising in real estate
- Venture capitalists focusing on PropTech investments

7.2.2.2 *Interview Themes*

The interviews aimed to collect key stakeholders' insights on the drivers and challenges of the PropTech ecosystem. Key themes explored in the interviews included:

- Factors driving PropTech adoption and innovation
- Technological, financial, and regulatory challenges in implementing PropTech solutions
- The role of various stakeholders in PropTech ecosystem development
- Future trends and potential innovations in PropTech

7.2.2.3 *Analysis Methods*

We employed thematic analysis to identify recurring themes and patterns in the interview data, particularly focusing on drivers and challenges. We used sentiment analysis to examine attitudes towards different aspects of PropTech adoption.

7.2.3 *Sampling Techniques*

To ensure a representative and diverse set of perspectives, we employed the following sampling techniques.

7.2.3.1 *Random Sampling*

We divided the population of PropTech stakeholders into categories based on:

- Company size (start-ups, small and medium sized enterprises, large corporations)
- Sector focus (residential, commercial, industrial)
- Geographic location (to capture regional variations in PropTech adoption)

7.2.3.2 *Purposive Sampling*

For in-depth case studies, we deliberately selected examples that represented:

- Innovative PropTech solutions addressing specific real estate challenges
- Diverse adoption scenarios (early adopters, laggards, etc.)
- Various stages of PropTech implementation (planning, execution, post-implementation)

7.2.3.3 *Snowball Sampling*

We leveraged industry networks and recommendations from initial participants to identify additional relevant stakeholders and case studies.

7.2.4 *Quantitative Data Analysis*

Our quantitative analysis employed rigorous statistical methods to derive meaningful insights about PropTech drivers and challenges.

7.2.4.1 *Data Sources*

- PropTech investment databases (e.g., CREtech, Unissu)
- Industry surveys on PropTech adoption rates
- Economic indicators related to real estate and technology sectors

7.2.4.2 *Analysis Methods*

We used descriptive statistics to summarise PropTech adoption trends and cluster analysis to group similar PropTech solutions or adoption patterns.

7.2.4.3 *Software Tools*

We utilised the statistical software package SPSS for data analysis and visualisation.

7.2.5 *Case Studies*

We developed detailed case studies of selected PropTech implementations to provide concrete examples of drivers and challenges. These case studies allowed us to:

- Examine PropTech solutions in real-world contexts
- Analyse the factors that drove adoption in specific cases
- Assess the challenges encountered during implementation and how they were addressed
- Identify best practices and lessons learned from both successful and unsuccessful implementations
- Explore the long-term impacts of PropTech adoption on real estate operations and strategy

Each case study involved a comprehensive analysis of project documentation, performance data, and the PropTech implementation process.

By combining these various research methods, we aim to provide a comprehensive and nuanced understanding of the drivers propelling the PropTech ecosystem's growth and its challenges in transforming the real estate industry.

7.3 Drivers for PropTech Growth

7.3.1 *Technological Advancements*

PropTech market data, including KPMG (2023)'s Global Tech Survey and JLL (2023)'s Global Real Estate Technology Report, show that technological advancement has demonstrated disruptive and enabling effects on the real estate industry. This section synthesises empirical evidence from these sources to examine how specific technologies have transformed traditional practices and created new opportunities within the sector.

7.3.1.1 Areas of Technological Impact

PropTech implementations have driven significant changes across major real estate markets, with data showing several primary areas of technological advancement. In the United States, data analytics, space and use design, and health and wellness technology have seen extensive adoption by over 50% of real estate companies in 2022 (shown in Table 7.5).

Other areas with notable adoption include property management, climate risk mitigation, and project management.

The PropTech sector has attracted substantial investment, with funding reaching over USD 24 billion in 2021, the highest on record. As shown in Figure 7.2, PropTech investments reached over USD 24 billion, the highest value on record. Technological innovation allows companies to improve the efficiency of existing processes and introduce new products and services. According to industry experts in the United States, some of the most

Table 7.5 PropTech Adoption by Companies in the United States in 2022, by Business Area

	Extensive adoption (%)	*Moderate adoption (%)*
Data analytics	63	37
Space and use design	57	43
Health and/or wellness	56	44
Property management	51	49
Climate risk and mitigation	43	57
Project management	43	57
Construction	39	61
Leasing/tenant transactions and relations	37	63
Transaction management	35	65
Portfolio management	30	70
Raising capital	30	70

Data source: PwC and ULI (2022)

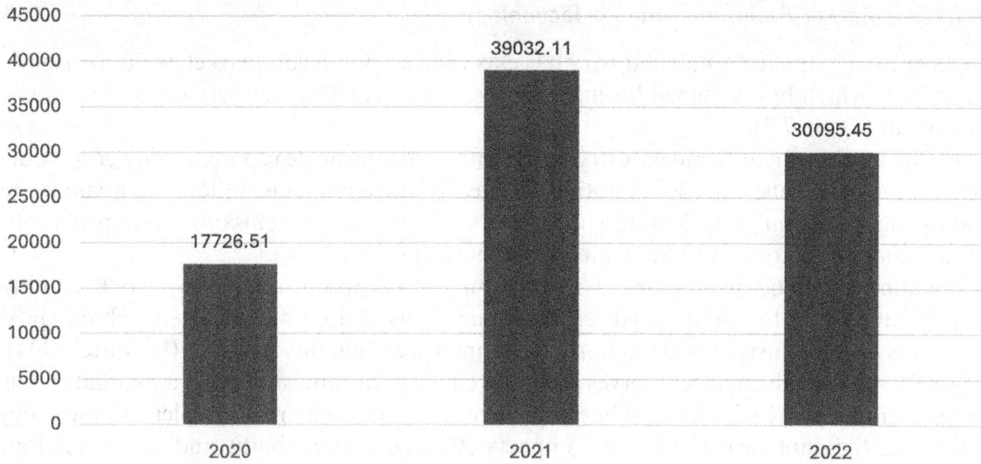

Figure 7.2 Top 100 PropTech Companies Worldwide Revenue (in Million USD)—End of Fiscal Year.

By author (2024); source: Market data by Nasdaq Data Link (2023).

potent disruptors in the real estate sector are cybersecurity, construction technology, big data, automation, and artificial intelligence.

Key disruptors in the industry include cybersecurity, construction technology, big data, automation, and artificial intelligence. Companies are increasingly incorporating innovative solutions like AI, digital twins, smart contracts, and tokenisation to improve efficiency and gain a competitive advantage.

In 2023, the largest PropTech funding rounds in the United States were led by EquipmentShare (USD 440 million), Palmetto (USD 150 million), and Juniper Square (USD 133 million), spanning construction, residential, and commercial real estate sectors (shown in Table 7.6).

Table 7.6 Largest PropTech Funding Rounds in the United States, 2023, by Company

	Subsector	*Equity Invested (in million USD)*
EquipmentShare	Construction	440
Palmetto	Residential real estate	150
Juniper Square	Commercial real estate	133
Eagle Eye Networks	Commercial real estate	100
Verkada	Commercial real estate	100
Span	Residential real estate	100
Jobber	Residential real estate	100
Measurabl	Commercial real estate	93
brivo	Commercial real estate	93
Kasa	Multifamily	70
Halio	Commercial real estate	70

Source: PitchBook (2024).

7.3.1.2 Market Evolution Through Technology

In recent years, the real estate industry has experienced significant market evolution due to PropTech, which has reshaped business processes and enabled new services and products (shown in Figure 7.3).

In 2023, 15% of real estate industry representatives worldwide reported using cryptocurrencies for payments, with 46% piloting or in early-stage implementation. Companies are increasingly adopting AI, digital twins, smart contracts, and tokenisation to improve efficiency and gain a competitive advantage (as shown in Figure 7.4).

Sustainability and technological innovation for decarbonisation are also becoming important focuses for commercial real estate investors. Adopting advanced technological solutions is the industry's most important planned sustainability measure (Deloitte, 2024).

Technological advancement serves as both a catalyst for innovation and a foundation for growth in the PropTech sector. While challenges exist, particularly in implementation and adoption, the data clearly positively impacts efficiency, accessibility, and value creation. The continued evolution of technology solutions and increasing market maturity suggest sustained growth potential for the sector.

7.3.2 Market Demand and Consumer Behaviour

Market data analysis from multiple industry sources reveals distinct patterns in consumer demands and behaviours driving PropTech adoption. In 2023, 41% of PropTech start-up CEOs targeted the multi-family residential sector, indicating a strong focus on this market segment (Metaprop, 2023). Figure 7.5 demonstrates the focus of PropTech start-ups on property asset types worldwide from 2022 to 2023.

Investor interest in PropTech innovations has shifted, with architecture, engineering, and construction becoming the most attractive area at 31% in 2023, followed by sustainability at 25% (Metaprop, 2023), as shown in Figure 7.6.

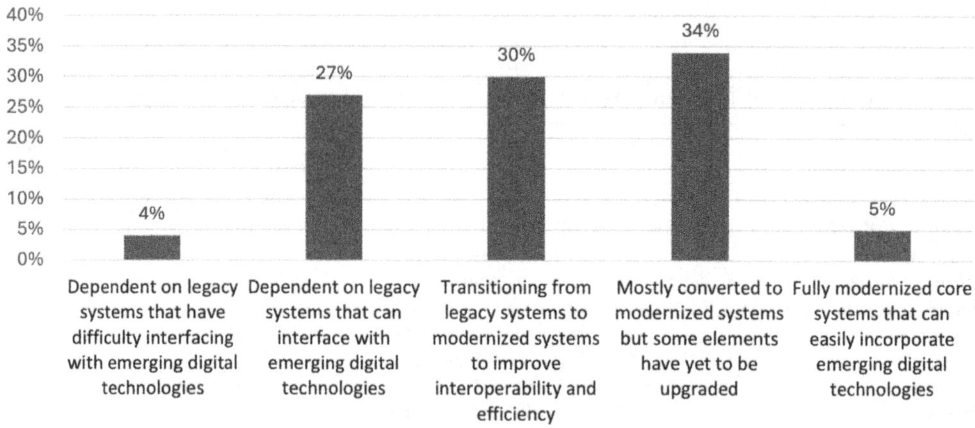

Figure 7.3 Dependence of Real Estate Firms on Legacy Technologies Globally 2023.
By author (2024); Source: Deloitte (2024).

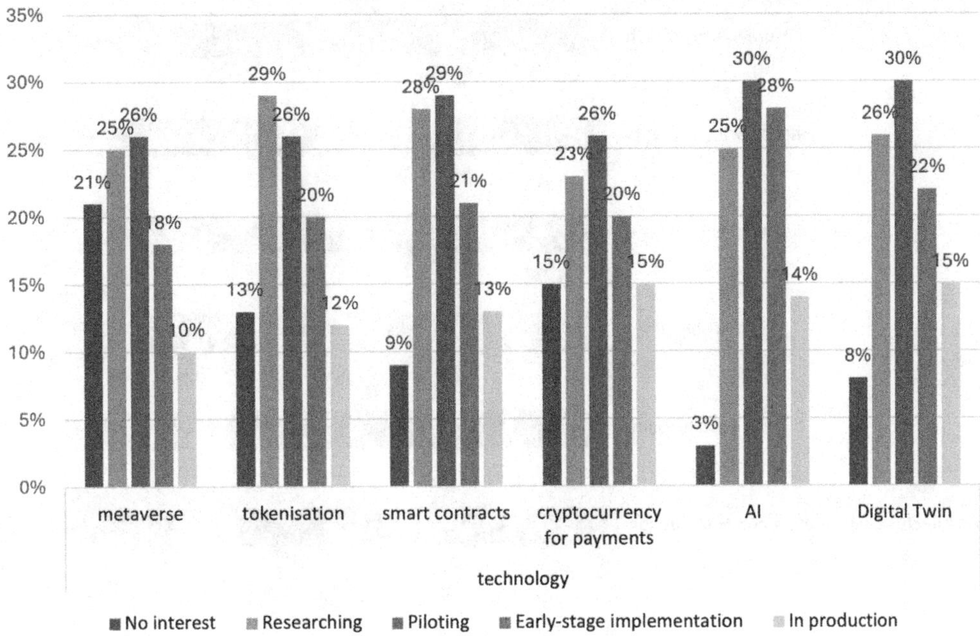

Figure 7.4 Adoption of PropTech by Real Estate Firms Worldwide According to Real Estate Industry Experts Worldwide in 2023.

By author (2024); source: Deloitte (2024).

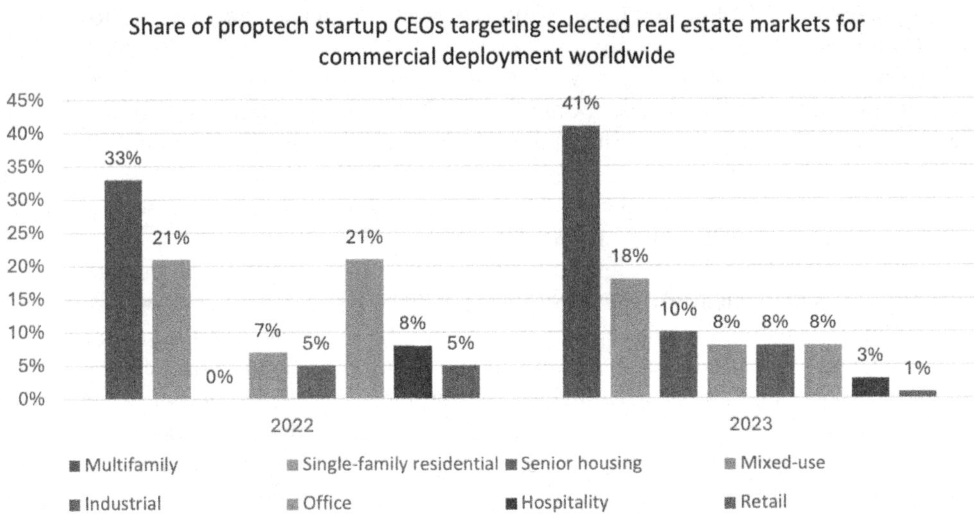

Figure 7.5 Focus of PropTech Start-Ups on Property Asset Types Worldwide, 2022–2023.

By author (2024); source: Metaprop (2023).

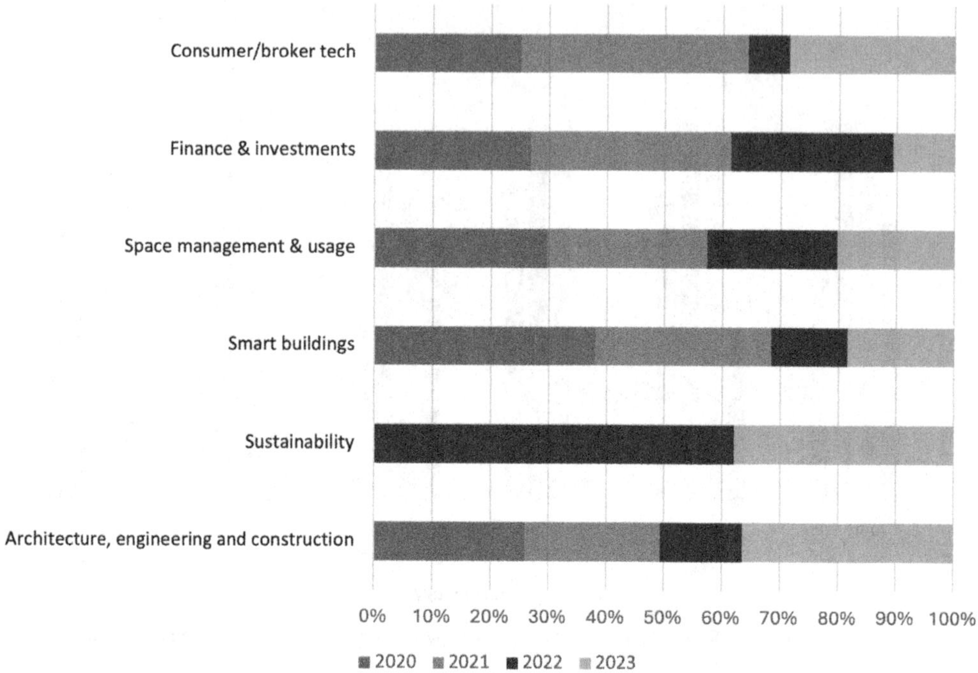

Figure 7.6 Investor Interest in Selected PropTech Innovations Worldwide, 2020–2023.
By author (2024); source: Metaprop (2023).

As shown in Figure 7.7, the global PropTech market is growing rapidly. Japan's market was estimated at 1.1 trillion yen in 2023 and is projected to reach nearly 2.4 trillion yen by 2030 (Yano, 2024).

Consumer behaviour is shifting towards digital solutions, with increasing interest in contactless online services like virtual room tours. In 2023, most commercial real estate occupiers were willing to pay a premium for a tech-enabled commercial rental. Globally, about 91% of companies were in favour of paying a premium for spaces that offer a range of technology innovations in collaboration, sustainability, employee experience, intelligent buildings, operations, and data and analytics. That percentage was even higher in the Americas, at 95% (JLL, 2023). Figure 7.8 demonstrates the occupiers' willingness to pay a premium for tech-enabled real estate worldwide in 2023.

Consumer behaviour and market demand significantly influence PropTech adoption patterns. The data shows clear trends towards digital solution preference, particularly among younger demographics, while also highlighting important considerations for technology development and implementation strategies. Understanding these patterns is crucial for PropTech companies in developing and deploying solutions that meet evolving consumer needs and preferences.

7.3.3 Investment and Funding Trends in PropTech

Venture capital investments in PropTech have experienced significant fluctuations in recent years. In 2023, the value of venture capital raised by PropTech companies worldwide

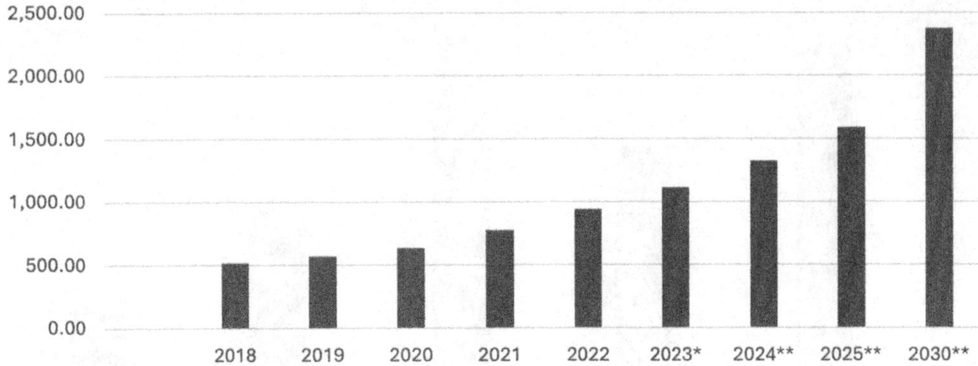

Figure 7.7 PropTech Market Size in Japan, FY 2018–2030.

By author (2024); source: Yano Research (2024).

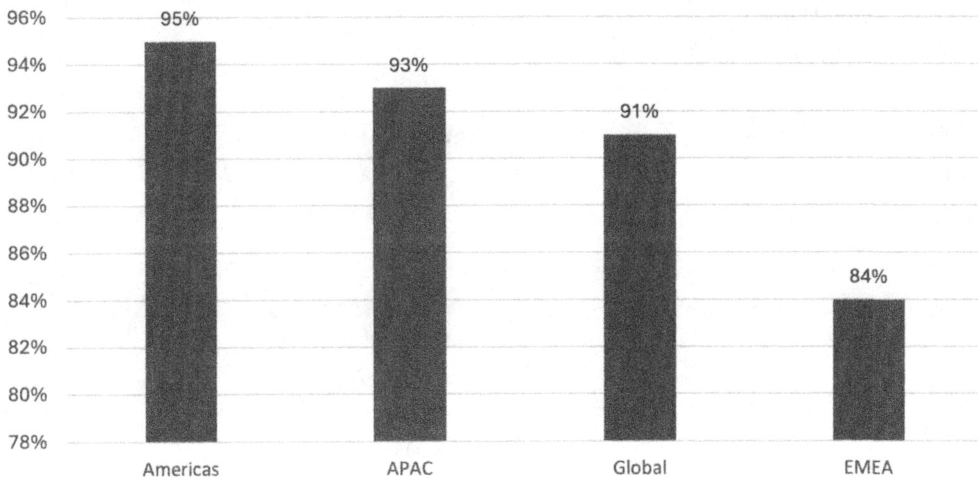

Figure 7.8 Share of Real Estate Occupiers Who Are Willing to Pay a Premium for a Tech-Enabled Space Worldwide in 2023, by Region.

By author (2024); source: JLL (2023).

declined to approximately USD 11.4 billion, marking the second consecutive year of decrease from the peak of USD 32 billion in 2021 (CRETI, 2023).

Figure 7.9 shows the value of venture capital raised by PropTech companies worldwide from 2019 to 2023. This decline over the past two years reflects a broader weakening in private equity markets and a reduction in investor confidence in the PropTech sector.

In the United States, PropTech deals followed a similar trend, reaching a record-breaking USD 19 billion in 2021 before falling to USD 4.7 billion in 2023 (PitchBook, 2024) (Figure 7.10).

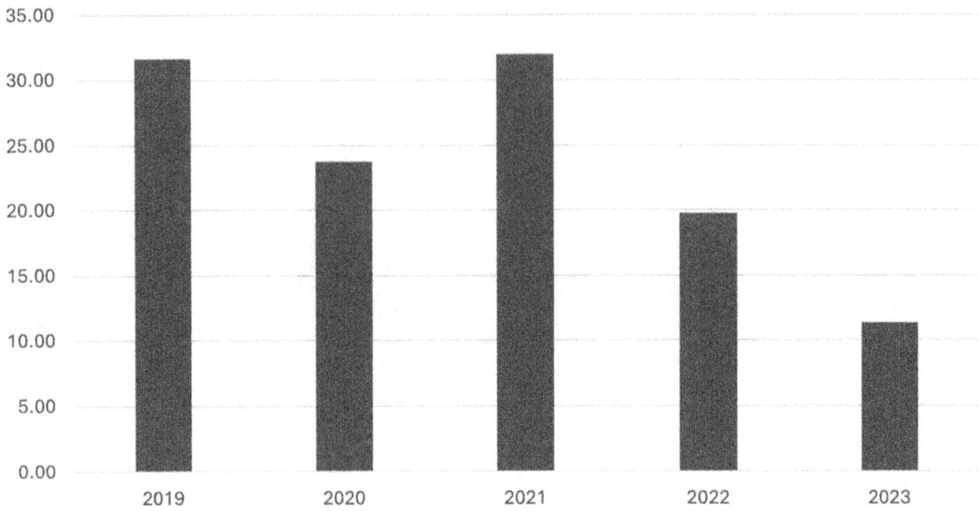

Figure 7.9 Value of Venture Capital Raised by Proptech Companies Worldwide from 2019 to 2023 (in Billion USD).

By author (2024); source: CRETI (2023).

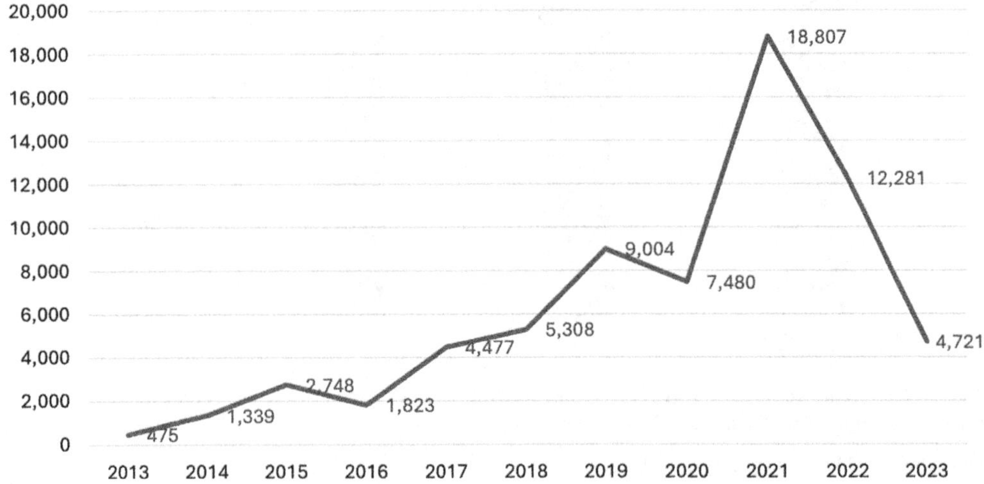

Figure 7.10 Volume of Proptech Deals in the United States from 2013 to 2023 (in Million USD).

By author (2024); source: PitchBook (2024).

Despite this decline, seed-stage investments accounted for about one-third of global PropTech investments in 2023, indicating strong early-stage investor confidence (CRETI, 2023) (Figure 7.11).

The PropTech confidence index, as shown in Figure 7.12, for investors and founders improved in 2023, indicating a potential recovery in the sector. Experts in the PropTech industry regained some of their confidence in the market this year, with the confidence

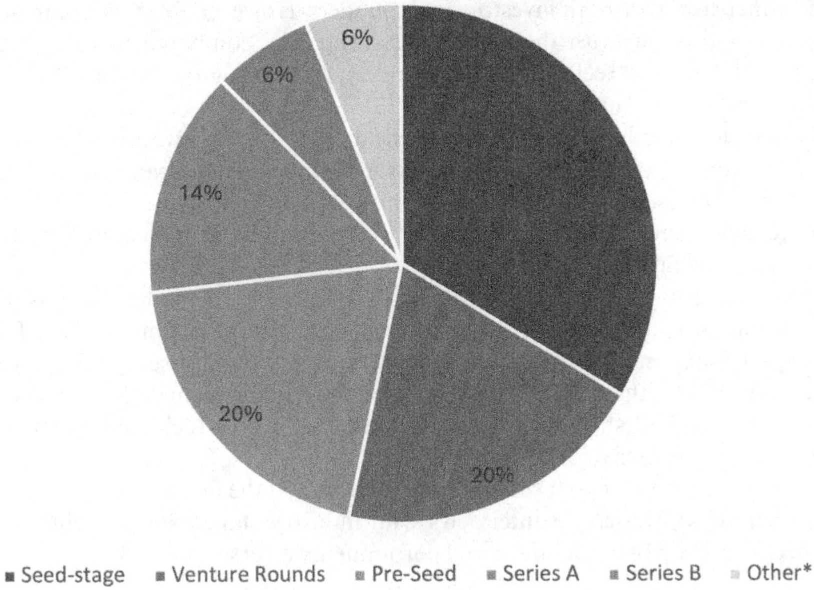

Figure 7.11 Distribution of Proptech Investment Worldwide in 2023, by Investment Stage.

By author (2024); source: CRETI (2023).

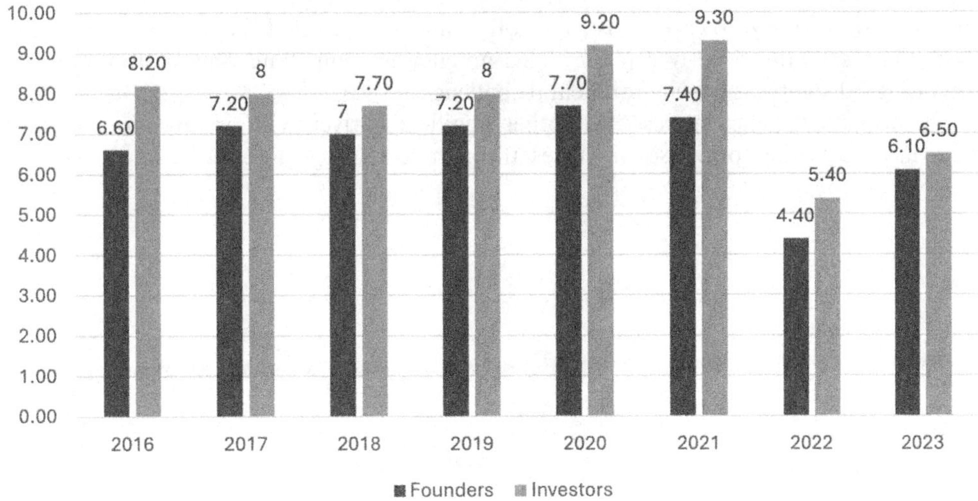

Figure 7.12 Proptech Start-Up Founder and Investor Confidence Index Worldwide from 2016 to 2023.

By author (2024); source: Metaprop (2023).

index score increasing for both investors and founders. However, PropTech start-up founders were less optimistic, with their score measuring 6.1, compared to 6.5 for investors. Overall, the global PropTech sector has been experiencing growth, with 2021 setting a record for investment volumes (Metaprop, 2023).

The PropTech sector has experienced significant growth and investment in recent years (Figure 7.13), and a complex financial ecosystem supports its expansion. In 2021, USD 24.3 billion were invested in PropTech companies globally. The unique investment amount has been steadily rising since 2012, except 2020, even though the number of PropTech companies founded has fallen since then.

The financial ecosystem supporting PropTech's growth includes various investors. Venture capital, private equity, and debt financing are crucial in funding PropTech start-ups. Though leasing and purchase-sales transactions have comprised the largest share of investment since 2015, the property development and management subsector has grown in importance. Figure 7.14 shows the investment volume in PropTech companies worldwide from 2010 to the first half of 2022, by subsector.

We also analyse venture capital investment patterns and the financial ecosystem supporting PropTech growth based on interviews with industry stakeholders, including venture capital investors, PropTech founders, and corporate investors.

7.3.3.1 Factors Driving PropTech Adoption and Innovation

The interview data reveals multiple significant forces driving PropTech adoption and innovation in the real estate sector. Market pressures emerge as a primary driver, with sustainability requirements playing an increasingly crucial role. As Participant 1 emphasises, "The 2030 deadline, which a lot of big institutions have and occupiers have within the built environment for net zero carbon, is so far off being achieved that they're going to have to start doing something fairly rapidly." This urgency is compelling organisations to seek technological solutions for environmental challenges.

Operational efficiency represents another significant driver, with organisations looking to reduce costs and optimise processes through technology adoption. Participant 2's

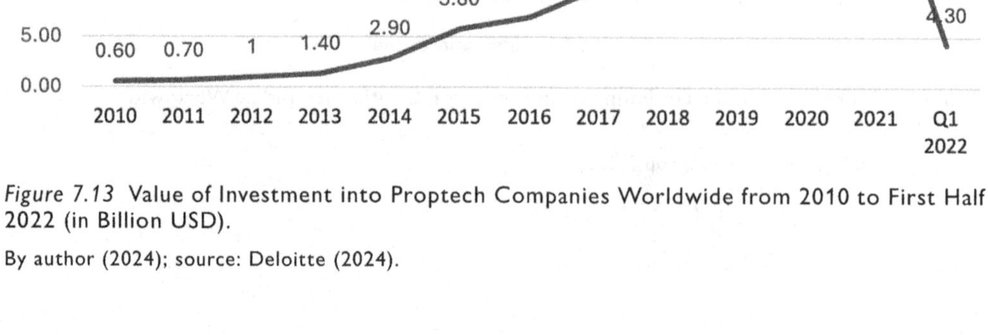

Figure 7.13 Value of Investment into Proptech Companies Worldwide from 2010 to First Half 2022 (in Billion USD).

By author (2024); source: Deloitte (2024).

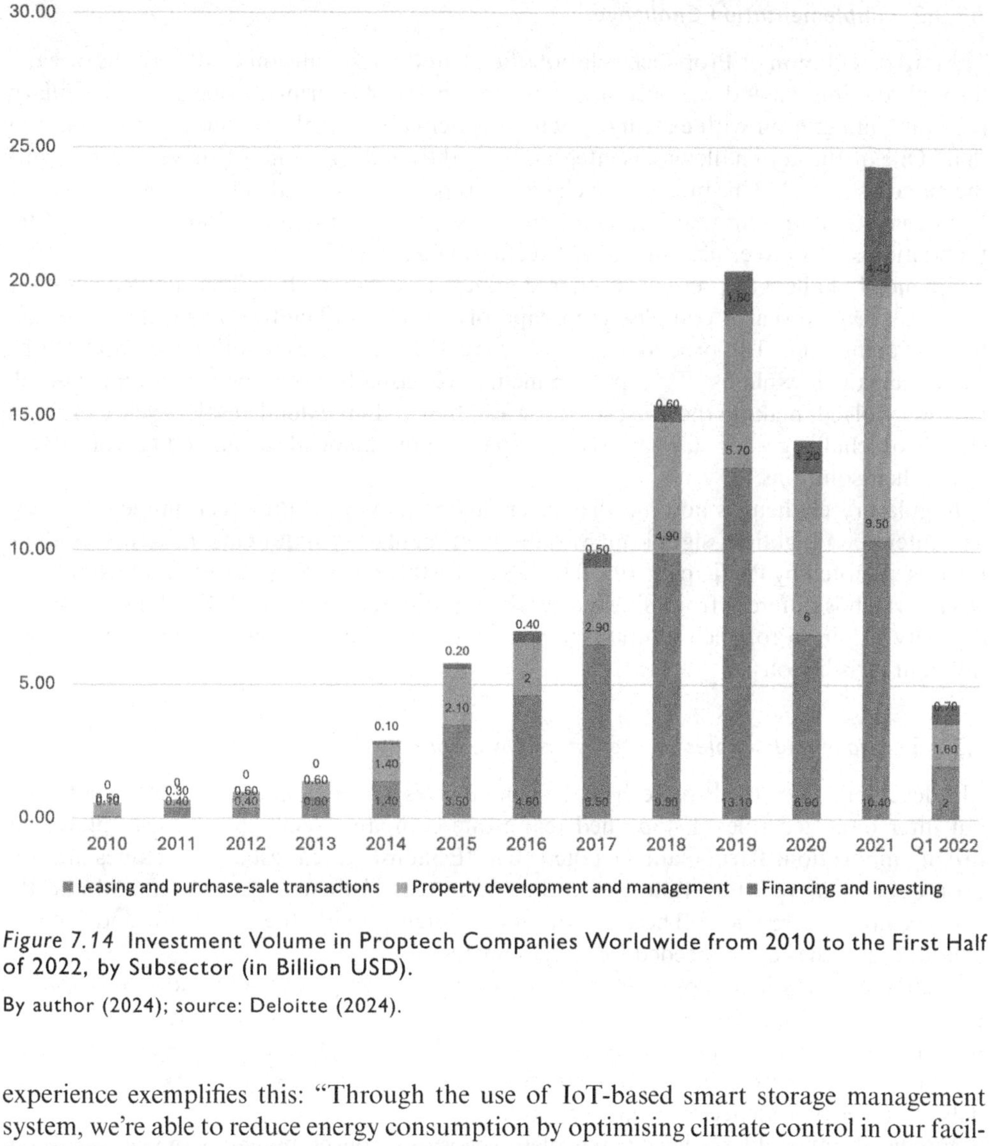

Figure 7.14 Investment Volume in Proptech Companies Worldwide from 2010 to the First Half of 2022, by Subsector (in Billion USD).

By author (2024); source: Deloitte (2024).

experience exemplifies this: "Through the use of IoT-based smart storage management system, we're able to reduce energy consumption by optimising climate control in our facilities." The focus on efficiency extends beyond energy management to encompass broader operational improvements and cost reductions across real estate operations.

External catalysts have also played a crucial role in accelerating PropTech adoption. The COVID-19 pandemic has fundamentally altered how organisations approach real estate technology. Participant 4 notes that "COVID-19 has definitely accelerated digitisation and how people work, live and play." This shift has created new imperatives for technology adoption, particularly in remote operation capabilities and space utilisation optimisation. Climate change concerns have further intensified the focus on sustainable technologies, with Participant 5 emphasising that "We can't get away from that. We can't pretend it's not happening because it is happening."

7.3.3.2 *Implementation Challenges*

The implementation of PropTech solutions faces multiple significant challenges, as revealed through the interview data. Technical barriers represent a primary obstacle, particularly regarding integration with existing systems. Participant 6 highlights this challenge, noting that "One of the key challenges is integration ... the challenge is how can we integrate with the air con system?" This integration challenge is particularly acute in older buildings, with Participant 7 observing that "Some of the sites are maybe really old and often lack the foundation system to easily integrate IT technologies."

Financial challenges present another significant barrier to PropTech implementation. The interviews reveal a complex landscape of funding difficulties, particularly for early-stage companies. The procurement cycle in real estate creates additional complications, as Participant 8 explains: "The procurement cycle usually takes long time and multiple parties involved, making the sale cycle long and hard." This extended sales cycle can create significant challenges for start-ups seeking to maintain financial sustainability while developing their solutions.

Regulatory challenges add another layer of complexity to PropTech implementation. The interviews highlight significant variations in regulatory requirements across different regions, as noted by Participant 10: "The US real estate market's quite different from Hong Kong, which is different from Singapore, which is different from the UK." This regulatory diversity requires PropTech solutions to be adaptable while maintaining compliance across different jurisdictions.

7.3.3.3 *Stakeholder Roles in Ecosystem Development*

The development of the PropTech ecosystem involves various stakeholders playing distinct but interconnected roles. Established real estate companies emerge as crucial players in driving innovation. Participant 11 noted that "Established real estate companies are well positioned to drive innovation due to their resources, their money, their capital and the existing market presence." These organisations often provide the testing ground for new solutions and the capital needed for development.

Technology providers approach the market with varying strategies and self-identifications. As Participant 12 explains, "We're not defining ourselves as a prop tech company ... we're really trying to be a light plug-and-play and help operations." This focus on practical solution delivery rather than sector identification reflects a maturing market where value delivery takes precedence over categorisation.

Financial stakeholders play diverse roles in ecosystem development, with venture capital firms and corporate investors bringing different perspectives and resources. Venture capital firms often focus on scalable business models, with Participant 13 noting that "VCs obviously love the whole software as a service model ... because software has huge margins." Corporate investors, meanwhile, often provide strategic value beyond capital, with their involvement lending credibility to emerging solutions.

7.3.3.4 *Future Trends and Innovations*

The interview data suggests several key trends shaping the future of PropTech. Technological evolution continues to drive innovation, particularly in AI and automation. Participant 13 highlights the growing importance of "generative AI when managing your documentation, managing your vendors trying to score tenants." This trend towards automation and intelligence is expected to accelerate as technologies mature and adoption increases.

Market evolution is another crucial trend, with industry consolidation emerging as a significant factor. Participant 17 observes, "There is a lot of merging of companies ... companies buying a lot of other companies to get that IP." This consolidation trend suggests a maturing market where scale and comprehensive solutions become increasingly important. Business model innovation is also expected to play a crucial role, with Participant 18 noting that "The industry cannot spend its way out of it ... they'll have to find innovative ways to keep people cool and comfortable, but without spending so much money."

The thematic analysis of interview data reveals a PropTech sector at a crucial point in its evolution. The combination of market pressures, technological advancement, and external catalysts drives significant change in how real estate organisations approach technology adoption. While implementation challenges remain significant, the diverse roles played by various stakeholders are helping to overcome these obstacles. The future outlook suggests continued innovation and market evolution, focusing on sustainable solutions and business model innovation. These insights provide valuable guidance for understanding the current state and future direction of the PropTech sector.

7.4 Regulatory Support and Policy Impacts

This section examines the empirical evidence of how government policies, regulatory frameworks, legal considerations, and compliance standards have shaped the development and adoption of PropTech solutions in the real estate industry.

7.4.1 *Government Initiatives*

Empirical evidence suggests that government policies and initiatives are crucial in shaping the PropTech ecosystem. By fostering a supportive environment, governments can enhance innovation and growth within the real estate sector. Conversely, restrictive policies can create barriers that slow down development.

Cities with favourable regulatory frameworks, such as London, New York, and Singapore, consistently rank among the top PropTech ecosystems. These cities have established clear regulations that ensure data security and consumer protection, fostering trust in digital platforms and enhancing user engagement. On the other hand, cities with stringent or outdated regulations, such as complex licensing processes or restrictions on digital transactions, tend to have less vibrant PropTech ecosystems.

Government funding and incentives are another critical driver of PropTech growth. The Smart Nation program (Singapore, n.d.) and the Smart Cities Challenge (U.S. Department of Transportation (Dof, n.d.)) have provided grants, subsidies, and tax incentives to support PropTech innovation and adoption.

Public–private partnerships have also emerged as a key mechanism for driving PropTech development. For example, the Sidewalk Toronto project, a collaboration between Alphabet's Sidewalk Labs[1] and the City of Toronto, aimed to create a smart city neighbourhood by leveraging PropTech solutions such as IoT sensors, AI-powered energy management, and autonomous vehicles. Although the project faced challenges with data privacy and community engagement, it demonstrates the potential for government–industry collaboration in deploying PropTech at scale.

Regulatory sandboxes have gained traction as a tool for fostering PropTech innovation while managing risks. The UK Financial Conduct Authority's (FCA) Regulatory Sandbox, launched in 2016, has supported several PropTech start-ups in testing innovative solutions, such as blockchain-based property transaction platforms and AI-powered mortgage underwriting tools (FCA, 2022).

Open data policies and digital infrastructure investments are critical enablers of PropTech's growth. Reynolds (2024) found that the release of public datasets, such as property transaction records and energy performance certificates, has stimulated the development of new PropTech applications and services, generating an estimated £100 million in economic value annually.

Government policies on housing affordability, sustainability, and urban development have created new opportunities for PropTech innovation. For example, the European Union's Green Deal and Renovation Wave initiatives have spurred the development of PropTech solutions focused on energy efficiency, renewable energy integration, and circular economy principles in the built environment. In the United States, the Biden administration's Infrastructure Investment and Jobs Act includes smart city technologies and digital infrastructure provisions, creating new avenues for PropTech deployment.

In addition to regulatory frameworks, funding programs, and infrastructure investments, government subsidies for PropTech adopters have emerged as a powerful tool for driving technology integration in the real estate sector. By providing financial support to users of PropTech solutions, governments can lower adoption barriers, increase market penetration, enhance technological capabilities, support sustainable practices, and facilitate digital transformation.

Subsidies can surge demand for PropTech solutions, increasing provider competition and innovation. Governments can use subsidies to promote adopting environmentally focused PropTech solutions, such as energy management systems and sustainable building materials.

7.4.2 *Regulatory Frameworks*

The evolving landscape of building codes, data privacy regulations, and financial regulations has profoundly impacted PropTech innovation and adoption. In the United States, the International Building Code (IBC) has progressively incorporated provisions for smart building technologies and sustainable construction practices (ICC, 2024).

Data privacy regulations, such as the EU's General Data Protection Regulation (GDPR) and California's Consumer Privacy Act (CCPA), have created challenges and opportunities for PropTech companies. They have also spurred the development of privacy-enhancing technologies (PETs) in the PropTech sector, such as the GDPR-compliant blockchain platform developed by the French start-up Propchain.[2]

Evolving securities regulations in the financial regulatory space have shaped the emergence of real estate crowdfunding and tokenisation platforms. For example, the US JOBS Act of 2012,[3] which eased crowdfunding and private placement restrictions, has been a key enabler of PropTech platforms like Fundrise[4] and RealtyMogul.[5]

7.4.3 *Legal Implications, Compliance, and Standards*

Adopting PropTech solutions in real estate transactions, property management, and construction has raised new legal questions and challenges. For example, using blockchain-based smart contracts for property sales and leases has sparked debates around their legal enforceability and compatibility with existing real estate laws (Garcia-Teruel, 2020). The development of PropTech-specific standards and certification programs, such as the OSCRE (Open Standards Consortium for Real Estate) and the WELL Building Standard, has also been crucial in promoting interoperability, data quality, and user trust in PropTech solutions.

7.5 Challenges Faced by the PropTech Ecosystem

While the PropTech ecosystem offers significant opportunities for innovation and transformation in the real estate sector, it faces numerous complex challenges that impact its growth, adoption, and implementation. This section examines these challenges through empirical analysis and industry insights, providing a comprehensive understanding of the hurdles stakeholders must navigate.

7.5.1 *Technology Implementation and Integration Challenges*

The implementation and integration of PropTech solutions present significant technical and operational challenges for organisations in the real estate sector. Legacy system integration remains one of the most pressing challenges, as many real estate organisations operate with outdated systems that are not designed with modern integration capabilities in mind. These legacy systems often utilise obsolete data formats and lack modern application programming interface (API) capabilities, making the integration of new PropTech solutions complex and costly. The challenge is compounded by the risk of operational disruption during integration processes, forcing organisations to carefully balance the need for modernisation with maintaining business continuity.

Technical infrastructure requirements pose another significant hurdle. PropTech implementation often demands substantial upgrades to existing technical infrastructure, including high-speed internet connectivity, cloud computing capabilities, and advanced data storage systems. Organisations must invest in robust security infrastructure and mobile device management systems to support modern PropTech solutions. These infrastructure requirements often represent significant capital expenditure and require effective implementation planning.

Data quality and standardisation challenges persist throughout the PropTech ecosystem. The real estate industry historically lacks standardised data formats and property identifiers, leading to data integration and analysis difficulties. Organisations must contend with variable data quality from different sources, requiring significant data cleaning and validation resources. Maintaining data accuracy over time becomes increasingly complex as organisations integrate more data sources and PropTech solutions.

7.5.2 *Financial and Resource Constraints*

Financial considerations and resource limitations represent significant barriers to PropTech adoption and implementation. The financial burden of PropTech adoption extends beyond initial software and hardware investments to include substantial system integration costs, staff training and development expenses, and ongoing maintenance requirements. Organisations must also account for recurring costs such as licensing fees and subscription payments, which can strain operational budgets.

Resource allocation presents particular challenges in the PropTech context. Traditional real estate organisations often lack the internal IT expertise to implement and maintain sophisticated technology solutions. This technical talent shortage is exacerbated by competition from other sectors that typically offer higher compensation and more attractive career paths for technology professionals. Organisations must invest significantly in training existing staff while recruiting new talent with the necessary technical skills. The need for dedicated implementation teams and ongoing support resources further strains organisational capabilities.

7.5.3 Organisational and Cultural Resistance

Cultural and organisational factors often present significant barriers to PropTech adoption, particularly in the traditionally conservative real estate sector. Change management challenges manifest in various ways, from direct resistance to new technologies by established staff to more subtle forms of organisational inertia. The disruption to existing workflows and processes can create anxiety and resistance among employees, requiring careful management and comprehensive training programs. Communication challenges across departments can impede effective implementation, as different stakeholders may have varying perspectives on the value and necessity of PropTech solutions.

The traditional nature of the real estate industry itself can impede PropTech adoption. The sector's conservative approach to technology adoption, rooted in decades of established practices, can make it difficult to implement innovative solutions. This conservatism often manifests in a preference for established methods over new technologies, even when the latter offers clear advantages. Risk-averse decision-making processes, common in real estate organisations, can further slow the adoption of innovative PropTech solutions.

7.5.4 Security and Privacy Concerns

Security and privacy considerations present increasingly critical challenges for PropTech implementation. Cybersecurity risks have become more sophisticated and frequent, requiring organisations to implement robust security measures to protect against potential threats. PropTech solutions must contend with various security challenges, including vulnerability to cyber attacks, data breach risks, and complex system access control requirements. The interconnected nature of modern PropTech solutions also introduces third-party security risks that must be carefully managed.

With the introduction of stringent data protection regulations across different jurisdictions, privacy compliance has become increasingly complex. Organisations must navigate requirements from various regulatory frameworks, such as GDPR in Europe and CCPA in California, while managing user consent and data-handling practices. Cross-border data compliance presents particular challenges for organisations operating in multiple jurisdictions, requiring careful attention to data storage and transfer restrictions. Privacy impact assessments have become necessary components of PropTech implementation, adding another layer of complexity to the adoption process.

7.5.5 *Market and Competition Challenges*

The PropTech ecosystem faces significant market-related challenges that affect its development and adoption. Market fragmentation remains a persistent issue, with multiple competing standards and platforms creating confusion and inefficiencies. The lack of interoperability between different PropTech solutions can limit their effectiveness and create barriers to widespread adoption. Regional variations in market needs and technological maturity levels further complicate the development and deployment of PropTech solutions.

Competitive dynamics in the PropTech sector present additional challenges. Rapid technological change requires organisations to update and adapt their solutions continuously to remain competitive. Market consolidation pressures can force smaller PropTech providers to merge with larger entities or risk becoming obsolete. The constant influx of new market entrants, while driving innovation, can create market uncertainty and make it difficult for organisations to commit to particular solutions. Evolving customer expectations and price competition further complicate the competitive landscape.

7.5.6 *Regulatory and Compliance Challenges*

As we discussed in Section 7.4, PropTech companies must navigate an increasingly complex regulatory environment that varies significantly by jurisdiction, industry, and market segment.

Legal considerations present additional challenges for PropTech organisations. Liability issues, contract enforcement, and intellectual property protection require careful management and often substantial legal resources. Data ownership rights and cross-border legal compliance add further complexity to PropTech operations. Organisations must balance innovation with legal risk management, often requiring sophisticated legal frameworks and expertise.

7.6 Insights from Industry Experiences

This section explores real cases of PropTech companies that have successfully navigated challenges and capitalised on real estate industry drivers. We've categorised these case studies into data and analytics platforms, real estate investment platforms, and property marketplaces.

7.6.1 *Data and Analytics Platforms*

In recent decades, some platforms have leveraged big data and AI to provide property valuations, market insights, and predictive analytics. In this section, we select three platforms to analyse their successes.

7.6.1.1 Zillow (USA)[6]

Zillow, founded in 2006, has become a major player in the online real estate marketplace by leveraging big data and artificial intelligence. It faced several challenges, including the accuracy of property valuations, competition from traditional real estate agents, and scaling data collection and analysis. Zillow adopted several strategies, such as developing the "Zestimate" algorithm, refining the AI models using machine learning, and acquiring broader data sources. Finally, it successfully transformed how consumers access real estate information, making it a go-to resource for property valuations and market trends.

7.6.1.2 Zoopla (UK)[7]

Founded in 2007, Zoopla has become one of the UK's leading property portals, offering property valuations and market insights. Zoopla applies the automated valuation model (AVM) for UK properties, integrates local area data (schools, transport, crime rates) into property listings, and develops tools for estate agents and property professionals. Therefore, it has become a crucial resource for UK property buyers, sellers, and renters, processing over 50 million property value estimates daily.

7.6.1.3 REalyse (UK)[8]

REalyse was founded in 2016 and is a data intelligence platform for the UK property market, focusing on the needs of property developers, investors, and lenders. It has an aggregation of over 300 datasets to provide comprehensive property market analysis, uses machine learning algorithms for predictive analytics and risk assessment, and also provides customised reporting tools for investment and development decision-making. REalyse has become a go-to platform for property professionals seeking in-depth market analysis and forecasting in the UK.

Data and analytics platforms have emerged as major players in the PropTech ecosystem, fundamentally changing how property is valued, analysed, and traded. Table 7.7 compares the three cases we selected for our further analysis.

These platforms have dramatically increased transparency in real estate markets by making previously inaccessible data available to the general public. This shift has empowered consumers and changed the dynamics between buyers, sellers, and real estate professionals. The development of sophisticated automated valuation models (AVMs) using machine learning and big data has improved the accuracy of property valuations. However, these models face challenges in unique or rapidly changing markets where historical data may need to be more relevant. Integrating diverse datasets (e.g., economic indicators, demographic trends, local amenities) allows these platforms to offer predictive insights, helping investors and developers make more informed decisions. This capability is particularly valuable in the commercial real estate sector. They face ongoing data collection, cleaning, and standardisation challenges, especially when operating across different regions with varying data availability and quality. As these platforms become more influential, they face increasing scrutiny regarding data privacy, the potential for market manipulation, and the accuracy of their valuations. The rise of these platforms is reshaping the role of real estate agents and appraisers, pushing them to adapt and provide value beyond mere information provision.

7.6.2 Real Estate Investment Platforms

In the last decade, there has been an emerging trend in real estate investment: democratising it and providing new ways for individuals and institutions to invest in property.

Table 7.7 Comparative Analysis of Data and Analytics Platforms

Aspect	*Zillow (USA)*	*Zoopla (UK)*	*REalyse (UK)*
Target audience	General public, real estate professionals	General public, real estate professionals	Property developers, investors, lenders
Primary innovation	Zestimate (automated home valuation)	Automated valuation model (AVM) for UK properties	Aggregation of 300+ datasets for comprehensive market analysis
Key features	- Property listings - Zestimate - Market trends - Mortgage marketplace	- Property listings - Automated valuations - Local area data integration - Tools for estate agents	- Predictive analytics - Risk assessment - Custom reporting tools - Development feasibility analysis
Data sources	MLS listings, public records, user-submitted data	Property transaction data, local area statistics, user-submitted data	Land registry, planning applications, census data, economic indicators
Geographic focus	United States	United Kingdom	United Kingdom
Business model	Advertising, lead generation for agents, iBuying	Advertising, software solutions for agents	Subscription-based access to analytics platform
Machine learning application	Predictive modeling for home values and market trends	Property valuation, personalized property recommendations	Predictive analytics for property values and market trends
User interface	Consumer-friendly, map-based search	Consumer-friendly, emphasis on local area insights	Professional-oriented, focus on data visualization and analysis tools
Mobile strategy	Strong mobile app presence	Mobile app with AR features	Web-based platform, mobile-responsive
Unique Selling Point	Comprehensive US property database with Zestimate	Integration of property data with local area insights	In-depth analysis tools for property professionals
Challenges	Accuracy of Zestimate in some markets, competition from traditional real estate services	Competing in a crowded UK property portal market	Acquiring and maintaining high-quality data across diverse sources

By author (2024)

7.6.2.1 *Cadre (USA)*[9]

Cadre, a technology-driven real estate investment platform founded in 2014, aims to democratise access to institutional-quality real estate investments using data science and machine learning. Cadre has implemented successful strategies despite challenges such as breaking into the closed-off world of institutional real estate investing, building trust with investors, developing accurate predictive models, and navigating complex regulations. These include developing a proprietary pricing model, creating a user-friendly digital platform, partnering with established firms, implementing a secondary market for liquidity, and focusing on transparency. As a result, Cadre has facilitated over USD 3 billion in real estate transactions, attracted individual and institutional investors, and delivered strong returns, demonstrating the potential of technology to transform real estate investment.

7.6.2.2 PeerStreet (USA)[10]

PeerStreet, an online platform for investing in real estate debt, was founded in 2013. The company has introduced several key innovations to the real estate investment space, including fractional investment in real estate loans, which has significantly lowered the barrier to entry for investors. PeerStreet also utilises algorithms for automated underwriting and offers integration with self-directed individual retirement accounts (IRAs), allowing for tax-advantaged investing. These innovations have proven successful, as PeerStreet has facilitated over USD 4 billion in transactions since its inception, effectively providing investors with access to a previously difficult-to-reach asset class.

7.6.2.3 BrickX (Australia)[11]

BrickX, launched in 2016, is a platform that revolutionised residential real estate investment in Australia by allowing investors to purchase fractional interests, known as "Bricks," in properties. The company's key innovations include dividing each property into 10,000 Bricks, representing fractional ownership; establishing a secondary market for trading these Bricks to provide liquidity; and implementing an automated system for distributing rental income to Brick owners. As a result of these innovations, BrickX has successfully made direct residential real estate investment more accessible to a wider range of investors in Australia, attracting over 70,000 members to its platform.

Real estate investment platforms are revolutionising access to property investment opportunities. Table 7.8 compares the three cases we selected for further analysis.

These platforms have significantly lowered the barriers to entry for real estate investing, allowing a broader range of individuals to participate in an asset class traditionally reserved for institutional investors or high-net-worth individuals. By creating secondary markets for traditionally illiquid real estate investments, these platforms address one of the key challenges of real estate as an asset class. This increased liquidity could have far-reaching implications for how real estate is valued and traded. Fractional ownership models allow investors to spread their capital across multiple properties or loans, potentially reducing risk through diversification. These platforms operate in a complex regulatory environment, often at the intersection of real estate and securities law. Navigating these regulations while innovating remains a key challenge. Using advanced analytics and machine learning in underwriting processes improves risk assessment and potentially leads to more efficient pricing of real estate assets and loans. Successful platforms in this category often emphasise investor education and intuitive user interfaces to help newcomers navigate the complexities of real estate investing. Many of these platforms have yet to be tested through a full real estate market cycle, raising questions about how they will perform during market downturns.

7.6.3 Property Marketplaces

These platforms connect buyers, sellers, renters, and agents, often expanding into additional services to create comprehensive real estate ecosystems.

Table 7.8 Comparative Analysis of Real Estate Investment Platforms

Aspect	*Cadre (USA)*	*PeerStreet (USA)*	*BrickX (Australia)*
Investment type	Commercial real estate equity	Real estate debt (loans)	Residential real estate equity
Minimum investment	USD 25,000	USD 1,000	As low as AUD 50
Target audience	Accredited investors, institutions	Accredited investors	Retail investors
Primary innovation	Data-driven investment selection	Fractional real estate debt investing	Fractional property ownership ("Bricks")
Key Features	- Institutional-quality deals - Secondary market - Data-driven due diligence	- Automated underwriting - Diverse loan types - IRA integration	- Property-specific investment - Secondary market for "Bricks" - Automated rental income distribution
Technology focus	Machine learning for deal sourcing and underwriting	Automated loan underwriting and risk assessment	Blockchain for transaction recording
Geographic focus	United States	United States	Australia
Liquidity mechanism	Secondary market for investments	Short duration of loans (6-24 months)	Secondary market for "Bricks"
Return distribution	Dividends and capital appreciation	Interest payments	Rental income and capital appreciation
Regulatory Framework	SEC Regulation D	SEC Regulation D	ASIC regulations
Unique Selling Point	Access to institutional-quality commercial real estate	Diversified exposure to real estate debt market	Micro-investment in specific properties
Challenges	High minimum investment, limited to accredited investors	Dependent on quality of underlying loans	Property selection, market liquidity

By author (2024)

7.6.3.1 *PropertyGuru (Southeast Asia)*[12]

PropertyGuru, founded in 2007 in Singapore, has grown into Southeast Asia's leading property technology company, operating digital property marketplaces across five countries and offering SaaS-based solutions for real estate developers. Despite challenges such as navigating diverse markets, building trust in online transactions, developing accurate valuation models, and addressing varied customer needs, PropertyGuru implemented successful strategies. These include creating localised platforms, leveraging AI and machine learning, introducing VR and AR tools for remote viewing, launching a FinTech arm, and acquiring regional competitors. As a result, PropertyGuru has digitised much of Southeast Asia's property search and transaction process, serving over 37 million property seekers monthly and 49,000 active agents across five countries. Its expansion into

FinTech and developer services has established a comprehensive ecosystem for regional property transactions.

7.6.3.2 *Rightmove (UK)*[13]

Rightmove, founded in 2000, is the UK's largest online real estate portal and property website. The company has introduced several key innovations to enhance the property search experience, including advanced search filters and map-based property search capabilities, property price comparison and market trend analysis tools, and a mobile app featuring augmented reality for property viewing. These innovations have propelled Rightmove to become the dominant platform for property search in the UK, attracting over 1.3 billion visits annually and solidifying its position as the go-to resource for property seekers and real estate professionals.

7.6.3.3 *Fang.com (China)*[14]

Fang Holdings Limited, established initially as SouFun in 1999, operates a leading real estate internet portal in China. The company has introduced several key innovations that have contributed to its success, including a comprehensive database of new and secondary properties spanning 658 Chinese cities, integrating VR technology for virtual property tours, and the development of a SaaS platform catering to real estate agencies and developers. As a result of these innovations, Fang.com has emerged as one of China's largest property advertising portals, playing a pivotal role in digitising the country's real estate market and significantly impacting how property transactions are conducted in the region.

Property marketplaces in many countries have become the primary interface between property seekers and the real estate market. Table 7.9 compares the three cases we selected for our further analysis.

One or two big players have emerged in many markets, benefiting from network effects. This consolidation raises questions about market power and the ability of new entrants to compete. Successful marketplaces are increasingly expanding their offerings to include additional services such as mortgage comparisons, insurance, and property management tools, creating more comprehensive real estate ecosystems. The shift to mobile-centric platforms has changed how people search for properties, with location-based services and app features playing a crucial role in user experience. The adoption of VR and AR technologies for virtual property tours has accelerated, particularly in light of recent global events with limited in-person viewings. These platforms increasingly leverage their vast datasets to create additional revenue streams, offering market insights and analytics to professionals and investors. Global players in this category face the challenge of adapting their platforms to local real estate practices, regulations, and cultural norms when expanding into new markets. The increasing power of these platforms has led to concerns about disintermediation among traditional real estate professionals, pushing agents and brokers to redefine their value proposition. Many of these platforms rely heavily on advertising revenue from real estate professionals. Balancing the needs of paying professionals with those of free users remains an ongoing challenge.

Table 7.9 Comparative Analysis of Property Marketplaces

Aspect	*PropertyGuru (Southeast Asia)*	*Rightmove (UK)*	*Fang.com (China)*
Geographic focus	Singapore, Malaysia, Thailand, Indonesia, Vietnam	United Kingdom	China
Primary audience	Property seekers, agents, developers	Property seekers, agents	Property seekers, agents, developers
Key features	- Property listings - Virtual tours - Agent directory - Market insights	- Advanced search filters - Price comparison tools - Market trend analysis	- New and secondary property listings - VR property tours - SaaS for agencies and developers
Revenue model	Subscription fees from agents, advertising, developer services	Advertising fees from agents and developers	Advertising, marketing services, e-commerce
Mobile strategy	Strong mobile app presence	AR-enabled mobile app	Mobile app with VR capabilities
Additional services	Mortgage marketplace, data analytics for developers	Property valuation tools	Home furnishing e-commerce, real estate financing services
Market position	Leading platform in Southeast Asia	Dominant player in the UK	One of the largest real estate portals in China
Technology innovation	AI for personalized recommendations, data analytics	Map-based search, market intelligence tools	VR technology, SaaS solutions
Unique challenge	Adapting to diverse markets across Southeast Asia	Maintaining market dominance in a mature market	Navigating China's complex and rapidly changing real estate market
Recent expansion	FinTech services (PropertyGuru Finance)	Expansion into commercial property listings	Development of blockchain-based services
Listing types	Residential, commercial, new developments	Residential, commercial, new developments	Residential, commercial, new developments

By author (2024)

7.6.4 *Insights from Industry Experiences*

As shown in Figure 7.15, key insights from the industry experiences can be categorised into users, challenges, collaboration, and innovation.

- *Users are the centre of success.* Companies prioritising user-friendly interfaces and streamlined processes are more likely to gain widespread adoption. Many successful PropTech companies have grown by expanding into adjacent services, creating more comprehensive solutions for their users. Companies that provide clear, transparent information about their processes and pricing are more likely to gain customer trust and market share. PropTech companies operating across diverse markets must adapt their solutions to local regulations, cultural norms, and market dynamics.

- *It's important to address industry-specific challenges.* Successful PropTech solutions are tailored to address the industry's unique pain points. PropTech can open up previously exclusive investment opportunities to more investors. Innovative PropTech solutions can also address traditional challenges in real estate, such as the lack of liquidity in investments.
- *Collaboration is critical, especially within the PropTech ecosystem.* Successful PropTech companies often form strategic partnerships and integrate with existing systems to provide comprehensive solutions. They also expand beyond their initial offerings to create integrated ecosystems that address multiple aspects of real estate transactions.
- *Innovation and technology are the keys to the success of PropTech companies.* The most successful companies in the PropTech ecosystem continually refine their technologies and business models to stay ahead of market demands and competition. Successful PropTech companies leverage big data and advanced analytics to provide valuable insights and improve operational efficiency. Successful PropTech companies often incorporate cutting-edge technologies like VR, AR, and blockchain to enhance their offerings and user experience.

By examining these diverse case studies and insights, we understand how PropTech companies navigate the complex landscape of drivers and challenges in the real estate industry

Figure 7.15 Key Insights from the Industry Experiences.

By author (2024).

across different global markets and business models. These insights provide valuable lessons for established players and newcomers in the PropTech ecosystem, particularly in investment, finance innovation, and market analysis.

7.7 Concluding Remarks

This chapter's comprehensive analysis of the PropTech ecosystem reveals significant insights into the drivers and challenges shaping the industry's evolution. Through empirical research and detailed case studies, we have uncovered patterns that have important implications for stakeholders across the real estate sector. The research demonstrates that PropTech's growth is driven by a convergence of factors, including technological advancement, changing consumer expectations, and supportive regulatory frameworks. Particularly significant is the role of data analytics and artificial intelligence in transforming traditional real estate practices, with successful PropTech implementations often resulting from a carefully balanced consideration of both technological capabilities and human factors.

Our analysis of leading PropTech platforms reveals that successful companies share several common characteristics, including a strong focus on user experience, leveraging data for competitive advantage, a flexible approach to regulatory compliance, strategic partnership development, and continuous innovation. However, despite the sector's growth, several challenges persist as significant barriers to PropTech adoption. These include integration complexity with legacy systems, resource constraints (particularly for smaller organisations), cultural resistance to technological change, data security and privacy concerns, and regulatory compliance across different jurisdictions.

Several trends are likely to shape the development of the PropTech ecosystem. The sector is moving towards consolidation, with successful platforms expanding their service offerings and creating more comprehensive solutions. Deep integration of emerging technologies like blockchain, IoT, and advanced AI will continue to transform real estate operations and transactions. The regulatory landscape will likely continue to evolve, requiring PropTech companies to maintain flexible and adaptable compliance strategies. Environmental considerations and sustainability requirements will also increasingly drive PropTech innovation and adoption.

Based on our analysis, different stakeholder groups should consider specific strategic approaches. PropTech companies should prioritise user experience while maintaining robust technical capabilities, develop scalable solutions that adapt to different market contexts, invest in strong data security and privacy measures, and build strategic partnerships within the ecosystem. Real estate organisations should develop comprehensive digital transformation strategies, invest in change management and staff training, focus on data quality and standardisation, and build internal technical capabilities. Regulators and policymakers should create balanced frameworks that promote innovation while protecting stakeholders, support standardisation efforts, and facilitate collaboration between traditional real estate and technology sectors.

The future of PropTech appears promising, yet it demands careful attention to both opportunities and challenges. Organisations that can successfully balance innovation with practical implementation while focusing on user needs and regulatory compliance will be best positioned to thrive in this evolving landscape. The continued evolution of the PropTech ecosystem will likely bring new challenges and opportunities, requiring stakeholders to remain adaptable and forward-thinking in their approaches to digital transformation in real estate.

Notes

1 https://www.google.com/earth/about/design/.
2 https://www.propchain.com/.
3 https://www.sec.gov/rules-regulations/jumpstart-our-business-startups-jobs-act.
4 https://fundrise.com/.
5 https://www.realtymogul.com/.
6 https://www.zillow.com/.
7 https://www.zoopla.co.uk/.
8 https://www.realyse.com/.
9 https://cadre.com/.
10 https://www.peerstreet.com/.
11 https://www.brickx.com/.
12 https://www.propertygurugroup.com/.
13 https://www.rightmove.co.uk/.
14 https://www1.fang.com/.

Recommended Reading

Battisto, J., Friedman, S., & Murillo, S. 2018. *Challenges and opportunities in cross-border real estate investment*. Federal Reserve Bank of New York. https://www.newyorkfed.org/medialibrary/media/outreach-and-education/community-development/constraints-and-opportunities-in-cross-border-real-estate-investment.pdf

Baum, A. 2017. *PropTech 3.0: The future of real estate*. Said Business School, University of Oxford. https://www.sbs.ox.ac.uk/sites/default/files/2018-07/PropTech3.0.pdf

Baum, A., Saull, A., & Braesemann, F. 2020. *PropTech 2020: The future of real estate*. Said Business School, University of Oxford. https://www.sbs.ox.ac.uk/sites/default/files/2020-02/proptech2020.pdf

Braesemann, F., & Baum, A. 2020. *PropTech: Turning real estate into a data-driven market?* Oxford Future of Real Estate Initiative, Saïd Business School, University of Oxford. https://www.sbs.ox.ac.uk/sites/default/files/2020-02/proptech-turning-real-estate-data-driven-market.pdf

Chegut, A., Eichholtz, P., & Kok, N. 2019. The price of innovation: An analysis of the marginal cost of green buildings. *Journal of Environmental Economics and Management*, 98, 102248. https://doi.org/10.1016/j.jeem.2019.07.003

Deloitte. 2020. 2020 commercial real estate outlook. https://www2.deloitte.com/us/en/insights/industry/financial-services/commercial-real-estate-outlook.html

JLL. 2021. Global real estate technology report 2021. https://www.us.jll.com/en/trends-and-insights/research/global-real-estate-technology-report

KPMG. 2021. Real estate in the digital age. https://home.kpmg/xx/en/home/insights/2021/01/real-estate-in-the-digital-age.html

Pärn, E. A., Edwards, D. J., & Sing, M. C. P. 2017. The building information modelling trajectory in facilities management: A review. *Automation in Construction*, 75, 45–55. https://doi.org/10.1016/j.autcon.2016.12.003

Shaw, J. 2018. Platform real estate: Theory and practice of new urban real estate markets. *Urban Geography*, 39(8), 1138–1156. https://doi.org/10.1080/02723638.2018.1524653

Tajani, F., Morano, P., Locurcio, M., & Torre, C. M. 2018. Data-driven techniques for mass appraisals. Applications to the residential market of the city of Bari (Italy). *International Journal of Business Intelligence and Data Mining*, 13(4), 421–450. https://doi.org/10.1504/IJBIDM.2016.081604

Ullah, F., Sepasgozar, S. M., & Wang, C. 2018. A systematic review of smart real estate technology: Drivers of, and barriers to, the use of digital disruptive technologies and online platforms. *Sustainability*, 10(9), 3142. https://doi.org/10.3390/su10093142

Źróbek, S., Manzhynski, S., Zysk, E., & Źróbek, R. 2020. Data analytics in the real estate market—A systematic literature review. *Sustainability*, 12(13), 5348. https://doi.org/10.3390/su12135348

References

Creti. 2023. 2023 Proptech venture capital report. https://creti.org/2023-proptech-venture-capital-report

Deloitte. 2024. 2024 commercial real estate outlook: Finding terra firma. https://www.deloitte.com/us/en/insights/industry/financial-services/financial-services-industry-outlooks/commercial-real-estate-outlook-2024.html

Dof, U. n.d. *Smart city challenge*. U.S. Department of Transportation. https://www.transportation.gov/tags/smart-city-challenge

Eisenmann, T. 2006. Strategies for two-sided markets. *Harvard Business Review*. https://hbr.org/2006/10/strategies-for-two-sided-markets

FCA. 2022. Regulatory sandbox cohort 7: Lessons learned and success stories. https://www.fca.org.uk/firms/innovation/regulatory-sandbox/accepted-firms

Garcia-Teruel, R. M. 2020. Legal challenges and opportunities of blockchain technology in the real estate sector. *Journal of Property, Planning and Environmental Law*, 12, 129–145.

ICC. 2024. *2024 international building code (IBC)*. International Code Council.

JLL. 2023. Global real estate technology survey 2023. https://www.jllt.com/research/global-real-estate-technology-survey-2023/

KPMG. 2023. KPMG global tech report 2023. https://kpmg.com/xx/en/our-insights/ai-and-technology/kpmg-global-tech-report-2023.html

Metaprop. 2023. Global PropTech confidence index. https://www.metaprop.com/research

Moore, J. F. 1993. Predators and prey: A new ecology of competition. *Harvard Business Review*, 71, 75–86.

Nasdaq Alternative Data. 2023. Nasdaq Data Link. https://www.nasdaq.com/solutions/data/nasdaq-data-link

North, D. C. 1990. *Institutions, institutional change and economic performance*. Cambridge University Press.

Parker, G. G. 2016. *Platform revolution: How networked markets are transforming the economy and how to make them work for you*. WW Norton & Company.

Pitchbook. 2024. https://get.pitchbook.com/investor-database/

PwC & ULI. 2022. Emerging trends in real estate. https://www.pwc.com/gx/en/industries/financial-services/real-estate/emerging-trends-real-estate/global-outlook-2022.html

Reynolds, J. 2024, August. The importance of open data for building a fully digital property market. *EstateAgentToday*. https://www.estateagenttoday.co.uk/features/2024/03/the-importance-of-open-data-for-building-a-fully-digital-property-market/

Singapore, G. n.d. Smart Nation Singapore. https://www.smartnation.gov.sg/

Yano, R. 2024. *PropTech market in Japan*. Yano Research Institute.

Chapter 8

The Future of PropTech for Sustainable Real Estate

8.1 Defining the Future Landscape

As we introduced in previous chapters, the PropTech landscape is rapidly evolving. This transformation is driven by technological advancements, changing consumer preferences, and an urgent need for sustainable solutions in the real estate sector. It is reshaping how we design, construct, manage, interact with, and invest in built environments, paving the way for a more sustainable and efficient future.

8.1.1 Convergence of Digital and Physical Realms

The future of PropTech lies in the seamless integration of digital technologies with physical spaces. This convergence blurs the lines between traditional real estate and technology, creating "smart" buildings and cities more responsive to environmental concerns and user needs. The rapid growth of Internet of Things (IoT) devices and sensors enables real-time monitoring and optimisation of building performance, from energy consumption to air quality. Virtual replicas (digital twins) of physical structures are becoming increasingly sophisticated, allowing for predictive maintenance and more efficient resource allocation.

8.1.2 Data-Driven Decision-Making

The ability to collect, analyse, and act upon large amounts of data is transforming real estate decision-making, with increasing focus on sustainability metrics. AI and machine learning technologies predict energy usage patterns, optimise building systems, and improve overall sustainability performance. Distributed ledger technology (DLT) revolutionises property transactions and energy trading, ensuring greater transparency and efficiency.

8.1.3 Sustainable Building Materials and Construction Methods

PropTech is not just about digital solutions; it's also driving innovation in physical materials and construction techniques. 3D printing and modular construction technologies are reducing waste, lowering carbon footprints, and accelerating the construction of sustainable buildings. Self-healing concrete, thermochromic materials, and other innovations enhance building longevity and energy efficiency.

DOI: 10.1201/9781003504436-8

8.1.4 Renewable Energy Integration

The future landscape of PropTech is inextricably linked with the renewable energy sector, as buildings become both consumers and producers of clean energy. Currently, solar technology is seamlessly integrated into building materials, transforming structures into power generators. Advanced energy management systems, such as microgrids and energy storage, enable buildings to operate more independently and resiliently.

8.1.5 User-Centric Design and Operation

PropTech increasingly focuses on the end-user experience, recognising that sustainable buildings must also be comfortable, healthy, and productive spaces. The technology incorporates nature into built environments, improving well-being and environmental performance. AI-driven systems tailor indoor environments to individual preferences while optimising energy use.

8.1.6 Regulatory Compliance and Reporting

PropTech is vital in helping real estate owners meet and exceed these requirements as sustainability regulations become more stringent. Software solutions, such as automated reporting tools, streamline the collection and reporting of sustainability data for regulatory compliance. Again, AI systems help property owners anticipate and prepare for future regulatory changes.

8.1.7 Real Estate Investment and Finance

PropTech is revolutionising the financial aspects of real estate, particularly sustainable investments. Digital platforms are emerging to connect investors with sustainable real estate projects and facilitate the flow of capital into green buildings and retrofits. Advanced sustainability analytics tools enable investors to assess the environmental, social, and governance (ESG) performance of real estate assets, informing informed investment decisions. Blockchain technology enables the fractional ownership of properties, potentially democratising access to sustainable real estate investments. Machine learning algorithms incorporate sustainability factors into property valuation, potentially reshaping how green buildings are valued in the market. PropTech solutions facilitate the measurement, verification, and trading of carbon credits generated by sustainable real estate projects, which incentivise the real estate industry to shift towards carbon neutrality.

8.1.8 Sustainable Property Management

PropTech is revolutionising the management and operation of sustainable properties. AI and IoT enable proactive maintenance of building systems, reducing energy waste and extending the life of assets. Mobile applications, especially tenant engagement apps, encourage sustainable behaviours among building occupants by gamifying energy conservation and waste reduction.

The future landscape of PropTech for sustainable real estate is one of integration, intelligence, and innovation. As these technologies continue to evolve and converge, they promise to create a built environment that is not only more sustainable but also more adaptive, efficient, and aligned with the needs of both occupants and the planet. Moreover, these advancements are reshaping the financial landscape of real estate, creating new opportunities for sustainable investment and value creation. This chapter proposes the future path PropTech will take to foster sustainability.

8.2 Integration of PropTech in Sustainable Development

PropTech is not only about software solutions. Hardware components are important in practically implementing sustainable and efficient building practices. These physical devices form the backbone of data collection, automation, and real-time monitoring systems that enable smart, sustainable buildings. PropTech's hardware integration has significantly improved building performance and sustainability metrics (WEF, 2021).

8.2.1 Key Hardware Components in Sustainable Building Practices

Integrating advanced hardware components is critical in transforming traditional buildings into sustainable, energy-efficient structures. These technologies form the backbone of smart building systems, enabling real-time monitoring, data collection, and automated control of various building functions. By leveraging these hardware solutions, building owners and managers can significantly reduce energy consumption, optimise resource usage, and enhance occupant comfort. The following key hardware components are at the forefront of sustainable building practices, each contributing to improved efficiency and environmental performance.

Smart meters measure and record energy consumption in real time, providing valuable data for energy management systems. Studies have shown that smart meters can lead to up to 15% energy savings in commercial buildings (Ahmad et al., 2016). *IoT sensors* monitor environmental conditions such as temperature, humidity, air quality, and occupancy, enabling precise control of building systems. The global market for IoT sensors in smart buildings is expected to grow from 45 million in 2022 to 115 million buildings in 2026 (Grant, 2022). This growth reflects the increasing demand for energy efficiency among property owners and tenants, including businesses and residents. *Smart HVAC systems* use sensor data to optimise energy usage while maintaining comfort. Compared to traditional systems, they can reduce energy consumption by up to 40% (Zhao et al., 2023; Chaouch et al., 2021; D'Silva, 2023). *LED lighting systems* with smart controls adjust based on occupancy and natural light levels. Adopting LED lighting in commercial buildings can save up to 75% of energy (Elliott et al., 2020) compared to traditional lighting systems. *Solar panels and energy storage* contribute to sustainable energy practices. Integrating solar panels with smart energy management systems can increase a building's energy efficiency by up to 30% (Bouckaert et al., 2021). *Water management systems* help conserve water resources. Advanced water management systems can reduce building water consumption by up to 30% (EPA, 2012).

8.2.2 Integration of Software and Hardware in PropTech Solutions

The true power of PropTech lies in the seamless integration of hardware and software components. This integration creates a complete ecosystem that enhances sustainability and efficiency. Hardware sensors collect real-time data, which is processed and analysed using sophisticated software algorithms to identify patterns and optimisation opportunities. Big data analytics in smart buildings can lead to 8–15% energy savings without significant hardware upgrades (Marinakis et al., 2020). By combining hardware monitoring with machine learning algorithms, buildings can predict and prevent equipment failures, reducing downtime and extending the lifespan of critical systems. Predictive maintenance can reduce maintenance costs by up to 40% and decrease breakdowns by 70% (Cerquitelli et al., 2021; Deloitte, 2017). Software platforms use smart meters and environmental sensor data to optimise energy usage across various building systems, reducing waste and costs. Advanced energy management systems have been shown to reduce energy consumption in both commercial and residential building scenarios (Mauser et al., 2016). Integrated systems balance energy efficiency with occupant comfort by adjusting lighting, temperature, and air quality based on real-time conditions and preferences. Studies have shown that optimised indoor environments can increase productivity by up to 11% (Allen et al., 2016). Building information modelling (BIM) software integrates with IoT devices to create digital twins of buildings, enabling better design, construction, and management practices. Nguyen and Adhikari (2023) discussed the integration of BIM and IoT technologies in the construction industry based on recent research and found that BIM and IoT integration creates a holistic ecosystem that enhances sustainability and efficiency in buildings. Real-time data from IoT sensors can be incorporated into BIM models, allowing for dynamic updates of building information. This integration enables better design, construction, and facility management decision-making by providing accurate, up-to-date information about building performance.

8.2.3 *Future Trends and Challenges*

As PropTech continues to evolve, several trends and challenges are emerging regarding hardware components for sustainable building practices:

8.2.3.1 *Advanced Sensor Technologies*

Developing more sophisticated, miniaturised, and energy-efficient sensors is a key trend in sustainable PropTech. These advanced sensors are becoming increasingly capable of collecting more diverse and accurate data while consuming less power (Plageras et al., 2018). The sensors can provide more accurate occupancy detection for optimised space utilisation and energy management, monitor indoor air quality precisely for improved health and well-being, and track resource consumption for better conservation strategies in property management (Minoli et al., 2017).

There have been some developments in sensor technologies. For example, nanosensors, which are tiny sensors at the nanoscale, are being developed. They allow for more granular and less intrusive data collection. These can be embedded in building materials to monitor structural health or air quality at a molecular level (Shahid et al., 2024). Instead of having multiple single-function sensors, there is a trend towards sensors that can measure multiple parameters simultaneously. For instance, a single device might measure temperature, humidity, CO_2 levels, and occupancy (Ahmad et al., 2016). Low-power wide-area network (LPWAN) sensors can transmit data over long distances while consuming minimal power, making them ideal for large commercial buildings or campuses (Mekki et al., 2019).

The small wireless microelectromechanical sensors (MEMS) can detect light, vibration, temperature, and chemicals. While still largely experimental, they hold promise for ubiquitous environmental sensing (Chong & Kumar, 2003).

While the future looks bright, there are some challenges in applying sensor technologies to building sustainability. Incorporating these advanced sensors into existing building structures without significant disruption or cost remains challenging (Minoli et al., 2017). The increased volume and variety of data from these sensors require robust data management and analysis systems (Plageras et al., 2018). More sophisticated sensors may raise privacy concerns, especially in residential or mixed-use buildings (Alaa et al., 2017). Ensuring the long-term reliability and ease of maintenance for these advanced sensors is important, particularly for hard-to-reach or embedded sensors. While sensor costs generally decrease, the most advanced technologies may still be prohibitively expensive for widespread adoption (Minoli et al., 2017).

Advancements in sensor technologies are expected to drive more precise, responsive, and efficient building management systems. As these sensors become more affordable and accessible to integrate, they will likely play a crucial role in achieving near-zero energy buildings and enhancing occupant comfort and well-being (Ahmad et al., 2016).

8.2.3.2 AI and Machine Learning Integration

The incorporation of AI and machine learning algorithms into hardware devices is enabling more intelligent and predictive building management systems. This integration transforms buildings' operations, leading to improved energy efficiency, occupant comfort, and overall sustainability.

AI algorithms can analyse data from sensors to predict equipment failures before they occur, reducing downtime and maintenance costs (Qiu et al., 2016); predict occupancy patterns and adjust HVAC, lighting, and other systems accordingly, significantly reducing energy waste (Gao, 2021); and learn individual preferences and adapt local environments to maximise occupant comfort while minimising energy use (Park et al., 2019). AI-driven building management systems that can balance energy efficiency, comfort, and air quality in real time.

ML models can learn from historical data and real-time inputs to optimise energy usage across various building systems (Ahmad et al., 2021). They can also identify unusual patterns in energy consumption or environmental conditions to alert facility managers of potential issues (Peña et al., 2016).

Although AI and machine learning have been increasingly discussed, several hurdles exist to their wide adoption. AI and ML models require large amounts of high-quality data to function effectively, which can be challenging to obtain in some building environments (Molina-Solana et al., 2017). The issue also lies in explainability and trust. The "black box" nature of some AI algorithms can make it difficult for building managers to understand and trust their decisions (Arrieta et al., 2020).

8.2.3.3 Blockchain for Secure Data Management

Blockchain technology is emerging as a powerful tool for secure data management in smart buildings. Its decentralised, immutable, and transparent nature makes it well-suited for managing sensitive building data, transactions, and operations. As we presented in Chapters

4 and 5, blockchain-based smart contracts automate and secure various building management processes, from lease agreements to maintenance schedules. These contracts can execute automatically when predefined conditions are met, reducing the need for intermediaries and increasing efficiency (Hewa et al., 2021). Blockchain enables secure and privacy-preserving identity management systems for building occupants and IoT devices, enhancing access control and personalisation of building services while protecting user privacy (Stokkink & Pouwelse, 2018). Blockchain is also used to track and verify the provenance of building materials and equipment, ensuring sustainability and quality standards are met (De la Cruz-Lovera et al., 2017).

At present, blockchain has been applied in a few aspects of real estate. For example, blockchain can provide a tamper-proof ledger for IoT sensor data, ensuring the integrity and authenticity of building performance data, which is important for energy management, predictive maintenance, and environmental monitoring (Panarello et al., 2018). Transparency and efficiency are the well-known strengths of blockchain. It can provide transparent building certifications (e.g., LEED, BREEAM, WELL) and make it easier to verify and maintain green building standards (Mahmudnia et al., 2022). Veuger (2018) found that blockchain-based platforms can streamline property transactions, reducing fraud and increasing transparency in real estate deals. Regarding property management and investment, blockchain enables more democratic decision-making processes in building management by allowing stakeholders to vote on and track building-related decisions (Kifokeris & Koch, 2019) and enables the tokenisation of real estate assets, potentially increasing liquidity and enabling fractional ownership of properties (Wouda & Opdenakker, 2019).

Blockchain challenges are increasingly discussed, such as regulatory uncertainty and energy consumption, scalability, integration with legacy systems, user adoption and education, and data privacy concerns (Kshetri, 2018; Reyna et al., 2018; Savelyev, 2018; Truby, 2018; Ølnes et al., 2017; Zheng et al., 2018).

Despite these challenges, the use of blockchain in smart buildings is expected to grow significantly. As the technology matures and more energy-efficient consensus mechanisms are developed, blockchain will likely become integral to secure and transparent building data management.

8.2.3.4 Data Privacy and Security

As PropTech solutions become increasingly prevalent in sustainable building practices, they bring significant data privacy and security concerns. Integrating IoT devices and smart systems generates vast amounts of sensitive data, ranging from energy consumption patterns to occupant behaviours and personal information. This raises several issues at different stages (in Figure 8.1).

- *Data Collection and Consent:* There are concerns about the extent and nature of data collected by smart building systems. Occupants may not be fully aware of what data is being gathered about them and how it's being used, raising questions about informed consent.
- *Data Storage and Protection:* The large volumes of data collected need to be stored securely. This includes protection against unauthorised access, data breaches, and cyber-attacks. According to a report by Gartner (2020), 75% of cyber-attacks against enterprises are expected to target IoT devices by 2025.

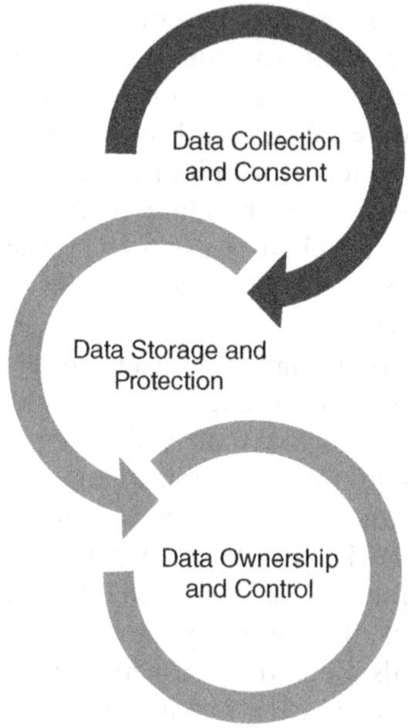

Figure 8.1 Data Privacy and Security Issues.
By author (2024).

- *Data Ownership and Control:* Questions arise about who owns the data collected by smart building systems – the building owner, the technology provider, or the occupants themselves. This has implications for data access, use, and deletion rights.

In commercial buildings or multi-tenant residences, there's a challenge in maintaining individual privacy while using data for overall building efficiency. Many PropTech solutions involve third-party service providers, raising concerns about how these entities access, use, and protect the data they handle. As buildings change ownership or management, there are questions about transferring, retaining, or deleting historical building data. To address these concerns, stakeholders in the PropTech ecosystem need to prioritise privacy-by-design principles, implement robust cybersecurity measures, and establish clear data governance policies. This includes using encryption, regular security audits, and transparent data practices. Integrating hardware and software in PropTech is critical for moving the industry towards being more environmentally friendly, efficient, comfortable, and adaptable to future needs.

8.3 The Role of PropTech in Reshaping Real Estate in the Future

PropTech plays a transformative role in reshaping the real estate industry in the coming years, driving cultural shifts, enabling new capabilities, and disrupting traditional processes. This section discusses how it will work in these three areas.

8.3.1 PropTech Contributes to Cultural Shifts Towards Sustainability

PropTech contributes to a few cultural shifts (in Figure 8.2) in real estate towards greater sustainability and environmental responsibility.

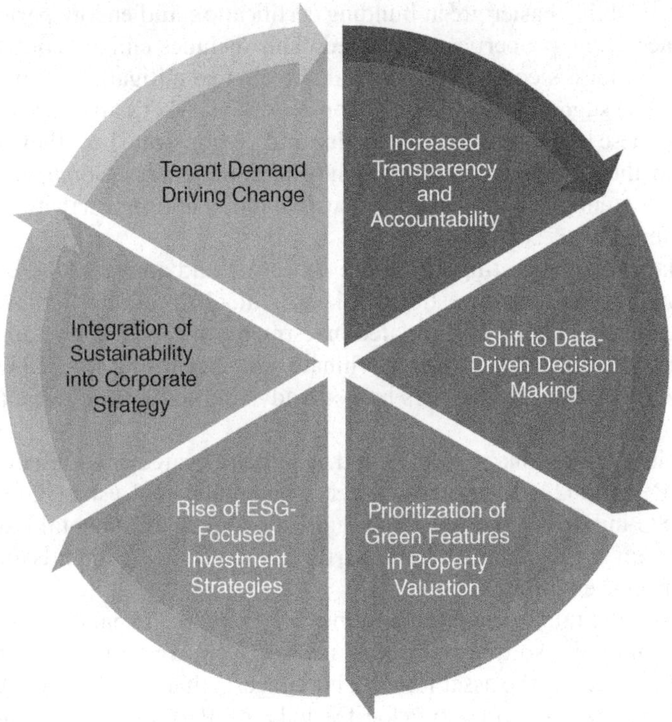

Figure 8.2 Cultural Shifts in Real Estate.
By author (2024).

The widespread adoption of sustainability rating systems like LEED, BREEAM, and GRESB, enabled by PropTech platforms, has created a culture of transparency (Ionaşcu & Anghel, 2020). For example, the GRESB (2024) survey showed that 96% of property companies and funds now disclose their ESG performance, up from just 45% in 2010.

The smart building technologies that use a measurable set of metrics, typically tracked through energy management software, to monitor and optimise building performance include features like real-time energy consumption monitoring, occupancy sensors, automated lighting controls, HVAC system optimisation based on weather data, and data analytics to identify inefficiencies and suggest adjustments, essentially allowing building managers to make data-driven decisions to reduce energy usage and improve overall building efficiency (Aguilar et al., 2021; Eini et al., 2021).

Key metrics tracked by energy management software in smart buildings include:

- *Energy Use Intensity (EUI)*: Total energy consumption per square foot of building area
- *Peak Demand*: Maximum power drawn by the building at any given time
- *Lighting Power Density (LPD)*: Lighting power per square foot
- *HVAC System Efficiency*: Heating, ventilation, and air conditioning performance
- *Occupancy Rates*: Percentage of space actively used by occupants
- *Environmental Data*: Outdoor temperature, humidity, and CO_2 levels

PropTech tools enabling easier green building certification and energy performance analysis have influenced how properties are valued. This includes climate contribution criteria that measure a business's contribution to climate change mitigation (Scherrenberg et al., 2024). RICS (2021) suggest that surveyors consider the market's attitudes towards sustainability and how these impact the property value and demonstrate how they have considered sustainability in their calculations, approach, and commentary. Some surveying services offer to identify potential improvements to sustainability and thermal efficiency as part of a property survey (Baker, 2024).

PropTech platforms facilitating ESG data collection and analysis have contributed to a surge in sustainable real estate investment. Biasin et al. (2024) found that investors increasingly recognise the benefits of ESG-related investments and realise the tangible impact of ESG real estate investment on mitigating climate risk. According to CBRE (2022)'s investor survey, 60% of real estate investors have already adopted ESG criteria as part of their investment strategies.

The accessibility of sustainability data and tools has elevated environmental concerns to C-suite level. Reiss (2022) interviewed three commercial real estate CEOs about their thoughts on ESG initiatives and indicated that ESG and environmental issues are becoming more important at the executive level, partly due to investor demands and the ability to set and track specific environmental goals.

Sonny Kalsi of BentallGreenOak mentions that they are "expanding our relationships with the tech community so that our access to PropTech can better support environment and social performance at the asset level." This suggests that technology is enabling better environmental performance tracking. Joe Derhake of Partner Engineering and Science, Inc. states that "ESG has become critically important because some of their investors are demanding it." He also mentions that they help firms with "property-specific ESG data collection, benchmarking, goal setting, and decarbonisation measures." Sara Armbruster of Steelcase notes that ESG is "a key pillar of our corporate strategy" and mentions setting science-based targets for emissions reduction.

PropTech-enabled green building features are increasingly influencing tenant decisions. Although traditional factors like location and basic building functionality still dominate the decision-making process for office tenants, there is a growing number of green-certified buildings in the market. The sensitivity analysis showed that even when the importance of these green/smart features was artificially increased to 100%, it did not significantly change the overall preference for different building types (Gluszak et al., 2019).

These shifts demonstrate how PropTech is providing tools and actively reshaping industry culture to prioritise sustainability as a core consideration in real estate operations and decision-making. PropTech is making sustainability initiatives more feasible, measurable, and economically viable.

8.3.2 PropTech Fosters Innovation and Sustainability Education

Education and awareness programmes enabled by PropTech are driving the tech-savvy and sustainability-focused real estate culture (Figure 8.3).

8.3.2.1 Education and Research Initiatives

Online platforms democratise access to PropTech and sustainability education for real estate professionals globally. These platforms offer flexible, on-demand learning

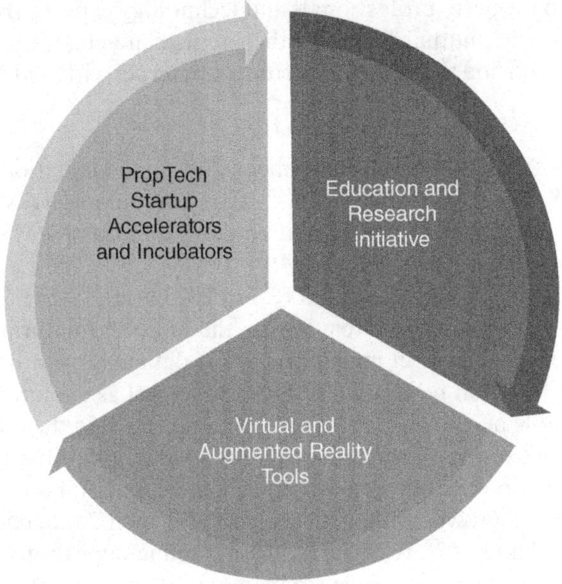

Figure 8.3 The Changes in Innovation and Sustainability Education.
By author (2024).

opportunities accessible from anywhere, allowing professionals to stay updated with the latest trends and technologies. Universities, corporations, and start-ups are increasingly collaborating on research projects focused on PropTech and sustainability.

For example:

- *MIT Real Estate Innovation Lab*: The MIT Real Estate Innovation Lab (REI Lab)[1] is a research initiative within MIT's School of Architecture and Planning that focuses on exploring the future of the built environment and how technology, design, and innovation will impact real estate. The lab conducts research on real estate innovation, including smart buildings and digital twins, data science and machine learning applications in real estate, sustainable and green building practices, and automation and AI in real estate. The lab collaborates with industry partners, including major real estate firms, technology companies, and other academic institutions. For example, they have worked with JLL and EY on various research projects. While being research-led, the lab is involved in education through courses and programs at MIT, including classes on data science in real estate and technological change in the built environment. The lab aims to be at the forefront of understanding and shaping how technology and innovation will transform the real estate industry and built environment in the coming years. Their work spans from theoretical research to practical applications and industry engagement.
- *RICS Tech Partner Programme*: The Royal Institution of Chartered Surveyors (RICS) Tech Partner Programme[2] is an innovative initiative designed to bridge the gap between traditional real estate practices and cutting-edge technology. This collaborative platform brings together forward-thinking firms from across the globe, fostering the development of thought leadership, content, and market insights that are reshaping the property profession. A key focus of the Tech Partner Programme is to highlight the

synergies between property professionals and technology experts. By encouraging ongoing collaboration and mutual learning, the programme creates a dynamic ecosystem where innovators outside the property sector can interact with and influence established industry practices.

The programme also supports the development and implementation of data standards, including the RICS Data Standard (RDS).[3] This comprehensive schema incorporates standards for International Property Measurement Standards (IPMS) and International Construction Measurement Standards (ICMS).

According to the survey conducted by RICS in 2023 and 2024, the program includes a wide range of sectors and practice areas across the built and natural environment, from planning and development to asset management and data analytics. It covers various asset types, with a strong focus on residential properties, as well as commercial, industrial, and alternative assets. While most tech partners are headquartered in the UK and Ireland, there is representation from Europe, North America, the Middle East, Africa, Asia, and Oceania. The program includes both established market participants (about 40%) and start-ups (about 60%), and it has grown, with 171 tech partners in 2023 increasing to 340 in 2024 (RICS, 2023, 2024). The RICS Tech Partner Programme appears to be a diverse, global initiative that brings together innovative companies to drive technological advancement in the built and natural environment sector, providing RICS with valuable insights to guide the profession's evolution.

Such collaborations are essential for bridging the gap between academic research and industry practice (Ankrah & Omar, 2015). Research has shown that online learning can be as effective as traditional classroom instruction, with the added benefits of flexibility and scalability (Means et al., 2013).

8.3.2.2 *Virtual and Augmented Reality Tools*

Immersive technologies are revolutionising sustainability training in real estate. These tools allow for experiential learning that can be more engaging and effective than traditional methods, including:

- *Virtual Green Building Tours*: Allowing learners to explore sustainable design features in a 3D environment
- *AR-Enhanced Maintenance Training*: Using augmented reality to provide real-time guidance on maintaining green building systems
- *VR-Based Energy Efficiency Simulations*: Enabling users to visualise and interact with building energy flows

Wang et al. (2018) documented the effectiveness of VR and AR in enhancing learning outcomes in fields like engineering and architecture.

8.3.2.3 *PropTech Start-Up Accelerators and Incubators*

Accelerators and incubators are crucial in nurturing innovation and knowledge sharing in the technology space, including PropTech. They provide mentorship, funding, and networking opportunities for start-ups working on sustainability-focused real estate technologies. Cohen et al. (2019) highlighted that accelerators can significantly improve start-up success rates and speed up innovation cycles. There are some notable examples around the world, such as MetaProp NYC,[4] which is based in the United States; Pi Labs,[5] which is focusing in Europe; and Colliers PropTech Accelerator,[6] which was launched in Canada.

All three accelerators have played significant roles in nurturing PropTech start-ups, but they each have unique characteristics that may appeal to different types of start-ups or focus on different aspects of the PropTech ecosystem:

- *Program Duration*: MetaProp has the longest program at 22 weeks, Colliers runs for 13 weeks, while Pi Labs' duration is not specified in the new information.
- *Investment Amount*: MetaProp offers the highest known investment at up to USD 250,000. Investment amounts for Pi Labs and Colliers are not specified in the new information.
- *Focus*: MetaProp emphasizes its industry connections and mentor network, Pi Labs stresses its global community and role in bridging tech and real estate, while Colliers highlights its partnership with Techstars.
- *Ecosystem*: Pi Labs particularly emphasizes its ecosystem and role in facilitating change in the built environment, which is not as prominently featured in the information about the other accelerators.

MetaProp NYC is known as "the world's premier PropTech accelerator program." It features:

- Access to free office space in midtown Manhattan at PropTech Place innovation hub;
- Participation in exclusive Pilot Days and Demo Days;
- Pairing with C-level mentors from MetaProp's RE200 industry mentor network;
- Access to industry decision-makers, technology pilots, and investment capital.

Pi Labs is famous for its unique ecosystem and relationships across the built environment. It builds a global community of change-makers and acts as a bridge between the tech and real estate worlds.

Colliers PropTech Accelerator is a partnership between a major real estate company (Colliers) and a leading start-up accelerator (Techstars). Therefore, it benefits from both sides, although it ran in 2019 only.

These educational and awareness initiatives are critical in bridging the knowledge gap and creating a workforce prepared to leverage PropTech for sustainability goals. PropTech is providing tools and shaping real estate professionals' mindsets towards a more sustainable and technologically advanced future by fostering a culture of continuous learning and innovation.

8.3.3 Disruptive Technologies Reshape Real Estate

The basis of the cultural shifts and educational changes is the disruptive technology in the real estate industry. Several emerging technologies have the potential to fundamentally disrupt and transform real estate processes across the value chain (Figure 8.4).

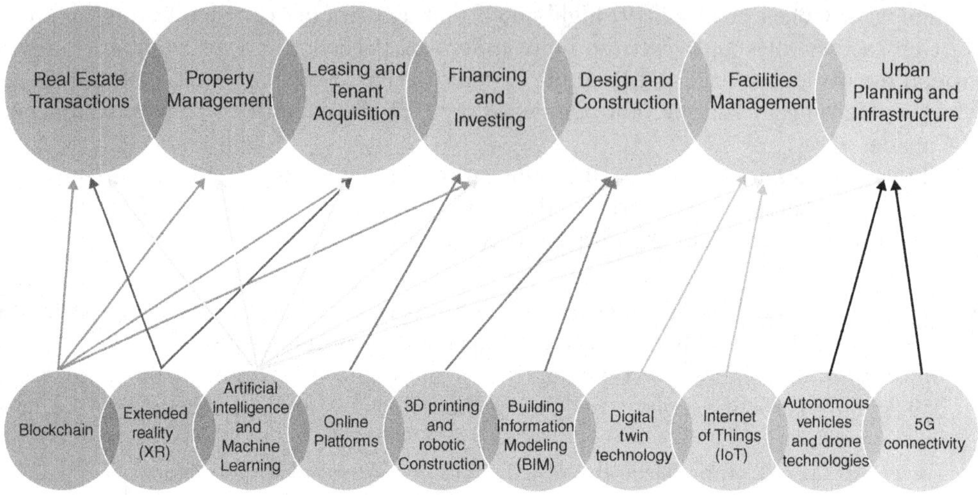

Figure 8.4 Disruptive Technologies Reshape Real Estate Value Chain.
By author (2024).

8.3.3.1 Real Estate Transactions

Blockchain and tokenisation enable fractional ownership, increased liquidity, and more efficient, secure asset transfers. Extended reality (XR) transforms how properties are visualised and marketed through virtual tours and augmented reality experiences. Artificial intelligence enhances automated valuation models and personalised recommendations for property searches.

8.3.3.2 Property Management

IoT and 5G connectivity enable smarter, more responsive built environments with real-time monitoring and control. Artificial intelligence and machine learning power predictive maintenance systems and optimise building operations. Blockchain improves lease management and automated payment systems.

8.3.3.3 Leasing and Tenant Acquisition

Artificial intelligence powers matching algorithms to connect tenants with ideal properties. Extended reality (XR) facilitates virtual property viewings and interactive space planning. Blockchain enables smart contracts for more efficient and transparent lease agreements.

8.3.3.4 Financing and Investing

Blockchain and tokenisation democratise real estate investment through fractional ownership and increased liquidity. Artificial intelligence enhances predictive analytics for investment decisions and risk assessment. Online platforms open up new avenues for real estate investment to a broader audience.

8.3.3.5 *Design and Construction*

3D printing and robotic construction technologies revolutionise how buildings are designed and constructed, potentially reducing costs and timeframes. BIM improves collaboration and efficiency in the design and construction process. Artificial intelligence powers generative design software to optimise building layouts and energy efficiency.

8.3.3.6 *Facilities Management*

Digital twin technology creates virtual replicas of buildings for improved monitoring and simulation of operations. The IoT builds comprehensive building management systems for energy optimisation and occupant comfort. Artificial intelligence provides predictive maintenance and intelligent resource allocation.

8.3.3.7 *Urban Planning and Infrastructure*

Autonomous vehicles and drone technologies will reshape transportation infrastructure and last-mile logistics, potentially altering property values and urban design. 5G connectivity will support smart city initiatives and improved urban services. Artificial intelligence will enhance urban planning decisions through data analytics and simulations.

As these PropTech solutions mature and gain wider adoption, they have the potential to make real estate processes more efficient, transparent, and accessible while supporting sustainability goals. However, successful integration will require continued education, evolving regulations, and cultural adaptation within the industry.

8.4 Sustainable Real Estate Through PropTech

PropTech is increasingly important in driving sustainability in the real estate sector. This section explores how PropTech enables more environmentally adaptive building practices, addresses affordable housing challenges, promotes socially inclusive developments, and shapes impact investing trends.

8.4.1 *Environmentally Adaptive Building Practices*

PropTech is revolutionising how buildings are designed, constructed, and operated to become more resilient to climate change and environmental challenges. PropTech solutions can help real estate become more resilient to climate change in several ways (Figure 8.5).

8.4.1.1 *Climate Resilience*

AI-powered climate modelling tools like Jupiter Intelligence[7] and ClimateAI[8] provide hyperlocal, long-term climate risk assessments. IoT-enabled flood sensors and predictive analytics platforms like FloodMapp[9] help buildings and cities prepare for and respond to flood events. PropTech solutions like Sensible Building Science[10] use AI to optimize HVAC systems based on weather forecasts and occupancy patterns. Drone technology and AI-powered image analysis, like those offered by Chooch AI,[11] can detect early signs of wildfires.

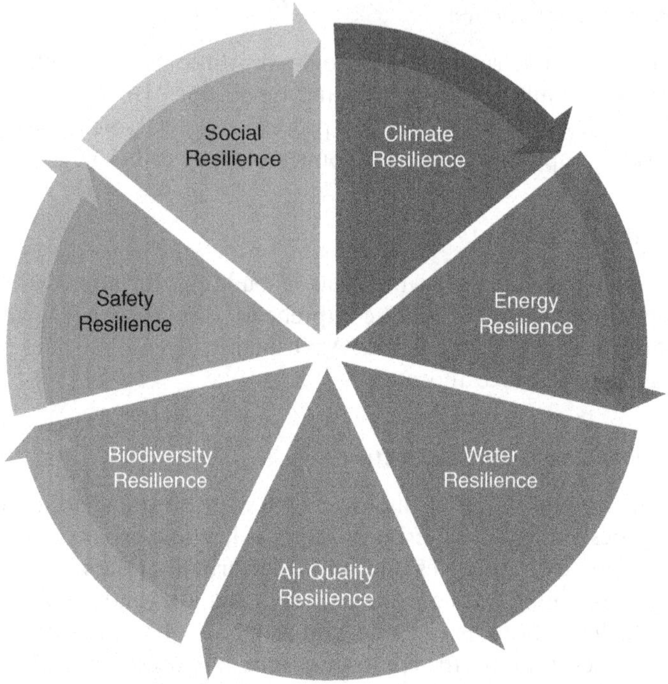

Figure 8.5 Increased Environmental Resilience Through PropTech.
By author (2024).

8.4.1.2 Energy Resilience

Companies like Schneider Electric[12] offer microgrid solutions that allow buildings to operate independently from the main power grid during outages. AI-optimized energy storage systems, such as those provided by Stem Inc.,[13] help buildings manage peak energy demands. IoT sensors and AI analytics from companies like Augury[14] detect potential failures in energy systems before they occur.

8.4.1.3 Water Resilience

IoT-enabled systems can use real-time data to optimise building water use and detect leaks instantly. Advanced water treatment systems and IoT monitoring enable efficient greywater recycling for non-potable uses in buildings.

8.4.1.4 Air Quality Resilience

AI-controlled ventilation systems adjust airflow based on outdoor air quality, occupancy, and indoor pollutant levels. IoT sensors from companies like Awair[15] provide real-time indoor air quality monitoring.

8.4.1.5 Biodiversity Resilience

Advanced sensor networks are being deployed on green roofs to monitor various parameters crucial for plant health and biodiversity (Gonsalves et al., 2022). The Green Roof Monitoring System uses a network of wireless sensors to continuously monitor these parameters, providing real-time data on green roof performance. The advantages of these technologies, such as the reduction of stormwater runoff, the reduction of cooling loads, and the reduction of the urban heat island, were assessed and confirmed by Bass et al. (2003). The company Sempergreen[16] has developed a smart blue-green roof system that uses sensor data to optimise water retention and release, supporting both biodiversity and stormwater management goals.

8.4.1.6 Safety Resilience

IoT sensors and AI analytics from companies like Worldsensing[17] continuously monitor building structural health. Cyber safety is now becoming more and more critical. Companies like Verkada[18] offer secure IoT platforms for buildings, protecting smart systems from cyber-attacks.

8.4.1.7 Social Resilience

As we introduced in Chapter 5, tools like coUrbanize[19] facilitate community input in development projects, fostering social cohesion and resilience.

PropTech also shapes the environmentally adaptive building designs. AI-powered tools like Autodesk's Revit[20] can generate multiple design options optimised for energy efficiency, natural light, and material use. Companies like Willow[21] create digital replicas of buildings, allowing for real-time monitoring and optimisation of environmental performance. PropTech firms like View Inc.[22] offer smart glass solutions that automatically adjust tint based on sunlight and temperature, optimising energy use and occupant comfort. PropTech solutions like Naava's smart green walls[23] incorporate nature into buildings, improving air quality and occupant well-being. By leveraging these technologies, the real estate industry can create buildings that are not only more resilient to climate change but also actively contribute to environmental sustainability.

8.4.2 Affordable Housing Solutions

Housing affordability has been a global issue for centuries. The real estate industry is using innovative technologies to reduce costs, increase efficiency, and improve access to affordable housing (Ferreri & Sanyal, 2022; Babcock, 2019). Figure 8.6 shows how PropTech contributes to affordable housing solutions.

8.4.2.1 Modular and Prefabricated Construction

PropTech is revolutionising construction methods to make housing more affordable. BIM tools enable the precise design of modular components, reducing waste and improving efficiency (Abanda et al., 2017). Companies like Katerra use robotics and automation in factory settings to produce building components more efficiently and at a lower cost

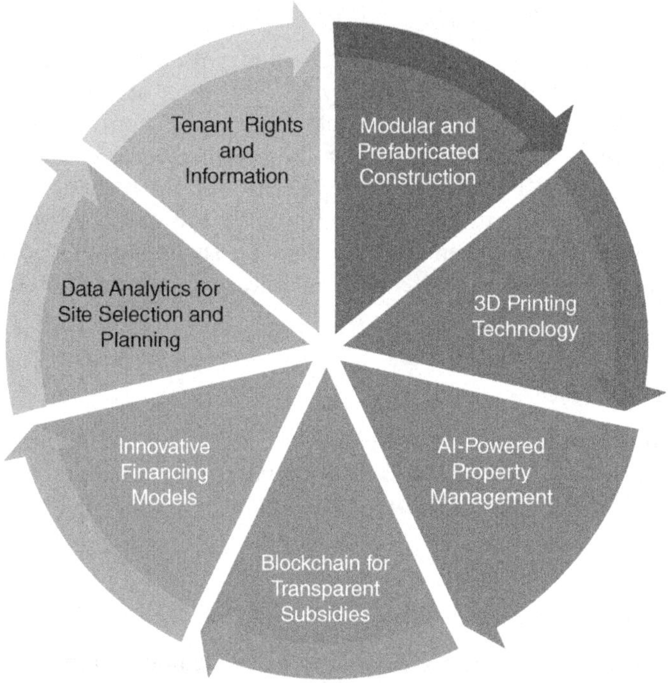

Figure 8.6 PropTech Solutions for Housing Affordability.
By author (2024).

(Barbosa et al., 2017). Drone technology and AI-powered robots assist in the on-site assembly of prefabricated components, reducing labour costs and construction time (Bogue, 2018). Module[24] offers an innovative approach to modular construction. Their homes are designed to be expandable, allowing buyers to start small and add on as their needs and finances change.

8.4.2.2 3D Printing Technology

3D printing is emerging as a game-changer for affordable housing. Companies like ICON[25] are using large-scale 3D printers to construct entire houses quickly and affordably (Hager et al., 2016). 3D printing of specific building components, such as walls or fixtures, can reduce costs and customise designs (Buchanan & Gardner, 2019). Research into new 3D-printable materials, including recycled plastics and sustainable composites, is making 3D-printed housing more environmentally friendly (Perrot et al., 2016).

8.4.2.3 AI-Powered Property Management

Operational costs in property management can be reduced by using AI-powered PropTech solutions. This is beneficial for affordable housing. AI algorithms analyse data from IoT sensors to predict maintenance needs, reducing repair costs and extending the lifespan of affordable housing units (Fraga-Lamas et al., 2016). Smart building systems use AI to optimise energy use, reducing utility costs for residents (Zhou et al., 2016). AI-powered platforms use machine learning to improve tenant screening processes, potentially reducing risks and costs for affordable housing providers.

8.4.2.4 *Blockchain for Transparent Subsidies*

Affordable housing is part of the social subsidies in many societies. Blockchain technology can improve the efficiency and transparency of housing subsidies. Blockchain-based smart contracts can automate the distribution of housing subsidies, reducing administrative costs and preventing fraud (Kshetri &Voas, 2018). As discussed in Chapters 4 and 5, real estate tokenisation platforms are exploring ways to make fractional property ownership more accessible, potentially opening up new models for affordable housing investment. Blockchain can provide a transparent, immutable record of how housing subsidies are allocated and used, improving accountability (Jaffer et al., 2020).

8.4.2.5 *Innovative Financing Models*

PropTech platforms are opening up new financing models for affordable housing. They offer homeowners the ability to sell fractional interests in their homes in exchange for a share in the property's future appreciation. This can provide homeowners with financial flexibility without taking on additional debt (Babcock, 2019). Platforms like Small Change[26] focus on crowdfunding for impactful urban development projects, including affordable housing. P2P lending platforms such as Groundfloor,[27] specifically for real estate, can provide alternative financing options for affordable housing developers.

8.4.2.6 *Data Analytics for Site Selection and Planning*

AI-powered geospatial analysis tools help identify optimal locations for affordable housing developments, considering factors like proximity to public transport, schools, and employment centres. Machine learning models analyse demographic and economic data to forecast affordable housing demand, helping planners make informed decisions (Nanda, 2019). PropTech tools such as Urban Footprint[28] help planners analyse the impact of zoning changes and policy decisions on affordable housing supply.

8.4.2.7 *Tenant Rights and Information*

PropTech is also being used to empower tenants and improve transparency in the rental market. Platforms can use publicly available data to generate grades for apartment buildings, giving tenants more information and bargaining power (Babcock, 2019). Companies like PadSplit[29] use technology to facilitate affordable co-living arrangements, particularly for low-income workers.

The real estate industry is finding innovative ways to address the affordable housing crisis. These technologies help reduce the cost of housing construction and management and create new models for financing and developing affordable housing projects.

8.4.3 *Socially Inclusive Developments*

As presented in Chapter 5, PropTech is creating more socially inclusive real estate developments. Tools like coUrbanize facilitate community input in development projects, ensuring diverse voices are heard in the planning process. PropTech solutions like AccessMap[30] provide detailed accessibility information for urban environments, promoting inclusive design. Platforms like Autodesk Forma[31] (previously known as Spacemaker[32]) use AI to optimise mixed-use developments, creating more diverse and inclusive urban spaces. Companies like Sidewalk Labs[33] integrate various technologies to create more inclusive and accessible urban environments. Platforms like Fractional[34] democratise real estate investment, allowing a more diverse range of individuals to participate in property ownership. With the help of these technologies, developers can create more inclusive spaces that cater to diverse populations and promote social cohesion.

8.4.4 *Impact Investing Trends*

The intersection of PropTech and impact investing is creating new opportunities for addressing housing affordability and sustainability challenges. Impact investing in real estate aims to generate measurable social and environmental benefits alongside financial returns. PropTech is playing a crucial role in facilitating and scaling these investments.

8.4.4.1 *ESG-Focused Real Estate Investments*

PropTech companies are developing sophisticated analytics tools to measure and report on ESG factors in real estate investments (Clayton et al., 2021). Artificial Intelligence can analyse vast amounts of data to assess the ESG performance of properties and portfolios (Gradillas et al., 2021; Walacik & Chmielewska, 2024). Blockchain technology can create immutable records of ESG credentials, enhancing transparency and trust in impact investments (Centobelli et al., 2022).

8.4.4.2 *Tokenisation of Impact Real Estate Assets*

Blockchain-based tokenisation platforms are making it possible to divide real estate assets into smaller, more accessible investment units (Wouda & Opdenakker, 2019). Tokenisation can potentially increase the liquidity of real estate impact investments (Veuger, 2018). Smart contracts associated with tokenized real estate can automatically track and report on impact metrics (Nijland, 2019).

8.4.4.3 *Green Bonds and Sustainability-Linked Loans*

Digital platforms are streamlining the issuance and trading of green bonds specifically for sustainable real estate projects (Lai, 2021). Machine learning algorithms help to assess the risk and potential impact of green bonds and sustainability-linked loans (Sembiyeva et al., 2024). IoT devices integrated into buildings can provide real-time data on energy efficiency and other sustainability metrics (Molina-Solana et al., 2017).

8.4.4.4 Crowdfunding for Social Impact Real Estate

PropTech platforms enable small-scale investors to participate in social impact real estate projects through crowdfunding models (Schweizer & Zhou, 2016). Again, AI and big data analytics can identify and evaluate potential social impact real estate projects (Himeur et al., 2023).

8.4.4.5 AI-Powered Impact Measurement and Reporting

PropTech solutions automate the collection and analysis of impact data (Jagannathan et al., 2017), and machine learning models are being developed to predict the potential impact of real estate investments (Christopoulos et al., 2017).

8.4.4.6 Sustainable Urban Development Platforms

Advanced digital twin technology is being used to model and simulate the impact of urban development projects (Shahat et al., 2021). PropTech solutions are increasingly integrated with smart city initiatives (Allam & Jones, 2021).

PropTech is a powerful enabler of sustainable real estate practices. Technology is reshaping the real estate industry to meet pressing environmental and social challenges, from enhancing climate resilience and promoting affordable housing to fostering social inclusion and facilitating impact investing. As these technologies continue to evolve and integrate, they promise to drive significant positive change in building, inhabiting, and investing in our built environment.

8.5 Global Adoption and Regulatory Support

The global adoption of PropTech in making real estate sustainable is accelerating, driven by the need for more efficient, sustainable, and technologically advanced built environments. However, the pace and nature of this adoption vary significantly across different regions, influenced by factors such as technological infrastructure, market maturity, and regulatory environments. This section examines the global landscape of PropTech adoption and the efforts to create supportive regulatory frameworks.

8.5.1 Global Adoption Trends

The adoption of PropTech varies across regions, influenced by factors such as technological infrastructure, market maturity, investment climate, and regulatory environments. Figure 8.7 shows the trends in key regions worldwide.

8.5.1.1 North America

North America, particularly the United States, continues to lead in PropTech innovation and adoption. The United States accounts for 58% of all PropTech companies globally, with a total of 1,287 companies. The United States attracts 64% of all investors in the PropTech sector, with 3,030 investors. Between 2000 and 2022, the U.S. PropTech sector raised USD 67.32 billion, accounting for 62% of all investments made in PropTech

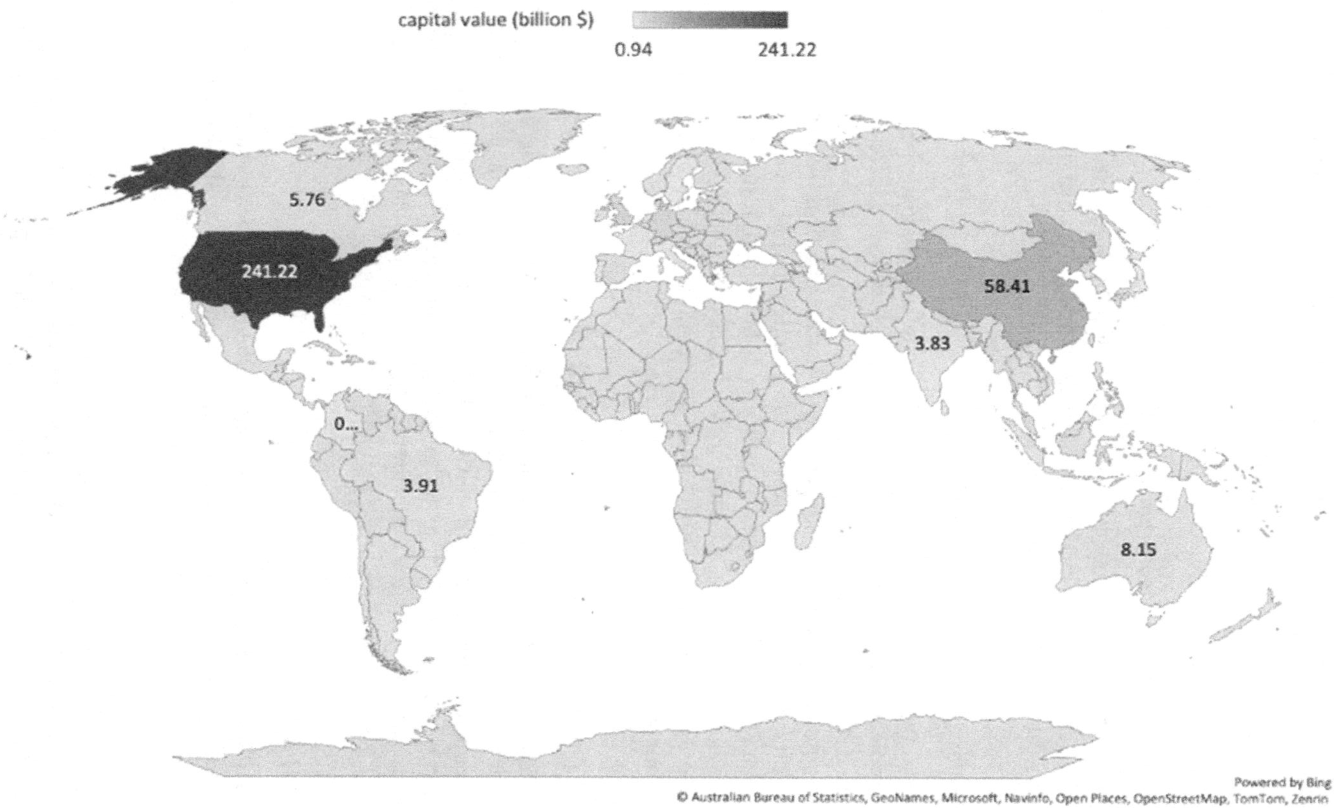

Figure 8.7 Value of Proptech Capital Raised Worldwide from 2012 to 2023, by Country (in Billion USD).
By author (2024), source: ESCP Europe (2023).

companies globally. Key PropTech hubs include California (36% of U.S. PropTech companies), New York (16.4%), Texas (6.1%), Washington (4.4%), and Massachusetts (4%) (Luque, 2022).

8.5.1.2 Europe

Europe has seen significant PropTech growth, with the UK, Germany, and France leading the way. The UK is the second-largest PropTech market globally after the United States. Germany, France, and Spain have become significant players in recent years. PropTech adoption in Europe is driven by the EU's push for sustainable development and smart cities, particularly in areas related to energy efficiency and green building technologies. Notable European PropTech companies include Colonies[35] in France (raised USD 213 million) and Deepki,[36] focusing on ESG strategy planning.

8.5.1.3 Asia-Pacific

The Asia-Pacific region presents a diverse landscape of PropTech adoption. India is emerging as a significant PropTech market, expecting to reach USD 1 trillion by 2030, up from USD 120 billion in 2017. China is a major player, with 75% of PropTech companies concentrated in Beijing, Shanghai, Shenzhen, and Hangzhou. South Korea has seen rapid growth, with companies like Yanolja[37] valued at USD 6.7 billion in 2021. Southeast Asian countries like Singapore and Indonesia are becoming PropTech hubs, with companies like 99 Group[38] attracting significant investments (Luque, 2023).

8.5.1.4 Middle East and Africa

According to Luque (2023), PropTech adoption in these regions is still in the early stages compared to other regions. Israel has seen significant growth, with PropTech companies increasing by 800% between 2018 and 2022, reaching almost 100 companies. Countries such as Saudi Arabia, Turkey, and the United Arab Emirates are exhibiting increased PropTech activity.

Globally, PropTech continues to attract significant investment despite the economic slowdown. As of the 2022 fiscal year, 2,209 PropTech firms were operating globally in 67 countries (Luque, 2022). The total investment in PropTech for the 2021–2022 fiscal year reached USD 20.41 billion, representing a 17% year-over-year increase. The Managing and Living categories of PropTech companies attract the most investment and have the highest number of companies globally.

PropTech solutions addressing sustainability and energy efficiency are increasingly adopted across all regions, driven by global climate concerns and regulations. The global shift towards remote work has accelerated the adoption of PropTech solutions for flexible workspace management and virtual property tours. The integration of AI, blockchain, and IoT in PropTech solutions is increasing across all regions, with varying degrees of adoption based on local technological infrastructure. Many successful PropTech companies are expanding globally. For example, VTS,[39] a leasing and asset management platform, has expanded from the United States to Europe and Asia.

These trends indicate a global movement towards increased technology integration in real estate. The pace of adoption is likely to accelerate as more countries recognise PropTech's potential to address urban challenges and promote sustainability.

8.5.2 *Integration Challenges and Opportunities*

The global integration of PropTech is a complex and multi-faceted process, shaped by diverse technological landscapes, varying market maturities, and distinct cultural contexts across different regions. As the PropTech sector grows, with investments reaching USD 20.41 billion in the 2021–2022 fiscal year (Luque, 2022), it faces challenges and opportunities. These range from infrastructural disparities and regulatory hurdles to the immense potential for cross-border collaboration and sustainable development.

As shown in Figure 8.8, the global integration of PropTech is facing five of the biggest challenges.

While the United States hosts 58% of all PropTech companies globally, emerging markets lag. For instance, the entire continent of Africa accounts for only a small fraction of global PropTech activity. In China, 75% of PropTech companies are concentrated in just four big cities, indicating infrastructure disparities even within a country.

According to Sophie Chateau from LBO France, "Most of the entrepreneurs [in PropTech] do not come from Real Estate." This knowledge gap necessitates significant

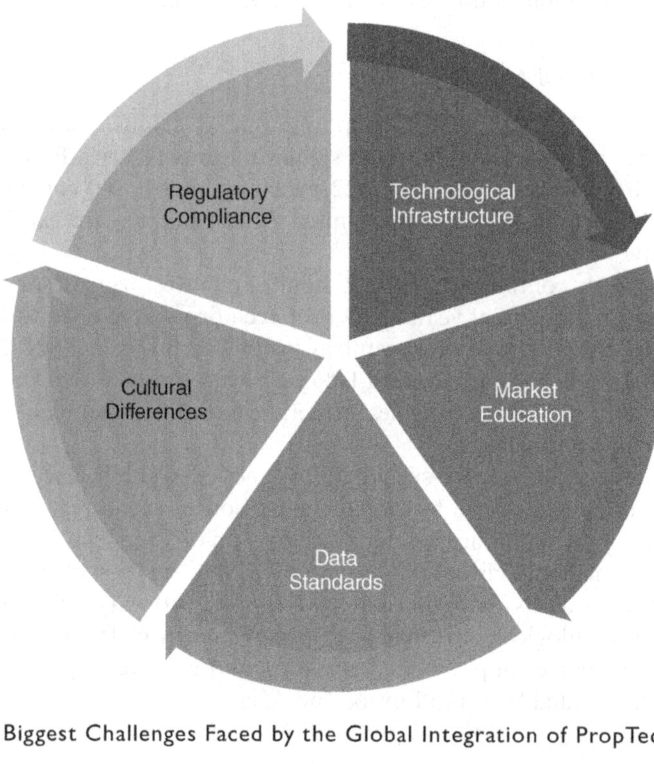

Figure 8.8 Five Biggest Challenges Faced by the Global Integration of PropTech.
By author (2024).

educational efforts (Luque, 2022). Traditional real estate professionals often lack understanding of new technologies, creating resistance to adoption.

The lack of unified data standards is evident in the disparity of PropTech categories across regions. For example, the "Managing" category accounts for 35.9% of U.S. PropTech companies but has different proportions in other markets. The adoption of shared economy models in real estate varies greatly. While companies like Airbnb have seen global success, similar models face regulatory challenges in countries like China. The regulatory pushback has already affected companies like Airbnb in cities like Barcelona and Berlin (Nieuwland & Van Melik, 2020).

While facing numerous challenges, PropTech also presents numerous opportunities for innovation, growth, and transformation in the real estate industry. These opportunities arise from the sector's rapid expansion, technological advancements, and the increasing demand for efficient, sustainable real estate solutions worldwide.

8.5.2.1 *Cross-Border Collaboration*

The global nature of PropTech is fostering unprecedented levels of cross-border collaboration, driving innovation and standardisation across the real estate industry. This trend is evident in the increasing global reach of PropTech, with the number of countries hosting PropTech companies rising from 64 in 2020 to 67 in 2022 (Luque, 2022). This expansion facilitates sharing best practices and technologies across different markets, leading to a more interconnected and efficient global real estate ecosystem. European companies are at the forefront of this international expansion. For instance, Colonies, a French PropTech firm, has raised significant funds (USD 213 million) with plans for international growth. This cross-border activity is about expansion, creating synergies, and learning from diverse market experiences. It's paving the way for international standards for property data and transactions, which could significantly streamline global real estate operations.

8.5.2.2 *Emerging Markets Focus*

Emerging markets represent a frontier of immense opportunity for PropTech, driven by rapid urbanisation and the potential for technological leapfrogging. India stands out as a prime example, with its PropTech market expected to grow from USD 120 billion in 2017 to an astounding USD 1 trillion by 2030 (Luque, 2023). This growth is fuelled by the country's large population, increasing urban migration, and the government's push for digital transformation. Southeast Asian countries are also becoming significant PropTech hubs. Singapore and Indonesia, in particular, are seeing a surge in PropTech activity. This influx of capital enables these companies to develop innovative solutions tailored to local needs and constraints. The rapid adoption of mobile-first PropTech solutions is a key characteristic of these emerging markets. With many consumers bypassing traditional computing devices and going straight to smartphones, PropTech companies are developing mobile-centric platforms for property search, management, and transactions.

8.5.2.3 *Sustainable Development Goals*

The industry is increasingly using technology to create more environmentally friendly buildings and cities, with promising results. This significant reduction highlights PropTech's power in addressing climate change challenges. Sustainable development opportunities in PropTech extend beyond reducing emissions. Innovative tools are being developed for measuring and reducing buildings' carbon footprints throughout their lifecycle. These technologies provide real-time data on energy consumption, water usage, and waste production, allowing property managers and owners to make informed decisions to improve their buildings' environmental performance.

Moreover, PropTech is revolutionising the retrofitting of existing buildings to improve energy efficiency. Given that a large portion of the building stock in many countries was constructed before modern energy efficiency standards, this represents a significant opportunity for reducing overall carbon emissions. Technologies such as smart HVAC systems, LED lighting with occupancy sensors, and advanced building envelope solutions are being integrated into older structures, dramatically improving their energy performance.

Another key area where PropTech is making strides is developing and implementing platforms for green building certification and compliance. These digital platforms streamline obtaining and maintaining green building certifications like LEED, BREEAM, or WELL, making it easier for property developers and owners to demonstrate their commitment to sustainability. These platforms often incorporate AI and machine learning to recommend improvements to a building's sustainability performance, ensuring ongoing optimisation.

The impact of these PropTech solutions extends beyond individual buildings to entire urban ecosystems. Smart city technologies, powered by IoT and AI, enable more efficient resource allocation and urban planning, contributing to more sustainable urban environments. From smart grids that optimise energy distribution to waste management systems that improve recycling rates, PropTech is helping cities reduce their environmental impact while improving the quality of life for residents.

As regulations around building energy performance and carbon emissions become more stringent globally, PropTech solutions for sustainability are likely to see increased adoption. This trend contributes to global efforts to combat climate change and presents significant business opportunities in the growing market for sustainable real estate technologies.

8.5.2.4 Emerging Technologies

The rapid advancement of technology is creating numerous opportunities in the PropTech sector, with several emerging technologies revolutionising various aspects of the real estate industry. At the forefront of this transformation is the increasing importance of data-driven decision-making. Data analytics' growing availability and sophistication in real estate enable more accurate property valuations, improved risk assessments, and optimised urban planning strategies. This trend is reflected in recent industry statistics, with the Managing category, which includes data analytics for property management, accounting for 32% of U.S. PropTech companies in 2021–2022, up from 28% in the previous year (Luque, 2022).

Blockchain and distributed ledger technology present another significant opportunity: They have the potential to revolutionise real estate transactions. By providing immutable digital records, blockchain can streamline property transactions and reduce fraud. It also enables fractional ownership of real estate assets, potentially increasing market liquidity and accessibility. Furthermore, blockchain facilitates the implementation of smart contracts for leases and property management, adding another layer of efficiency and transparency to real estate operations.

AI and machine learning are set to transform real estate decision-making processes. These technologies offer enhanced predictive analytics for property valuation and investment decisions, improved property management through AI-driven maintenance scheduling and energy optimisation, and personalised property search and recommendations for buyers and renters. The application of AI in real estate is expected to lead to more informed decisions and improved operational efficiency across the industry.

The IoT and smart building technologies create more sustainable and efficient real estate opportunities. IoT integration allows for the creation of energy-efficient buildings, enhances occupant comfort and productivity through smart environmental controls, and enables predictive maintenance to reduce downtime and operational costs. As sustainability becomes an increasingly important consideration in real estate, these technologies are likely to play a crucial role in the future of building design and management.

Finally, VR/AR technologies are opening up new possibilities for property marketing and management. VR enables virtual property tours and remote inspections, potentially revolutionising how properties are viewed and assessed. AR can provide immersive architectural visualisations for off-plan properties, enhancing the pre-construction sales process. Additionally, these technologies offer new avenues for facilities management and maintenance training, improving operational efficiency and reducing costs.

The convergence of these emerging technologies is creating a new paradigm in real estate, where data-driven insights, automation, and immersive experiences are reshaping traditional processes. As these technologies mature and become more widely adopted, they have the potential to address long-standing challenges in the industry, from improving transaction efficiency to enhancing sustainability and user experience. The PropTech sector stands at the edge of these technological advancements, driving significant innovation and transformation in the real estate industry.

8.5.3 Regulatory Frameworks and Standardisation Efforts

Standardised regulatory frameworks are required to ensure responsible innovation and adoption. The challenge is to create flexible regulations that accommodate rapid technological advancements while providing adequate protection for consumers and maintaining market stability.

8.5.3.1 Efforts Towards Global Standards

Some international organisations and initiatives (in Figure 8.9) have worked towards setting global standards for the development of PropTech.

The International Organization for Standardization (ISO) has been working on developing standards relevant to smart cities and buildings that intersect with PropTech. The Royal Institution of Chartered Surveyors (RICS) has been developing global guidelines for adopting PropTech in various aspects of real estate. The WEF has launched initiatives to promote responsible development and use of PropTech globally.

8.5.3.2 Regulatory Harmonisation Challenges

The global adoption of PropTech faces significant challenges due to the diverse regulatory landscapes across different regions. One of the primary obstacles is the varying approach

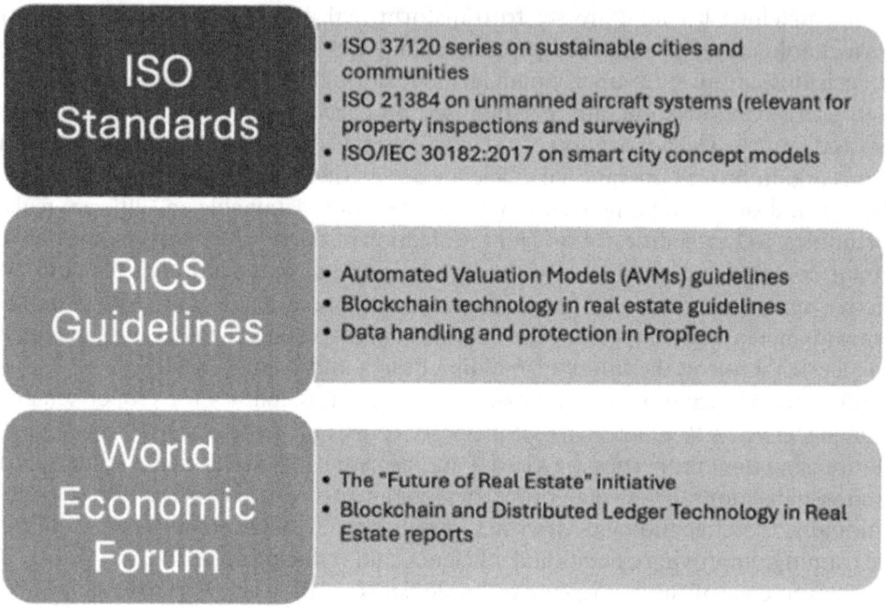

Figure 8.9 International Organisations and Initiatives' Efforts Towards Global Standards. By author (2024).

to data privacy and security. The European Union leads with its comprehensive General Data Protection Regulation (GDPR), setting a high standard for data protection. In contrast, the United States lacks a federal equivalent, relying instead on a patchwork of state laws, such as the California Consumer Privacy Act (CCPA). Meanwhile, China has implemented its Personal Information Protection Law (PIPL). These disparate regulations create complexities for PropTech companies aiming to operate on a global scale, as they must navigate and comply with multiple, often conflicting, data protection regimes.

The diversity of property laws across nations presents another significant hurdle for PropTech solutions, particularly in property transactions and management. The fundamental differences between common law and civil law systems result in varied property rights and transaction approaches. Furthermore, concepts such as leasehold and freehold may have different interpretations or may not exist in some jurisdictions. Property registration systems also differ widely, from highly digitised and centralised systems in some countries to fragmented, paper-based systems in others. These variations make it challenging to develop universally applicable PropTech solutions for property transactions and management.

Building codes and sustainability standards present a third major challenge to global PropTech adoption. In the United States, while the International Building Code (IBC) provides a baseline, individual states often have variations, leading to a complex patchwork of regulations. The European Union uses Eurocodes, but these, too, are subject to national annexes that can introduce significant differences between member states. China operates with its own national GB standards, which can differ substantially from Western norms. These variations in building codes and sustainability standards can significantly impact the applicability and effectiveness of PropTech solutions across different markets, and particularly those focused on construction, building management, and energy efficiency.

8.5.3.3 Emerging Regulatory Approaches

As the PropTech industry develops, regulators are adopting innovative strategies to encourage progress while ensuring proper oversight. One such strategy is the use of regulatory sandboxes, which offer controlled environments for testing new PropTech solutions. The United Kingdom leads in this area with the Financial Conduct Authority's (FCA) regulatory sandbox, enabling firms to trial innovative products and services in a live market setting. Likewise, Singapore's Monetary Authority has launched a FinTech Regulatory Sandbox, allowing experimentation with financial technologies that could benefit the property sector. Australia has also taken similar steps, with the Australian Securities and Investments Commission (ASIC) providing a regulatory sandbox for FinTech, including PropTech innovations. These sandboxes enable regulators to observe new technologies in action and develop suitable regulatory frameworks without hindering innovation.

Another emerging trend is the formation of public–private partnerships to shape PropTech regulations. Governments are increasingly recognising the value of collaborating with industry experts to develop regulatory frameworks that are both effective and practical. In the United States, the Department of Housing and Urban Development has established an Innovation Lab to explore and support new technologies in the housing sector. Singapore has adopted a comprehensive approach with its Real Estate Industry Transformation Map, a collaborative effort between the government and private sector to drive innovation and productivity in the real estate industry. The United Arab Emirates, particularly Dubai, has proactively partnered with PropTech firms to modernise its land department operations and regulations. These partnerships enable the development of regulations informed by real-world industry needs and technological capabilities.

Recognising the global nature of PropTech, there are growing efforts towards international cooperation in developing harmonised regulations. This is particularly evident in data standards and cybersecurity, where consistency across borders is crucial. The EU–US Privacy Shield, although currently invalidated, represented an attempt to reconcile different approaches to data protection. The Asia-Pacific Economic Cooperation (APEC) has developed the Cross-Border Privacy Rules System to create consistent data privacy standards across its member economies. On a broader scale, the G20 AI Principles provide a framework for the responsible development of artificial intelligence, which has significant implications for PropTech. These international efforts reflect a growing recognition that the challenges posed by PropTech often transcend national borders and require coordinated global responses.

When examining regulatory approaches across different regions, several distinct patterns emerge in how various jurisdictions address the challenges and opportunities PropTech presents. In data protection, the European Union stands out as a global leader with its comprehensive GDPR. This far-reaching legislation has established a new standard for data privacy and security, influencing regulations worldwide. In contrast, the United States has adopted a more fragmented approach, with a patchwork of state-level laws rather than a unified federal framework. This disparity is exemplified by regulations like the CCPA. Meanwhile, Asian countries, including China and Singapore, are developing comprehensive data protection laws, often drawing inspiration from the GDPR while tailoring regulations to their specific national contexts.

As shown in the foregoing analysis, support for innovation in PropTech varies significantly across regions. The United Kingdom and Singapore have positioned themselves at the forefront of regulatory innovation by implementing regulatory sandboxes. These controlled environments enable the testing of new PropTech solutions under regulatory supervision, fostering innovation while mitigating risks. The United States, in keeping with its traditionally market-driven approach, tends to adopt a more hands-off stance, allowing market forces to largely drive innovation in the PropTech sector. China presents a unique case, blending strong government direction with market-driven innovation, resulting in a rapidly evolving PropTech landscape shaped by state priorities and entrepreneurial dynamism.

Standardisation efforts also differ markedly across regions. The European Union, emphasising harmonisation across member states, tends to adopt more standardised approaches to PropTech regulation. This consistency facilitates the development of PropTech solutions that can operate across multiple European markets. In contrast, the United States exhibits greater variation between states, reflecting its federal system and the significant autonomy granted to state governments in areas like property law and building regulations. Asian countries often develop national standards for PropTech, but these may diverge significantly from Western norms, reflecting different priorities and market structures.

Regarding PropTech-specific regulation, most countries are still adapting existing regulatory frameworks to accommodate PropTech innovations rather than creating entirely new, PropTech-specific laws. This approach allows for flexibility but can also lead to regulatory uncertainty. Singapore and the United Arab Emirates stand out as exceptions, having launched targeted PropTech initiatives that directly address the unique challenges and opportunities presented by these new technologies.

The regulation of blockchain technology and smart contracts in real estate transactions represents another area of divergence. Some jurisdictions, such as the state of Arizona in the United States, have taken proactive steps by passing laws that explicitly recognise smart contracts. The European Union is working towards a more comprehensive approach through the European Blockchain Partnership, aiming to create a harmonised framework for blockchain applications across member states. China has adopted a more controlled approach, banning cryptocurrency transactions while simultaneously promoting the development and application of blockchain technology in other areas, including real estate.

The global regulatory landscape for PropTech is still evolving, with different regions taking varied approaches based on their existing legal frameworks and policy priorities. As the industry matures, we expect to see more targeted PropTech regulations and increased efforts towards international harmonisation.

8.5.3.4 Outlook

The global adoption of PropTech in sustainable real estate is set for continued growth and evolution. As nations increasingly recognise PropTech's potential to address urban challenges and promote sustainability, we can anticipate a surge in investment and the development of more supportive regulatory frameworks. This trajectory will likely shape the future of real estate markets worldwide, driving innovation and efficiency across the industry.

One key area of future development is the creation of global data platforms. These international platforms will facilitate sharing of real estate data and best practices in PropTech, fostering greater transparency and efficiency in global real estate markets. Such platforms could revolutionise how property information is accessed and utilised, enabling more informed decision-making and promoting standardisation across borders.

Alongside this, we can expect the emergence of more cross-border PropTech solutions designed with global applicability in mind from their inception. These solutions will likely be built to navigate the complexities of different regulatory environments and market conditions, offering scalable technologies that can be deployed across multiple countries and regions. This trend could significantly accelerate the global adoption of PropTech and drive greater uniformity in real estate practices worldwide.

Progress towards harmonised international regulations for PropTech is another crucial area for future development. This is particularly pertinent in areas such as data protection and PropTech-enabled transactions. As the industry matures, efforts will likely be increased to create consistent regulatory frameworks that can accommodate technological innovation while ensuring adequate consumer protection and market stability across different jurisdictions.

Furthermore, we notice an increased global focus on PropTech solutions that directly address climate change and support sustainable development goals. As environmental concerns become increasingly pressing, PropTech will play a vital role in creating more sustainable built environments, from energy-efficient smart buildings to urban planning tools that promote greener cities.

As the PropTech ecosystem matures globally, collaboration will be key to its success. The interplay between governments, industry stakeholders, and technology innovators will create a supportive environment for responsible innovation and adoption. This collaborative approach will be essential in navigating the complex challenges of implementing PropTech solutions across diverse regulatory landscapes and market conditions.

8.6 Concluding Remarks

As we've explored throughout this chapter, PropTech plays a transformative role in shaping the future of sustainable real estate. Technology is becoming integral to the real estate industry's response to pressing environmental and social issues, from revolutionising building practices to addressing affordability challenges and fostering social inclusion. Our analysis includes cultural shifts, innovation and education, disruptive technologies, sustainability solutions, affordability and inclusion, global adoption, and regulatory evolution.

We can expect PropTech to continue its rapid evolution, with several key trends likely to shape its future. One of these is increased focus on ESG-driven solutions as sustainability becomes a top priority for investors and occupiers alike. Greater technology integration creates more comprehensive and powerful real estate management systems. Expansion of PropTech solutions to address urban-scale challenges will contribute to the development of smart, sustainable cities. There is growing emphasis on data standardisation and interoperability to facilitate more efficient real estate transactions and management globally.

However, realising PropTech's full potential in creating a more sustainable real estate industry will require ongoing collaboration between technology innovators, real estate professionals, policymakers, and academics. It will also necessitate a commitment to responsible innovation that balances the drive for efficiency and profitability with broader societal and environmental goals.

As we move forward, PropTech will undoubtedly play a crucial role in helping the real estate industry meet the complex challenges of the 21st century, from climate change to urbanisation and social equity. By embracing these technological innovations responsibly and strategically, the real estate sector can transform itself into a powerful force for sustainable development, creating more efficient, profitable, liveable, inclusive, and environmentally sound environments.

8.7 Post-Chapter Engagement

Activities are designed to encourage readers to actively engage with and apply the knowledge gained from this chapter.

1 Reflection Questions

- What do you think are the most significant ways PropTech is reshaping the real estate industry towards greater sustainability?
- Which emerging PropTech technologies or applications do you think have the most potential to transform real estate in the next 5–10 years? Why?
- What challenges do you foresee in the widespread adoption of PropTech solutions for sustainability? How might these be overcome?

2 Design Challenge

- Please try to conceptualize an innovative PropTech solution that could help address a specific sustainability challenge in real estate (e.g., reducing energy consumption in existing buildings). Outline the concept, technology involved, and potential impact.

Notes

1 https://realestateinnovationlab.mit.edu/.
2 https://www.rics.org/networking/rics-tech-partner-programme.
3 https://github.com/RICS-Data-Standard/RDS/releases.
4 https://www.metaprop.com/.
5 https://pilabs.vc/.
6 https://www.colliers.com/en-xe/news/20190929-colliers-proptech-accelerator-powered-by-techstars-welcomes-class-of-2019-startups.
7 https://www.jupiterintel.com/industry/real-estate.
8 https://climate.ai/.
9 https://www.floodmapp.com/.
10 https://sensiblebuildingscience.com/.
11 https://www.chooch.com/.
12 https://www.se.com/uk/en/.
13 https://www.stem.com/.
14 https://www.augury.com/.
15 https://uk.getawair.com/.
16 https://www.sempergreen.com/en.
17 https://www.worldsensing.com/.
18 https://www.verkada.com/uk/.
19 https://www.courbanize.com/.
20 https://www.autodesk.com/uk/products/revit/overview?term=1-YEAR&tab=subscription.
21 https://developers.willowinc.com/docs/resource-guide-willow-digital-twin/.
22 https://view.com/.
23 https://www.naava.io/en/.
24 https://www.modulehousing.com/.
25 https://www.iconbuild.com/.

26 https://www.smallchange.co/.
27 https://groundfloor.com/.
28 https://urbanfootprint.com/.
29 https://www.padsplit.com/.
30 https://www.accessmap.app/.
31 https://www.autodesk.com/products/forma/overview?term=1-YEAR&tab=subscription.
32 https://techcrunch.com/2020/11/17/spacemaker-ai-software-for-urban-development-is-acquired-by-autodesk-for-240m/.
33 https://www.google.com/earth/about/design/.
34 https://www.fractional.app/.
35 https://www.livecolonies.com/en/paris.
36 https://www.deepki.com/.
37 https://www.yanoljagroup.com/.
38 https://www.99.co/about-us.
39 https://www.vts.com/.

References

Abanda, F., Tah, J. & Cheung, F. 2017. BIM in off-site manufacturing for buildings. *Journal of building engineering*, 14, 89–102.

Aguilar, J., Garces-Jimenez, A., R-Moreno, M. D., & García, R. 2021. A systematic literature review on the use of artificial intelligence in energy self-management in smart buildings. *Renewable and Sustainable Energy Reviews*, 151, 111530.

Ahmad, M. W., Mourshed, M., Mundow, D., Sisinni, M., & Rezgui, Y. 2016. Building energy metering and environmental monitoring: A state-of-the-art review and directions for future research. *Energy and Buildings*, 120, 85–102.

Ahmad, T., Zhang, D., Huang, C., Zhang, H., Dai, N., Song, Y., & Chen, H. 2021. Artificial intelligence in sustainable energy industry: Status quo, challenges and opportunities. *Journal of Cleaner Production*, 289, 125834.

Alaa, M., Zaidan, A. A., Zaidan, B. B., Talal, M., & Kiah, M. L. M. 2017. A review of smart home applications based on Internet of Things. *Journal of Network and Computer Applications*, 97, 48–65.

Allam, Z., & Jones, D. S. 2021. Future (post-COVID) digital, smart and sustainable cities in the wake of 6G: Digital twins, immersive realities and new urban economies. *Land Use Policy*, 101, 105201.

Allen, J. G., MacNaughton, P., Satish, U., Santanam, S., Vallarino, J., & Spengler, J. D. 2016. Associations of cognitive function scores with carbon dioxide, ventilation, and volatile organic compound exposures in office workers: A controlled exposure study of green and conventional office environments. *Environmental Health Perspectives*, 124, 805–812.

Ankrah, S., & Omar, A.-T. 2015. Universities–industry collaboration: A systematic review. *Scandinavian Journal of Management*, 31, 387–408.

Arrieta, A. B., Díaz-Rodríguez, N., Del Ser, J., Bennetot, A., Tabik, S., Barbado, A., García, S., Gil-López, S., Molina, D., & Benjamins, R. 2020. Explainable artificial intelligence (XAI): Concepts, taxonomies, opportunities and challenges toward responsible AI. *Information Fusion*, 58, 82–115.

Babcock, P. 2019. Tech start-ups look to disrupt the affordable housing industry. *The Washington Post*. https://www.washingtonpost.com/realestate/tech-start-ups-look-to-disrupt-the-affordable-housing-industry/2019/07/11/6ca0a678-9f30-11e9-b27f-ed2942f73d70_story.html

Baker, J. 2024. Why sustainability must go beyond ticking boxes. *Built Environment Journal*. RICS.https://ww3.rics.org/uk/en/journals/built-environment-journal/sustainability-building-surveying-skills.html

Barbosa, F., Woetzel, J., & Mischke, J. 2017. *Reinventing construction: A route of higher productivity*. McKinsey Global Institute.

Bass, B., Liu, K. K. Y., & Baskaran, B. A. 2003. *Evaluating rooftop and vertical gardens as an adaptation strategy for urban areas*. National Research Council of Canada.

Biasin, M., Delle Foglie, A., & Giacomini, E. 2024. Addressing climate challenges through ESG-real estate investment strategies: An asset allocation perspective. *Finance Research Letters*, 63, 105381.

Bogue, R. 2018. What are the prospects for robots in the construction industry? *Industrial Robot: An International Journal*, 45, 1–6.
Bouckaert, S., Pales, A. F., McGlade, C., Remme, U., Wanner, B., Varro, L., D'Ambrosio, D., & Spencer, T. 2021. *Net zero by 2050: A roadmap for the global energy sector*. International Energy Agency (IEA). https://iea.blob.core.windows.net/assets/deebef5d-0c34-4539-9d0c-10b13d840027/NetZeroby2050-ARoadmapfortheGlobalEnergySector_CORR.pdf
Buchanan, C., & Gardner, L. 2019. Metal 3D printing in construction: A review of methods, research, applications, opportunities and challenges. *Engineering Structures*, 180, 332–348.
CBRE. 2022. *ESG and real estate: The top 10 things investors need to know*. https://cbre.com.kh/research-center/downloads/top-10-things-investors-need-to-know-about-esg/
Centobelli, P., Cerchione, R., Del Vecchio, P., Oropallo, E., & Secundo, G. 2022. Blockchain technology for bridging trust, traceability and transparency in circular supply chain. *Information & Management*, 59, 103508.
Cerquitelli, T., Nikolaris, N., O'Mahony, N., Macii, E., Ippolito, M., & Makris, S. 2021. *Predictive maintenance in smart factories*. Springer.
Chaouch, H., Çeken, C., & Ari, S. 2021. Energy management of HVAC systems in smart buildings by using fuzzy logic and M2M communication. *Journal of Building Engineering*, 44, 102606.
Chong, C.-Y., & Kumar, S. P. 2003. Sensor networks: Evolution, opportunities, and challenges. *Proceedings of the IEEE*, 91, 1247–1256.
Christopoulos, G. I., Liu, X.-X., & Hong, Y.-Y. 2017. Toward an understanding of dynamic moral decision making: Model-free and model-based learning. *Journal of Business Ethics*, 144, 699–715.
Clayton, J., Fabozzi, F. J., Giliberto, S. M., Gordon, J. N., Liang, Y., MacKinnon, G., & Mansour, A. 2021. Introduction: Resilient real estate. *Journal of Portfolio Management*, 47, 11–24.
Cohen, S., Fehder, D. C., Hochberg, Y. V., & Murray, F. 2019. The design of startup accelerators. *Research Policy*, 48, 1781–1797.
D'Silva, M. 2023. Energy saving in smart buildings. https://smartbuildingmag.com/market-overview/68658-energy-saving-in-smart-buildings
De La Cruz-Lovera, C., Perea-Moreno, A.-J., De La Cruz-Fernández, J.-L., Alvarez-Bermejo, J. A., & Manzano-Agugliaro, F. 2017. Worldwide research on energy efficiency and sustainability in public buildings. *Sustainability*, 9, 1294.
Deloitte. 2017. *Predictive maintenance: Taking pro-active measures based on advanced data analytics to predict and avoid machine failure*. Deloitte Analytics Institute.
Eini, R., Linkous, L., Zohrabi, N., & Abdelwahed, S. 2021. Smart building management system: Performance specifications and design requirements. *Journal of Building Engineering*, 39, 102222.
Elliott, C., Lee, K., & Guidehouse, I. 2020. *Adoption of light-emitting diodes in common lighting applications*. U.S. Department of Energy.
EPA. 2012. *WaterSense at work: Best management practices for commercial and institutional facilities*. Office of Water, U.S. Environmental Protection Agency.
ESCP Europe. 2023. PropTech global trends, 2023. https://escp.eu/sites/default/files/PDF/news-events/proptech-global-trends-2023.pdf
Ferreri, M., & Sanyal, R. 2022. Digital informalisation: Rental housing, platforms, and the management of risk. *Housing Studies*, 37, 1035–1053.
Fraga-Lamas, P., Fernández-Caramés, T. M., Suárez-Albela, M., Castedo, L., & González-López, M. 2016. A review on Internet of Things for defense and public safety. *Sensors*, 16, 1644.
Gao, Y. 2021. *Artificial intelligence for building energy management in the electricity market and transmission power flow planning*. The University of Alabama.
Gartner. 2020. *IoT security primer: Challenges and emerging practices*. https://www.gartner.com/en/doc/iot-security-primer-challenges-and-emerging-practices
Gluszak, M., Gawlik, R., & Zieba, M. 2019. Smart and green buildings features in the decision-making hierarchy of office space tenants: An analytic hierarchy process study. *Administrative Sciences*, 9, 52.
Gonsalves, S., Starry, O., Szallies, A., & Brenneisen, S. 2022. The effect of urban green roof design on beetle biodiversity. *Urban Ecosystems*, 25, 205–219.
Gradillas, M., Castro, A., & Salandra, R. 2021. *Understanding the adoption of voluntary environmental, social and governance (ESG) standards in the real estate sector*. UCL. https://www.ucl.ac.uk/bartlett/news/2021/nov/new-report-explores-voluntary-adoption-esg-standards-real-estate-sector

Grant, D. 2022. *Smart buildings and the battle for sustainability*. Juniper Research.

GRESB. 2024. *2023 real estate assessment results*. GRESB.

Hager, I., Golonka, A., & Putanowicz, R. 2016. 3D printing of buildings and building components as the future of sustainable construction? *Procedia Engineering*, 151, 292–299.

Hewa, T., Ylianttilla, M., & Liyanage, M. 2021. Survey on blockchain-based smart contracts: Applications, opportunities and challenges. *Journal of Network and Computer Applications*, 177, 102857.

Himeur, Y., Elnour, M., Fadli, F., Meskin, N., Petri, I., Rezgui, Y., Bensaali, F., & Amira, A. 2023. AI-big data analytics for building automation and management systems: A survey, actual challenges and future perspectives. *Artificial Intelligence Review*, 56, 4929–5021.

Ionaşcu, E., & Anghel, I. 2020. Improvement of the real estate transparency through digitalisation. *Proceedings of the International Conference on Business Excellence*, Bucharest, Romania, 371–384.

Jaffer, S. A., Pandey, S., Mehta, R., & Bhavathankar, P. 2020. Blockchain-based direct benefit transfer system for subsidy delivery. In *2020 International Conference for Emerging Technology (INCET)*. IEEE, 1–6.

Jagannathan, R., Ravikumar, A., & Sammon, M. 2017. *Environmental, social, and governance criteria: Why investors are paying attention*. National Bureau of Economic Research.

Kifokeris, D., & Koch, C. 2019. Blockchain in building logistics: emerging knowledge, and related actors in Sweden. In *Proceedings of the 35th Annual ARCOM Conference*, Leeds, UK, 426–435.

Kshetri, N. 2018. 1 Blockchain's roles in meeting key supply chain management objectives. *International Journal of Information Management*, 39, 80–89.

Kshetri, N., & Voas, J. 2018. Blockchain in developing countries. *IT Professional*, 20, 11–14.

Lai, T. 2021. Blockchain, law and governance: General conclusion. In B. Cappiello & G. Carullo (eds.), *Blockchain, law and governance*. Springer, 289–304.

Luque, J. 2022. *PropTech global trends 2022 annual barometer*. ESCP Business School.

Luque, J. 2023. *PropTech global trends 2023 annual barometer*. ESCP Business School.

Mahmudnia, D., Arashpour, M., & Yang, R. 2022. Blockchain in construction management: Applications, advantages and limitations. *Automation in Construction*, 140, 104379.

Marinakis, V., Doukas, H., Tsapelas, J., Mouzakitis, S., Sicilia, Á., Madrazo, L., & Sgouridis, S. 2020. From big data to smart energy services: An application for intelligent energy management. *Future Generation Computer Systems*, 110, 572–586.

Mauser, I., Müller, J., Allerding, F., & Schmeck, H. 2016. Adaptive building energy management with multiple commodities and flexible evolutionary optimization. *Renewable Energy*, 87, 911–921.

Means, B., Toyama, Y., Murphy, R., & Baki, M. 2013. The effectiveness of online and blended learning: A meta-analysis of the empirical literature. *Teachers College Record*, 115, 1–47.

Mekki, K., Bajic, E., Chaxel, F., & Meyer, F. 2019. A comparative study of LPWAN technologies for large-scale IoT deployment. *ICT Express*, 5, 1–7.

Minoli, D., Sohraby, K., & Occhiogrosso, B. 2017. IoT considerations, requirements, and architectures for smart buildings—Energy optimization and next-generation building management systems. *IEEE Internet of Things Journal*, 4, 269–283.

Molina-Solana, M., Ros, M., Ruiz, M. D., Gómez-Romero, J., & Martín-Bautista, M. J. 2017. Data science for building energy management: A review. *Renewable and Sustainable Energy Reviews*, 70, 598–609.

Nanda, A. 2019. *Residential real estate: Urban & regional economic analysis*. Routledge.

Nguyen, T. D., & Adhikari, S. 2023. The role of BIM in integrating digital twin in building construction: A literature review. *Sustainability*, 15, 10462.

Nieuwland, S., & Van Melik, R. 2020. Regulating Airbnb: How cities deal with perceived negative externalities of short-term rentals. *Current Issues in Tourism*, 23, 811–825.

Nijland, M. 2019. Influence of blockchain in the real estate sector. *International Journal of Applied Science*, 2, 22–22.

Ølnes, S., Ubacht, J., & Janssen, M. 2017. *Blockchain in government: Benefits and implications of distributed ledger technology for information sharing*. Elsevier.

Panarello, A., Tapas, N., Merlino, G., Longo, F., & Puliafito, A. 2018. Blockchain and IoT integration: A systematic survey. *Sensors*, 18, 2575.

Park, J. Y., Ouf, M. M., Gunay, B., Peng, Y., O'Brien, W., Kjærgaard, M. B., & Nagy, Z. 2019. A critical review of field implementations of occupant-centric building controls. *Building and Environment*, 165, 106351.

Peña, M., Biscarri, F., Guerrero, J. I., Monedero, I., & León, C. 2016. Rule-based system to detect energy efficiency anomalies in smart buildings, a data mining approach. *Expert Systems with Applications*, 56, 242–255.

Perrot, A., Rangeard, D., & Pierre, A. 2016. Structural built-up of cement-based materials used for 3D-printing extrusion techniques. *Materials and Structures*, 49, 1213–1220.

Plageras, A. P., Psannis, K. E., Stergiou, C., Wang, H., & Gupta, B. B. 2018. Efficient IoT-based sensor BIG Data collection–processing and analysis in smart buildings. *Future Generation Computer Systems*, 82, 349–357.

Qiu, J., Wu, Q., Ding, G., Xu, Y., & Feng, S. 2016. A survey of machine learning for big data processing. *EURASIP Journal on Advances in Signal Processing*, 2016, 1–16.

Reiss, R. 2022. *Three commercial real estate CEOs committed to ESG. Forbes.*

Reyna, A., Martín, C., Chen, J., Soler, E., & Díaz, M. 2018. On blockchain and its integration with IoT: Challenges and opportunities. *Future Generation Computer Systems*, 88, 173–190.

RICS. 2021. *Sustainability and ESG in commercial property valuation and strategic advice*. RICS.

RICS. 2023. *RICS tech partner programme survey*. Royal Institution of Chartered Surveyors.

RICS. 2024. *RICS tech partner programme survey*. Royal Institution of Chartered Surveyors.

Savelyev, A. 2018. Copyright in the blockchain era: Promises and challenges. *Computer Law & Security Review*, 34, 550–561.

Scherrenberg, J., Wessels, P., & Nelisse, P. 2024. The future of real estate valuations: The impact of ESG. Royal Institution of Chartered Surveyors. https://www.rics.org/content/dam/ricsglobal/documents/latest-news/WBEF-ESG-and-valuation-2024.pdf

Schweizer, D., & Zhou, T. 2016, October 2. Do principles pay in real estate crowdfunding? *Journal of Portfolio Management*, 43(6), 2017. http://dx.doi.org/10.2139/ssrn.2846892

Sembiyeva, L., Zhagyparova, A., Zhumadillayeva, A., Zholamanova, M., Bekbolsynova, A., & Zhanabergenova, M. 2024. Applying advanced artificial intelligence to predict the green bond market in Kazakhstan: Fostering sustainable financial instruments and environmental objectives. *Montenegrin Journal of Economics*, 20, 237–250.

Shahat, E., Hyun, C. T., & Yeom, C. 2021. City digital twin potentials: A review and research agenda. *Sustainability*, 13, 3386.

Shahid, I., Shahzad, M. I., Tutsak, E., Mahfouz, M. M., Al Adba, M. S., Abbasi, S. A., Rathore, H. A., Asif, Z., & Chen, Z. 2024. Carbon based sensors for air quality monitoring networks: Middle East perspective. *Frontiers in Chemistry*, 12, 1391409.

Stokkink, Q., & Pouwelse, J. 2018. Deployment of a blockchain-based self-sovereign identity. In *2018 IEEE International Conference on Internet of Things (iThings) and IEEE Green Computing and Communications (GreenCom) and IEEE Cyber, Physical and Social Computing (CPSCom) and IEEE Smart Data (SmartData)*, IEEE, 1336–1342.

Truby, J. 2018. Decarbonizing Bitcoin: Law and policy choices for reducing the energy consumption of blockchain technologies and digital currencies. *Energy Research & Social Science*, 44, 399–410.

Veuger, J. 2018. Trust in a viable real estate economy with disruption and blockchain. *Facilities*, 36, 103–120.

Walacik, M., & Chmielewska, A. 2024. Real estate industry sustainable solution (environmental, social, and governance) significance assessment—AI-powered algorithm implementation. *Sustainability*, 16, 1079.

Wang, P., Wu, P., Wang, J., Chi, H.-L., & Wang, X. 2018. A critical review of the use of virtual reality in construction engineering education and training. *International Journal of Environmental Research and Public Health*, 15, 1204.

WEF. 2021. *A framework for the future of real estate*. World Economic Forum.

Wouda, H. P., & Opdenakker, R. 2019. Blockchain technology in commercial real estate transactions. *Journal of Property Investment & Finance*, 37, 570–579.

Zhao, D., Watari, D., Ozawa, Y., Taniguchi, I., Suzuki, T., Shimoda, Y., & Onoye, T. 2023. Data-driven online energy management framework for HVAC systems: An experimental study. *Applied Energy*, 352, 121921.

Zheng, Z., Xie, S., Dai, H.-N., Chen, X., & Wang, H. 2018. Blockchain challenges and opportunities: A survey. *International Journal of Web and Grid Services*, 14, 352–375.
Zhou, K., Fu, C., & Yang, S. 2016. Big data driven smart energy management: From big data to big insights. *Renewable and Sustainable Energy Reviews*, 56, 215–225.

Chapter 9

Conclusions

9.1 Key Findings and Insights

The comprehensive research presented throughout this book reveals several fundamental insights about real estate transformation through PropTech innovation. These findings demonstrate how technological advancement reshapes traditional practices while creating new opportunities for sustainable development and improved operational efficiency. The evidence points to a profound shift in how real estate assets are managed, traded, and valued, with implications extending across the industry.

9.1.1 The Evolution of PropTech and Its Impact

PropTech has evolved from simple digital tools to sophisticated platforms transforming real estate operations. This evolution is most evident in integrating advanced technologies such as artificial intelligence, blockchain, and Internet of Things (IoT) systems. The research demonstrates that AI and machine learning have significantly enhanced decision-making processes, with automated valuation models achieving accuracy rates above 90% in markets like Singapore. Blockchain technology has enabled new transaction models, as evidenced by platforms like Propy successfully facilitating property transfers through smart contracts. Meanwhile, IoT sensors and smart building systems have dramatically improved operational efficiency, with documented energy savings of 15–25% in buildings using advanced building management systems.

The advancement in data analytics capabilities has transformed market analysis and investment strategies. Big data analytics now enable more sophisticated approaches to market assessment, while predictive analytics have improved risk assessment and property management. The development of enhanced visualisation tools, including virtual and augmented reality, has revolutionised property marketing and evaluation methods. These technological improvements have facilitated the evolution from single-purpose tools to integrated ecosystems, leading to the development of comprehensive property management solutions and new investment platforms that enable broader market participation.

PropTech has catalysed significant changes in market dynamics, particularly on transparency and efficiency. Online platforms have greatly enhanced access to market data, providing improved visibility into property performance metrics and enabling more efficient price discovery mechanisms. This increased transparency has been accompanied by considerable improvements in operational efficiency, featuring reduced transaction times and costs, streamlined property management processes, and automated compliance and

DOI: 10.1201/9781003504436-9

reporting systems. The democratisation of real estate investment through crowdfunding platforms has lowered entry barriers for smaller investors and expanded opportunities for cross-border investment.

9.1.2 *Impact on Investment and Finance*

PropTech has profoundly transformed real estate investment and finance practices. Advanced analytical tools have fundamentally changed investment decisions, enabling more sophisticated risk assessment and enhanced portfolio management capabilities. The emergence of new investment vehicles, including real estate crowdfunding platforms and tokenised assets, has created novel opportunities for market participation. These innovations have been accompanied by significant improvements in transaction efficiency, reduced processing times and costs, and improved transparency in financial dealings.

Financial innovation has transcended traditional investment models to encompass new strategies for risk management and value creation. PropTech solutions have enabled more sophisticated risk assessment tools, enhanced monitoring capabilities, and improved predictive analytics. These advancements are vital for institutional investors and property managers, who can now leverage technology to optimise portfolio performance and manage risk more effectively.

9.1.3 *Sustainability Outcomes*

The research provides compelling evidence of PropTech's impact on sustainability outcomes in the real estate sector. Regarding environmental performance, smart building systems have significantly reduced energy consumption through AI-driven controls and improved HVAC optimisation. The integration of IoT technology has enabled better resource management, from waste reduction to water conservation, while facilitating the integration of renewable energy systems.

The social impact of PropTech has been equally significant. Digital platforms have enhanced community engagement and transparency in development projects while improving access to sustainable housing options. Smart building systems have contributed to occupant well-being through improved indoor environmental quality and personalised environmental controls. Automation of service delivery systems has enhanced the occupant experience while improving operational efficiency.

Beyond direct environmental and social impacts, PropTech has enabled more precise measurement and verification of sustainability performance. This has improved the industry's ability to meet increasingly stringent regulatory requirements while providing better data for sustainability reporting and environmental, social, and governance (ESG) compliance. The enhanced measurement capabilities have also facilitated the development of more effective sustainability strategies and improved accountability for environmental performance.

These findings demonstrate that PropTech is not merely enhancing existing practices but is fundamentally reshaping how real estate assets are developed, managed, and traded. The evidence suggests that this transformation is particularly significant in advancing sustainable practices, where technology has enabled more precise measurement, better management, and improved outcomes across environmental and social dimensions. As the industry continues to evolve, the integration of PropTech solutions appears increasingly central to achieving operational efficiency and sustainability goals.

9.2 Emerging Trends and Future Directions

The research presented throughout this book points to several significant trends that will likely shape PropTech's future and its impact on the real estate industry. These emerging patterns suggest continuing the technological transformation documented in previous chapters and indicate new directions for innovation and development.

9.2.1 *Technology Trends*

Artificial intelligence and machine learning technologies are set to play an increasingly central role in real estate operations and decision-making. Research indicates a clear trajectory towards more sophisticated predictive analytics capabilities, with AI systems becoming more proficient at forecasting market trends, property values, and maintenance needs. This evolution is particularly evident in the development of automated valuation models, which continue to enhance in accuracy and scope. Integrating machine learning with operational systems facilitates more advanced automation of property management tasks, from maintenance scheduling to tenant communication, resulting in significant efficiency improvements and cost reductions.

Blockchain technology is emerging as a transformative force in real estate transactions and ownership structures. The increasing adoption of smart contracts is streamlining property transactions and reducing the need for intermediaries whilst enhancing transparency and security. The tokenisation of real estate assets continues to gain momentum, with new platforms enabling fractional ownership and improved liquidity in property investments. These developments are particularly significant for cross-border transactions and international investment, where blockchain's capacity to provide transparent, immutable records offers substantial advantages over traditional systems.

The IoT is rapidly expanding within real estate applications, as smart building systems become increasingly sophisticated and interconnected. Advanced sensor networks facilitate more precise monitoring and control of building systems, resulting in enhanced energy efficiency and occupant comfort. The integration of IoT devices with building management platforms creates more comprehensive and responsive property management systems. This enhanced connectivity generates richer data streams that support improved decision-making and operational optimisation, while also enabling more effective sustainability initiatives.

9.2.2 *Market Evolution*

The real estate market is undergoing significant structural changes driven by technological innovation and user demands. Industry consolidation is becoming more prevalent, with increasing integration of PropTech solutions across different aspects of real estate operations. Strategic partnerships between traditional real estate firms and technology providers are growing, leading to more comprehensive and sophisticated service offerings. This consolidation is accompanied by efforts to standardise technology protocols and data formats, facilitating better interoperability between different systems and platforms.

New business models that challenge traditional approaches to real estate ownership and management are emerging. Property-as-a-service is gaining traction, with flexible space solutions and enhanced service offerings becoming more common. Digital-first real estate platforms are proliferating, offering new ways to access and interact with property markets. These platforms are particularly significant in developing sustainable investment vehicles, enabling better tracking and verifying of environmental performance while facilitating green investment strategies.

The evolution of financing mechanisms creates new opportunities for investors and property owners. Crowdfunding platforms continue to democratise real estate investment, while blockchain-based systems facilitate new forms of property ownership and trading. Integrating sustainability metrics into investment decisions is becoming increasingly sophisticated, with an enhanced ability to measure and verify environmental performance influencing financing terms and investment strategies.

9.2.3 Sustainability and Technology Convergence

The convergence of sustainability initiatives and technological innovation creates new opportunities to improve environmental performance. Smart building technologies are becoming more sophisticated in optimising resource use and reducing environmental impact. Integrating renewable energy systems with building management platforms enables more effective energy management, while advanced analytics improve the ability to measure and verify sustainability performance.

Data analytics is becoming increasingly vital in sustainability management, with new tools facilitating enhanced tracking and reporting of environmental metrics. The development of more sophisticated environmental performance indicators, along with improved measurement capabilities, supports more informed decision-making regarding sustainability initiatives. This convergence is particularly apparent in the rising integration of ESG considerations into property investment and management decisions.

Social sustainability is garnering increased attention, with technology facilitating improved community engagement and measurement of social impact. Digital platforms encourage more effective communication and participation among stakeholders in property development decisions. Enhanced data collection and analysis capabilities are boosting the ability to quantify and manage social impacts whilst also enabling better reporting and verification of social sustainability initiatives.

These emerging trends suggest that the transformation of real estate through PropTech will continue to accelerate, with technology taking on an increasingly central role in addressing industry challenges and creating new opportunities. The convergence of various technological innovations, along with a growing emphasis on sustainability and evolving market demands, is likely to drive ongoing development in how real estate assets are created, managed, and traded. This continual transformation presents both challenges and opportunities for industry stakeholders, requiring careful attention to emerging trends and strategic adaptation to shifting market conditions.

9.3 Challenges and Opportunities

While PropTech presents significant potential for transforming the real estate industry, the research reveals substantial implementation challenges and compelling opportunities for future development. Understanding these challenges and opportunities is crucial for stakeholders seeking to navigate the evolving landscape of real estate technology and sustainability.

Integrating PropTech solutions into existing real estate operations presents several significant technical challenges. One of the most pressing issues is the integration with legacy systems, as many real estate organisations are equipped with outdated technological infrastructures that were not designed for contemporary digital solutions. Research indicates that successful integration often requires substantial investment in both technology and change management within organisations. Data quality and standardisation also emerge as critical challenges, with organisations struggling to maintain consistent data formats and ensure accuracy across various systems and platforms. This issue is particularly acute in larger organisations managing multiple properties across different regions.

Cybersecurity concerns present an increasingly significant challenge as real estate operations become more digitised. Research indicates a growing vulnerability to cyber threats as more building systems become connected and property data transitions to cloud-based platforms. These security challenges are exacerbated by the necessity to protect sensitive financial and personal information while maintaining the accessibility and transparency that make PropTech solutions valuable. Scalability issues also pose considerable technical hurdles, especially for organisations attempting to implement solutions across large portfolios or multiple jurisdictions.

Market challenges present another set of significant hurdles for PropTech adoption. Regulatory compliance emerges as a major concern, requiring organisations to navigate complex and often changing regulatory requirements across various jurisdictions. This is particularly challenging for cross-border operations and innovative financial products enabled by new technologies. Rates of industry adoption continue to vary significantly, with some sectors exhibiting resistance to technological change despite clear evidence of its benefits. This resistance often arises from concerns about implementation costs, disruptions to existing processes, and uncertainty regarding returns on investment.

The skills gap in the workforce represents another substantial challenge, as organisations struggle to find and retain talent with the necessary combination of real estate and technology expertise. This challenge is particularly acute in areas demanding specialised knowledge, such as data analytics and artificial intelligence applications in real estate. The significant investment requirements for comprehensive PropTech implementation also pose barriers, especially for smaller organisations and those operating in markets with lower profit margins.

Despite these challenges, the research highlights numerous compelling opportunities for the development and implementation of PropTech. One of the most significant opportunities is the enhancement of sustainability through technology, which could improve environmental performance via smart building systems, superior resource management, and more effective monitoring of sustainability metrics. The research shows that organisations successfully implementing PropTech solutions can achieve considerable reductions in energy consumption and operational costs while elevating their environmental performance.

Enhanced operational efficiency presents another significant area of opportunity, with the potential for substantial cost savings and performance improvements via automation and optimisation of property management processes. Research indicates that organisations adopting comprehensive PropTech solutions can realise operational cost reductions of 15–30% while enhancing service delivery and tenant satisfaction. The creation of new revenue streams through technology-enabled services and business models offers further prospects for value generation.

Market opportunities continue to expand as technology facilitates access to new markets and customer segments. Global market expansion becomes increasingly feasible through digital platforms and enhanced data analytics capabilities, whilst new investment vehicles enabled by blockchain and other technologies generate opportunities for market growth and increased liquidity. Improved service offerings enabled by technology present opportunities for differentiation and value addition, particularly in competitive markets.

Risk management presents another significant opportunity, as advanced analytics and monitoring capabilities enable improved risk assessment and mitigation strategies. Research indicates that organisations utilising sophisticated PropTech solutions can achieve greater predictability in property performance and a reduced exposure to operational risks. This enhanced risk management capability becomes particularly valuable in the face of increasing environmental and market uncertainties.

The increasing emphasis on ESG factors offers opportunities for organisations that can effectively utilise technology to assess and improve their sustainability performance. Research shows a growing demand for properties and organisations that can demonstrate strong ESG credentials, offering competitive advantages for those implementing comprehensive PropTech solutions. This trend is particularly significant in markets with strict regulatory frameworks relating to sustainability and environmental performance.

Enhancing customer experience offers a significant opportunity, as technology facilitates more personalised and responsive service delivery. Research indicates a growing demand for technology-enabled services and amenities, particularly among younger demographics and within premium market segments. Organisations that effectively utilise technology to enhance tenant and user experiences will position themselves for a competitive advantage in increasingly demanding markets.

These challenges and opportunities indicate that although the journey to comprehensive PropTech implementation may be intricate, the potential rewards render it a worthwhile endeavour for organisations dedicated to long-term success in the real estate sector. The key to success seems to rest in judiciously balanced approaches that tackle technical and organisational hurdles while strategically seizing opportunities aligned with organisational capabilities and market demands.

9.4 Recommendations for Stakeholders

Drawing from the research and analysis presented throughout this book, several critical recommendations emerge for different stakeholder groups in the PropTech ecosystem. These recommendations aim to address current challenges while capitalising on emerging opportunities for sustainable development and improved operational efficiency in real estate.

For industry practitioners, implementing PropTech solutions requires a comprehensive and strategic approach. Real estate companies should develop robust digital transformation strategies that align technological innovation with their broader business objectives. As demonstrated in Chapter 7, organisations must invest significantly in workforce development to bridge the growing skills gap in PropTech implementation. This includes training existing staff and recruiting new talent with relevant technical expertise. Data quality and security should be prioritised, as shown by the case studies in Chapter 4, where robust data management practices underpinned successful PropTech implementations.

Furthermore, sustainability initiatives should be integrated into core business strategies rather than treated as separate initiatives, following the frameworks outlined in Chapter 6.

Technology providers face distinct challenges in effectively serving the real estate sector. Integration capabilities should be a primary focus, ensuring that PropTech solutions seamlessly connect with existing systems and workflows. As evidenced by the analysis in Chapter 5, user experience is crucial for successful adoption; solutions should be intuitive and accessible while maintaining sophisticated functionality. Security concerns must be proactively addressed, particularly given the sensitive nature of real estate data and transactions. Technology providers should also ensure their solutions actively support sustainability goals, incorporating features for environmental monitoring, reporting, and optimisation, as discussed in Chapter 3.

The research suggests several crucial areas for policymakers' attention. Developing supportive regulatory frameworks is essential to encouraging innovation while protecting stakeholder interests. As shown in Chapter 2, successful PropTech implementation often depends on clear regulatory guidance, particularly in areas like tokenisation and automated valuations. Policymakers should work to promote standardisation across the industry, facilitating interoperability and data sharing while maintaining security and privacy standards. Supporting sustainable development through appropriate incentives and requirements, as outlined in Chapter 6, remains crucial for achieving broader environmental and social goals.

These stakeholder-specific recommendations are interconnected and mutually reinforcing. Success in the evolving PropTech landscape requires collaboration and coordination among all stakeholder groups. The evidence presented throughout this book suggests that organisations that embrace these recommendations while maintaining flexibility to adapt to changing conditions will be best positioned to thrive in the future real estate industry.

Beyond these specific recommendations, stakeholders should remain mindful of the rapid pace of technological change and the need for continuous adaptation. Regular reassessment of strategies and approaches, informed by emerging technologies and evolving market conditions, will be crucial for long-term success in the PropTech ecosystem.

9.5 Future Research Directions

The swift evolution of PropTech and its increasing influence on real estate investment, finance, and sustainability indicate several critical areas for future research. These research directions arise from both the findings and limitations noted throughout this book, highlighting opportunities for greater understanding and practical application in both technological and market domains.

In the realm of technology research, several promising areas warrant further exploration. Advanced AI applications in real estate present numerous research opportunities, particularly in the development of more sophisticated predictive models and automated decision-making systems. As exemplified in Chapter 4, although current AI applications show promise in areas such as automated valuation and risk assessment, significant potential remains for the expansion of these capabilities. Blockchain implementation strategies require additional scrutiny, especially concerning scalability, energy efficiency, and integration with existing systems. As highlighted in Chapters 2 and 4, while blockchain technology offers compelling benefits for real estate transactions and tokenisation, practical implementation challenges still need thorough resolution. IoT integration methods represent another crucial research area, particularly in the development of more effective approaches to data collection, analysis, and system integration in smart buildings. The findings in Chapter 5 suggest that while IoT technologies can significantly enhance building performance and sustainability, optimal implementation strategies are still evolving. Furthermore, sustainability measurement techniques necessitate further refinement to establish more standardised and effective methods for assessing the environmental impact of real estate assets and PropTech solutions.

Market research directions also present several significant avenues for investigation. The impact of PropTech on market efficiency remains a crucial area of study, particularly in understanding how technological innovation affects property values, transaction speeds, and market liquidity. As discussed in Chapter 7, while PropTech has demonstrably influenced market dynamics, the full extent and implications of these changes necessitate further analysis. The evolution of investment vehicles, especially those enabled by new technologies, offers another important research direction. Chapter 4's examination of digital platforms and tokenisation suggests numerous opportunities for exploring how these innovations may reshape real estate investment structures and accessibility. Sustainability performance metrics represent a critical area for future research, especially in devising more comprehensive and standardised approaches to measuring and comparing the environmental impact of real estate assets. As highlighted in Chapter 6, while various sustainability metrics exist, there remains a need for more unified and widely accepted measurement frameworks.

Social impact assessment emerges as another crucial area for future research, particularly in understanding how PropTech innovations affect various stakeholder groups and communities. This entails investigating the democratisation of real estate investment, the effect of digital platforms on housing accessibility, and the role of technology in fostering social equity in real estate development. The findings in Chapters 5 and 6 indicate that while PropTech can contribute to positive social outcomes, further research is necessary to comprehend and optimise these impacts fully.

Future research should also explore the intersection of these various domains, investigating how technological innovations interact with market dynamics and sustainability objectives. This includes examining the role of regulatory frameworks in shaping PropTech development, the relationship between technological innovation and market efficiency, and the potential for PropTech to drive sustainable transformation in the real estate sector. As demonstrated throughout this book, these areas are intricately interconnected, and their relationships merit further study to inform effective policy and practice.

Advancing these research directions will necessitate collaboration among academia, industry practitioners, and technology providers. Mixed-method approaches that integrate quantitative analysis with qualitative insights from case studies and stakeholder experiences will be crucial.

9.6 Concluding Remarks

This comprehensive examination of PropTech's role in transforming real estate investment, finance, and sustainability has revealed several fundamental insights into the industry's evolution and future direction. Throughout this book, we have illustrated how technological innovation is not merely altering operational practices but is fundamentally reshaping the real estate landscape. From democratising investment opportunities through digital platforms to enhancing sustainability via smart building technologies, PropTech is driving a paradigm shift in how real estate assets are developed, managed, and traded.

Integrating advanced technologies with traditional real estate practices has become a crucial driver of industry transformation. As evidenced by the case studies and empirical research presented throughout this book, successful PropTech implementation enables more informed decision-making through advanced analytics, enhances transparency via blockchain and digital platforms, and improves sustainability outcomes through smart building systems and IoT integration. These technological advances are not occurring in isolation but are part of a broader evolution towards more efficient, sustainable, and accessible real estate markets.

Looking ahead, technological innovation will increasingly shape the future of real estate, with PropTech playing a central role in tackling industry challenges and creating new opportunities. The research presented in this book indicates that success in this evolving landscape will require stakeholders to embrace technological change while maintaining a strong emphasis on sustainable development goals. This necessitates the adoption of new technologies along with the cultivation of organisational capabilities and a cultural mindset necessary to leverage these innovations effectively.

The relationship between technology and sustainability has emerged as a critical theme throughout this investigation. PropTech solutions are vital in enabling the real estate sector to fulfil its environmental responsibilities while ensuring economic viability. From smart building systems that optimise energy consumption to blockchain platforms that enhance transparency in green building certification, technology is providing the necessary tools for sustainable transformation. This synergy between technological innovation and sustainability objectives represents both a challenge and an opportunity for all stakeholders in the real estate ecosystem.

As our analysis has demonstrated, the success of PropTech initiatives relies significantly on effective collaboration among industry practitioners, technology providers, and policymakers. The recommendations of this book highlight the necessity of developing comprehensive digital strategies, investing in workforce development, and establishing supportive regulatory frameworks. These elements must function in unison to foster an environment conducive to innovation while ensuring that technological advancement aligns with the broader objectives of sustainable development and social responsibility.

The path forward will necessitate ongoing research and development in PropTech, particularly in artificial intelligence, blockchain implementation, and sustainability measurement. As highlighted in our discussion of future research directions, significant opportunities remain for enhancing our understanding of how technology can effectively serve the needs of the real estate industry while promoting sustainable practices. This continual evolution will require flexibility and adaptability from all stakeholders as new technologies emerge and market conditions change.

In conclusion, this book emphasises that PropTech is more than just a technological advancement; it represents a significant shift in how we approach real estate investment, finance, and sustainability. The findings suggest that organisations that successfully integrate technology with sustainable practices, while prioritising the needs of their stakeholders, are most likely to thrive in the evolving real estate landscape. As we move forward, PropTech will undoubtedly present new challenges and opportunities, but the innovations we are witnessing today establish a strong foundation for a more efficient, sustainable, and inclusive real estate sector.

Index

Pages in *italics* refer to figures and pages in **bold** refer to tables.

Made in the USA
Las Vegas, NV
16 May 2026

47035719R00149